CW01621610

Astronomer, Cartographer and Naturalist of the New World

VOLUME 1

Life, Work and Legacy

Astronomer, Cartographer and Naturalist of the New World

The Life and Scholarly Achievements of Georg Marggrafe (1610-1643) in Colonial Dutch Brazil

VOLUME 1

Life, Work and Legacy

Huib J. Zuidervaart & Oscar T. Matsuura

Routledge
Taylor & Francis Group
LONDON AND NEW YORK

The publication of this book has been made possible by grants from the Stichting Dr. Henrik Muller's Vanderlandsch Fonds; the Dr. C. Louise Thijssen-Schoute Stichting; the Huygens Instituut (KNAW) and the Louwman Collection of Historic Telescopes.

This book highlights the scientific achievements of the astronomer, cartographer and naturalist Georg Marggrafe (1601-1643).
It consists of two volumes:
Vol. 1. Discusses his biography and legacy.
Vol. 2. Presents his previously unpublished astronomical observations, collected in colonial Dutch Brazil between 1638 and 1643. This volume contains the earliest known series of observations of the Southern Hemisphere, collected by scientific instruments made according to the European standards of the time. (ISBN 978 94 6372 228 5)

First published in 2022 by Amsterdam University Press Ltd.

Published 2025 by Routledge
4 Park Square, Milton Park, Abingdon, Oxon OX14 4RN
605 Third Avenue, New York, NY 10158

Routledge is an imprint of the Taylor & Francis Group, an informa business

ISBN: 9789463722186 (hbk)
ISBN: 9781041175957 (pbk)
ISBN: 9781003691297 (ebk)
NUR 685

Cover illustration: Former Portuguese house at Antonio Vaz. First residence of the Dutch governor-general Johan Maurits von Nassau-Siegen, depicted in 1639 by Zacharias Wagener. On top of the house stands the observatory constructed on behalf of Georg Marggrafe. (Thierbuch, Ca 226, fol. 107. Kupferstich-Kabinett, Staatliche Kunstsammlungen Dresden, Photo: Herbert Boswank.

Cover design: Marijke Maarleveld, ViaMare, Almen

DOI 10.5117/9789463722186

CONTENTS

Volume 1
Life, Work and Legacy

Preface 11

PART I – CONTEXT

CHAPTER 1 INTRODUCTION 17
A forgotten astronomer
The aim of this book
A publication postponed by more than three centuries
The correct spelling of Georg Marggrafe's surname

CHAPTER 2 CIRCULATION OF KNOWLEDGE IN EARLY MODERN DUTCH BRAZIL? 27
The Dutch trading companies and the emergence of a pre-modern knowledge society
Dutch Brazil and its first and only governor general
Johan Maurits's achievements in Brazil
The background of a governor general
A colony built on slavery
A princely court in Dutch Brazil: Vrijburg and Boa Vista
Dutch Brazil as a seventeenth-century Tower of Babel

PART II – BIOGRAPHY

CHAPTER 3 YOUTH IN GERMANY 51
Childhood in Liebstadt
Wandering years in Germany
Becoming a Medical Candidate at Wittenberg University (1633-1634)
A horoscope for a Wittenberg professor of astronomy (1634)
Constructing astronomical tables in Stettin (1634-1635)

CHAPTER 4 A CRUCIAL YEAR AT LEIDEN UNIVERSITY 61
Arrival in Leiden
Leiden University in the seventeenth century
The university's brand new observatory
Marggrafe's activities at Leiden University
Horoscopes for Leiden students
Friends of Marggrafe's roommate Kechel
A special friendship with Jørgen From
Marggrafe as preceptor for Estonian students?
The Leiden Lutheran Church
Preparations for Brazil
Hired as assistant to the court physician Willem Piso
The job description of the court physician and his assistants

CHAPTER 5 LIFE IN BRAZIL 93
Travel to and arriving in Brazil
The siege of São Salvador in Bahia
The decor of Marggrafe's life in Brazil: the fast-changing island of Antonio Vaz
The count's first residence on Antonio Vaz
A change of profession
A courtier for a ruler with imperial ambitions
Horoscopes for local officials
Striving for the fulfilment of an astronomical dream
Expeditions into the interior of North-Eastern Brazil
The emergence of an additional goal
Setback in March 1640: the collapse of part of the count's residence
Collector of naturalia for a German prince
Gathering cartographic information
Processing geographical information, adding some from others
A map with astonishing qualities
Astronomical and meteorological achievements
Dutch strongholds on the West African coast
The inevitable end

CHAPTER 6 LEGACY 151
The spread of Marggrafe's paper legacy
The genesis of the Historia Naturalis Brasiliae
Processing the raw material
Selecting the proper images
Ad vivum?
Transmission of visual evidence into woodcuts
Framing a frontispiece
De Laet's additions
The last phase of editing and printing
Innovative aspects of the Historia Naturalis Brasiliae
The reception of the Historia Naturalis Brasiliae
Piso's 1658 rewriting of the Historia Naturalis Brasiliae
Christian Marggrafe claiming the inheritance of his late brother Georg
After the verdict
The fate of Georg Marggrafe's astronomical manuscripts
The lost mathematical notebooks and the accompanying planisphaerium
Surviving manuscripts in the Leiden archive
The fate of a press-ready manuscript, lost in Spain
Surviving manuscripts in Paris and Lisbon

PART III – ASTRONOMY

CHAPTER 7 BACKGROUND 193
Early astronomical observations of the southern heavens
The emergence of astronomical aspirations
Astronomical observations in Leiden before Marggrafe's arrival

CHAPTER 8 SETTING AND EQUIPMENT 203
Observatories in Leiden and Recife: the design, organization and instruments compared
Location and building: 1. Leiden
Location and building: 2. Recife
Platform and dome
Quadrant
Pinacidium Tychonicum, or Tychonic aligning sight
Other angular measuring instruments
Telescopes
Helioscope
Clock, sandglass and pendulum
Marggrafe's concise library

CHAPTER 9 OBSERVATIONS IN LEIDEN AND RECIFE 221

1. Training in Leiden
   - *An eager student in practical astronomy*
   - *Witnessing a learning curve: Marggrafe's observations of sunspots*
   - *Characteristics of the Leiden telescope*
   - *Early observations of Jupiter's satellites*
   - *Marggrafe and Mercury*
   - *Measurements with the Tychonic quadrant made by Willem Jansz Blaeu*
   - *Measurements with the sextant*
   - *Measuring time*
   - *The accuracy of Marggrafe's timing*
   - *The personal equation: Marggrafe's accuracy as calculator*
   - *What has Marggrafe achieved during his training at Leiden Observatory?*
2. Observing in Brazil
   - *A leap forward at the very start*
   - *The aspired Progymnastica Mathematica Americana*
     - *- Mercury revisited*
     - *- An excellent observation with a dubious result: the determination of Recife's longitude*
     - *- Other eclipses*
     - *- The determination of the Sun's angular diameter*
     - *- Charting the Southern Stars with the Tychonic quadrant*
     - *- The obliquity of the ecliptic*
     - *- The determination of the Earth's radius*
     - *- Accidental data and other astronomical rarities*
     - *- What part of the Progymnastica Mathematica Americana has been achieved?*
   - *Missing observations*
     - *- The famous 1639 Transit of Venus*
     - *- Sunspots*
     - *- Some planetary riddles*
     - *- Comets and Supernovae*
   - *A meeting with Casper Schmalkalden?*
   - *Vrijburg, the observatory that never was*
   - *The fate of the Governor's old residence*
   - *The quality of the instruments used in Brazil*
     - *- The 5–foot quadrant*
     - *- Marggrafe's telescope*
     - *- The pendulum*
3. Is Marggrafe's astronomical legacy ever applied?

PART IV – CONCLUSION AND EPILOGUE

CHAPTER 10 RETROSPECTIVE 277
Marggrafe's importance as an early modern scholar
Marggrafe's importance for natural history
Marggrafe's importance for cartography
Marggrafe's importance for early modern astronomy
The preconditions for Marggrafe's scholarly work in colonial Dutch Brazil
Intercultural cooperation across several language barriers: the missing Go-betweens
The transmission of gathered facts into canonized knowledge
Epilogue

APPENDICES 289
1. Chronology
2. Marggrafe's horoscopes
3. A false attribution of an 'astronomical' manuscript

NOTES 307
CONSULTED ARCHIVES 369
ABBREVIATIONS 371
LITERATURE 373
INDEX 393

Volume 2
Text edition

PART V – PROGYMNASTICA ASTRONOMICA AMERICANA

TRANSCRIPTION AND ENGLISH TRANSLATION OF MARGGRAFE'S ASTRONOMICAL OBSERVATIONS MADE IN COLONIAL DUTCH BRAZIL IN THE YEARS 1638-1643

Introduction
The Paris manuscript: a note about the transcription
A note about the English translation
Transcription and Translation of Marggrafe's astronomical observations in Dutch Brazil
Index of of personal names, celestial bodies and observed phenomena

FIG. 1. Both authors at the 'Instituto Ricardo Brennand', Recife, in September 2010, investigating one of the eight surviving coloured copies of the *Historiae Naturalis Brasiliae* (Leiden, 1648), in which Marggrafe's 'Historiae Rerum Naturalium Brasiliae' is published.
Right: Oscar Matsuura. *Middle*: Curator 'Instituto Ricardo Brennand'.
Left: Huib Zuidervaart. Photo: the authors.

PREFACE

This book has a long history. It started in 2009, in Budapest, Hungary, when the two authors of this book, Oscar Matsuura and Huib Zuidervaart, met for the first time during the XXIII International Congress of History of Science and Technology. Oscar had already done a lot of research into Marggrafe's astronomical legacy, both in Paris and Leiden. From the Budapest program, he learned about Huib's presentation on the use of the telescope in the Dutch Republic, and approached him with the question whether he could tell him something more about Leiden Observatory in the seventeenth century. Huib was very surprised by this question. Why on earth would an astronomer from Brazil be interested in specifically that observatory, at this early period of its existence? Like most Dutch citizens, Huib was then barely aware of the fact that a Dutch colony had once existed in Northern Brazil. Oscar was able to quickly fill that gap in Huib's knowledge, and concisely told what he knew about Georg Marggrafe's pioneering astronomical work in seventeenth-century colonial Dutch Brazil. It was a very inspiring meeting, which led to both researchers joining forces. Oscar's archival research could be supplemented with some additional research by Huib. This led to joint presentations, first, in September 2010, in Recife, Brazil, during a symposium celebrating 'Georg Marcgrave's fourth centennial' (*see photo*), and in October 2012, during the XXXI Symposium of the Scientific Instrument Commission in Rio de Janeiro, Brazil.

In the proceedings of this last symposium a joint paper was published, but long before that the idea had already arisen to extend this piece into a larger biography of Georg Marggrafe in the English language; this to supplement Oscar's own book on Marggrafe's astronomical work, published in the Portuguese language in Recife in 2011.

Busy work on both sides hampered any progress for a long time, which in hindsight was actually a fortunate circumstance, as it allowed the study of a previously unknown second Marggrafe manuscript in Lisbon, as well as a series of Marggrafe draft letters, which in 2017 – by a happy coincidence – were found again in the Leiden City Archives (ELO), after being missing for more than a century. The ongoing digitalisation of libraries and archives – such as the fully digitized archive of the Dutch West India Company at the Dutch National Archive in The Hague – has further largely expanded the possibilities of research into Marggrafe's life. This book therefore presents much more information than would have been possible a decade ago. The long time of preparation and research for this book implies that many people in libraries, archives and other institutions have contributed to this project; far too many to mention here. We would like to make an exception for the following persons, who each contributed in their own way to the completion of this book.

On Huib's account, he desires to honour Luís Tirapicos, who alerted him to the existence of Marggrafe's Lisbon Manuscript, and Ingrid Pot-Noordman, librarian of *Erfgoed Leiden en Omstreken*, who, after the unexpectedly trouvaille of Marggrafe's long-lost draft letters, remembered that we had searched for these seven years earlier, and was keen to inform us of this recovered treasure. Unfortunately, efforts to recover a missing Marggrafe manuscript in Spain, among others through the network of Marta Lourenço, have yielded nothing. Nevertheless, she is greatly thanked for her efforts. Further, Huib pays gratitude to his colleague and friend Ilja Nieuwland, who, together his spouse Marieke van der Duin, has constantly listened, stimulated and helped him; to Marika Keblusek, who inspired the writing process with perseverance and enthusiasm; to Albert van Helden, who was always ready to answer questions; to Peter Louwman, who in word and deed has supported Huib's historical research of astronomy for decades; to Milo van de Pol and David Baneke, who both commented meticulously on the manuscript, and to Douglas Anderson, who, as a native speaker, patiently corrected our English.

Oscar wishes to thank Christoph Ostendorf, director of the Brazil-Germany Cultural Center in Recife, for providing him with key contacts, enabling him to organize in Brazil the international meetings promoting Marggrafe's scientific work. Without those meetings, this book would never have come into existence; to Carlos Ziller Camenietski, then head of the Research Department of the Museum of Astronomy and Related Sciences (MAST/MCTI) in Rio de Janeiro, for providing xerographic copies of Marggrafe's manuscripts deposited at the Paris Observatory; to D. José Rodrigues Leandro da Costa, OSB (in memoriam), of the Monastery of São Bento in São Paulo, for reviewing the preliminary translation of Marggrafe's Latin manuscripts; to André Luiz da Silva, at the time intern at the Planetarium of Ibirapuera, São Paulo, for the positional identification of the Southern stars observed by Marggrafe; to the architect Filipe Jacopucci dos Reis for the suggestion to use SketchUp for the 3D modeling of Marggrafe's observatory with its astronomical instruments and to Companhia Editora de Pernambuco (CEPE) for the publication of *O observatório no telhado,* celebrating the 400th anniversary of Marggrafe's birth, the forerunner of this book.

In addition to these dear supporters, over the years there have been many unnamed individuals, who also deserve the authors credit. We are of course also very grateful to them.

Leiden / São Paulo, May 2022

HUIB ZUIDERVAART / OSCAR MATSUURA

PART I

CONTEXT

CHAPTER 1

INTRODUCTION

A FORGOTTEN ASTRONOMER

Maybe it's still there ... the seventeenth-century astronomical manuscript that gave cause to this book. Maybe it is still there ..., somewhere in Spain. Perhaps in the attic of a house from one of the descendants of the Spanish admiral ANTONIO DE ULLOA. After all, he was the last known owner, a man who after his death in 1795 left a huge load of books and documents. The manuscript was last seen when DE ULLOA bought it around 1770, so will it ever be found again? Fortunately, in the eighteenth century two accurate copies of this extremely interesting manuscript were made. These are now preserved and cherished, one in a Paris library, and the other (partly preserved) in Lisbon.[1] They contain nothing more or less than the astronomical legacy of GEORG MARGGRAFE (1610-1643), the first person who attempted to map the heavenly bodies in the southern sky in a systematic way. He did this, using contemporary state-of-the-art instruments on an European-style astronomical observatory, built specifically for that purpose.

MARGGRAFE had the ambition to become famous in astronomy, and he indeed became famous. But he achieved his fame in a scholarly field he originally did not seek. GEORG MARGGRAFE (erroneously mostly named *Marcgraf, Markgraf, or Marcgrave*), worked more than five years in colonial Dutch Brazil; to be exact from the spring of 1638 to mid-August 1643. Posthumously he obtained many credits for his important and innovative contributions to the *Historia Naturalis Brasiliae* (1648). This book was the influential first account of Brazil's zoology and botany, in which also an early form of ethnography is presented. It became the standard work for knowledge about Brazilian nature for more than two centuries. In it MARGGRAFE also presented the first daily meteorological records made in Brazil, as well as one of the first microscopical surveys of insects made in the Dutch scholarly community. MARGGRAFE is also hailed for his construction of four maps of Dutch Brazil, which are regarded as one of the most accurate and elegant products of seventeenth century Dutch cartography.

But MARGGRAFE's first and foremost desire was to become a famous astronomer. He came to Brazil with the firm intention to chart the Southern Hemisphere in a similar systematic way to what TYCHO BRAHE (1546-1601) had done for the northern part of the heavens at the end of the sixteenth century. BRAHE was a Danish nobleman, who in the years 1576-1596 had founded and used two astronomical observatories (*Uraniborg* and *Stjerneborg*) on the Island Hven in the strait Øresund, today situated between Denmark and Sweden. There he developed several new measuring instruments for the naked eye, including a large quadrant and sextants with aligning sights that enabled him to measure stellar positions with an accuracy of one arc minute. With these instruments TYCHO systematically mapped around a thousand stars in the Northern Hemisphere for the first time. His *Astronomiae Instauratae Progymnasmata* with his star catalogue, including an elaborate description of his observatories and instruments, was published in the years 1598-1602. The title that MARGGRAFE had outlined for his intended – but never finalized – book about the southern stars was *Progymnastica Mathematica Americana*.[2] This title was evidently inspired by TYCHO BRAHE's monumental book, which also showed that he wanted to follow in TYCHO's footsteps.

To fulfil his dream, the governor of Dutch Brazil, JOHAN MAURITS VAN NASSAU-SIEGEN, enabled MARGGRAFE to build on top of his residence the first European-style observatory on the American continents. There, on the island of Antonio Vaz in present-day Recife, the capital of the Brazilian province of Pernambuco, MARGGRAFE pursued an ambitious astronomical observation program. However, this program was interrupted several times when he was assigned to other tasks, for instance when he was sent on an expedition to the inland of Brazil, where he devoted himself to botanical, zoological, cartographic and ethnographic researches. In August 1643, shortly before he would return to the Netherlands, MARGGRAFE received the order to map the recently conquered Dutch territories in Africa (Sao Thomé and Angola), where, unfortunately, he fell ill and died a few weeks later, aged only thirty-three.

The aim of this book

If Marggrafe had succeeded in completing his observation program and published its results, he could have become the southern Tycho Brahe, following in the footsteps of his great role model. Unfortunately, due to bad luck, virtually all the observations he completed before his sudden death have remained unpublished to this very day. Therefore, the aim of this book is as follows:

(1) First, we provide some context by presenting a brief outline of the current historiography of seventeenth-century scholarship in a Dutch context and an introduction to colonial Dutch Brazil during Marggrafe's stay.

(2) Then we give a new biography of Georg Marggrafe, based on the current literature, supplemented by various previously unpublished archival finds. We also pay extensive attention to the way in which his scholarly legacy with regard to his findings on Brazilian botany, zoology, ethnography, meteorology, cartography and astronomy was processed after his death.

(3) Further, we give a description and virtual reconstruction of Marggrafe's astronomical observatory and the equipment he used, followed by an analysis of his astronomical achievements, both during his training in Leiden (1637), as well as during his research in Recife (1638-1643).

(4) Then we use Marggrafe's case to investigate questions with regard to knowledge formation, organisation and transfer in the Early Modern Period. His case, touching divergent fields like astronomy, meteorology, natural history and cartography, is very interesting, because his work predates the time in which the institutional foundations were laid for the development of modern science. In his day, a scientific infrastructure was still lacking. After all, scholarly societies and scholarly journals would emerge only in the 1660s. For the dissemination of new knowledge, a scholar had to rely on newly published books or on written correspondence distributed through the informal channels of the 'Republic of Letters'. And, of course, in order to be able to do any scholarly work, one needed time and funds. So in most cases a form of financial support – royal or otherwise – was essential. So, what exactly made Marggrafe's work in Brazil possible? Was it only the support of a semi-royal court that enabled him to pursue his scientific curiosity, or were other factors also relevant? For instance, what was the role of the West

Indian Company, a trading company active in the whole Atlantic area?

(5) Finally, in a number of appendices, we present a new chronology of MARGGRAFE's life, an overview of the horoscopes he made for various persons, and a correction of some false attributions made in the past. And we finally present in a supplemented (digitally published) Volume 2, MARGGRAFE's astronomical observations, both in the original Latin text and in English translation. In this translation each observed star is identified according to modern astronomical nomenclature. The Latin text was already prepared for the printing press by – or under the supervision of – the Leiden professor of astronomy and oriental languages JACOB GOLIUS. Due to the sudden death, in 1667 and 1668 respectively, of both this professor and his assistant SAMUEL KECHEL, its publication was unfortunately prevented.

A publication postponed by more than three centuries

Thanks to GEORG's youngest brother CHRISTIAN MARGGRAFE, it is known what GOLIUS intended to do with the rare Brazilian observations he had received. In a letter to the astronomer JOHANNES HEVELIUS in Danzig, CHRISTIAN wrote in July 1652:

> At last, I saw the astronomical observations of my brother GEORGE. They concern a new theory of the planets, especially Mercury, which in the place where he lived, could be better seen than with us. These [observations] will be published shortly by the honorable GOLIUS, together with other astronomical observations transmitted [to him] from Arabia.[3]

So, strangely enough, GOLIUS planned to publish MARGGRAFE's observations together with another astronomical text he was working on. This concerned the Latin translation of a manuscript by the Persian astronomer AHMAD IBN MUHAMMAD IBN KATHĪR AL-FARGHĀNĪ, a man who lived in the ninth (!) century. GOLIUS had obtained this manuscript during a tour through the Middle East in the 1620s, and desired to publish it in Latin, together with elaborate annotations. GOLIUS's unfortunate decision to combine both astronomical texts not only delayed the publication of MARGGRAFE's observations, but in the end also forestalled it.

When GOLIUS suddenly died in 1667, his work on AL-FARGHĀNĪ was not yet published, and the unknown person who eventually led the book through the printing press understandably decided *not* to include contemporary observations from Brazil in a book on ninth-century Persian astronomy.[4] As a consequence, MARGGRAFE's astronomical work never became known to his contemporaries. Although he had been the first to map the southern sky with state-of-the-art instruments, *nota bene* from a newly built European-style astronomical observatory, this made in fact his efforts worthless. Had MARGGRAFE's observations been published in the 1650s, they would have greatly increased his fame. He would have been remembered as an important astronomer who had contributed substantially to the knowledge of the Southern Hemisphere. But as it turned out, his gathering of astronomical data did not contribute to the contemporary corpus of astronomical knowledge.

So, why do we now want to finish this publishing project that got stuck in the middle of the seventeenth century? Here, two reasons are relevant. Firstly, MARGGRAFE's astronomical observations represent the earliest systematic research of the southern sky performed with state-of-the-art scientific instruments, on a Tychonic observatory especially constructed for that purpose. As such, this corpus of early modern observations deserves to be known to the present-day astronomical community. Further, this concerns a unique document that testifies about early modern science in action in a colonial setting. It shows in detail how science was done in these early days of scientific endeavour. Reasons why – in our view – this manuscript that was made ready for the printing press in the late 1650s still deserves to be published.

MARGGRAFE's surviving astronomical documents in Leiden and Paris were first touched upon in 1979 by the historian of astronomy JOHN NORTH (1934–2008).[5] NORTH rejected the exaggerated thesis of EUGENE GUDGER (1866-1956), one of MARGGRAFE's early biographers, that, had MARGGRAFE survived, he would have developed himself into 'possibly the greatest natural historian since ARISTOTLE'.[6] But having said that, NORTH, too, concluded that indeed it had been MARGGRAFE's ambition to become a second TYCHO BRAHE, but that by an unfortunate twist of fate 'his astronomical efforts were to be without significant consequence for the advancement of the subject'.[7] NORTH also concluded that MARGGRAFE 'simply did not live long enough to accumulate or assemble the materials for the promised book'.[8]

So, what results did MARGGRAFE want to deliver? The answer to this question is provided by his editor JOHANNES DE LAET, who wrote – based on documents he had found in MARGGRAFE's chest – that MARGGRAFE intended to include the following in his astronomical book.

I. The first section would be on astronomy and optics. It would contain:
   i. An overview of all the southern stars found between the Tropic of Cancer and the Antarctic Pole;
   ii. Observations of all the planets and of eclipses of the sun and moon, 'worked out in an original way';
   iii. New and true theories of the inferior planets Venus and Mercury, based on his own observations;
   iv. A theory of astronomical refraction and parallax;
   v. A new figure for the obliquity of the ecliptic;
   vi. Accidental data, not only on sunspots, but also on other astronomical rarities.
II. The second part would be geographic and geodetic in nature. It would discuss:
   i. A theory of the longitude of the earth and a manner of computing the same;
   ii. A demonstration of the true dimensions of the earth, based on his own observations;
   iii. A discussion of previous geographical errors.
III. The third section would be based on the two preceding ones and would contain the *Tabulae Mauritii Astronomicae*.[9]

The name of the tables in this last section was again a reference to MARGGRAFE's ambition to parallel TYCHO BRAHE's work for the Southern Hemisphere. In 1627 BRAHE's assistant in Prague, JOHANNES KEPLER, published the so-called *Tabulae Rudolphinae*, a catalogue of observational data collected by BRAHE, listing the positions of stars and planets, named in memory of RUDOLPH II, the Holy Roman Emperor in whose service they both had worked. GEORG MARGGRAFE evidently had a similar role in mind for his patron and employer, the governor of Dutch Brazil, Count JOHAN MAURITS VAN NASSAU-SIEGEN.

Without running ahead of the details, it can already be said that NORTH was correct in his remark that MARGGRAFE's ambitious program had not been completed in its entirety. However, as we will demonstrate in part III of this book, a large part of MARGGRAFE's ambitions formulated in the first section have been fulfilled. In his paper, NORTH gave an excellent first review of MARGGRAFE's astronomical achievements. But unfortunately, NORTH also misinterpreted a few things. One of these mistakes was that NORTH thought that MARGGRAFE's observatory, which was built in 1639 on top of the first residence of JOHAN MAURITS, was relocated to one of the two towers of the count's new residence *Vrijburg*, built during the years 1641-1642. This erroneous supposition is repeated as a fact in numerous publications. In 1984 the Brazilian amateur astronomer and teacher of astronomy JORGE POLMAN already argued that NORTH's analysis was wrong and that the observatory stayed on top of JOHAN MAURITS's first house.[10] NORTH also thought that MARGGRAFE's prime inspiration for his observatory was the description of TYCHO BRAHE's *Uraniborg*, as published in the *Astronomiae Instauratae*. However, as we have demonstrated in an earlier article, in Brazil, MARGGRAFE almost exactly copied the site, as well as the equipment and working methods, of Leiden observatory, the place where he had been trained in practical astronomy.[11]

A more thorough assessment of MARGGRAFE's astronomical achievements was published in 2011, by one of the present authors – OSCAR MATSUURA – but as this book was written in Portuguese, it did not receive the attention it deserved outside the domain of this language.[12] Our joint publication, published in Brazil in 2014, also touched only a small part of what deserves a deeper attention. Our promise made in that article to return to MARGGRAFE's life and work more in detail is hereby fulfilled.

THE CORRECT SPELLING OF GEORG MARGGRAFE'S SURNAME

Over the years, the spelling of MARGGRAFE's surname has been extremely polymorphic. In 1648, JOHANNES DE LAET, the editor of the *Historia Naturalis Brasiliae*, used the Latin spelling 'Marcgravi' and 'Marcgravius', reason why, in 1912, his biographer the fish biologist EUGENE GUDGER used the German 'Marcgrave'; THEO MEIJER (1972) wrote 'Marggraf' and JOHN NORTH (1979) noted 'Markgraf'. But the same year, when there still was no autograph available, PETER WHITEHEAD introduced the spelling 'Marcgraf', arguing that the German family name would have been written in this way, overlooking the fact that in the birth register of Liebstadt, published by

VICTOR HANTZSCH in 1896, the surname was spelled as 'Marggravij' and that the first biography, published in 1685 by Marggrafe's youngest brother, was titled 'Vitae Georgii Marggravii' (both with the letter 'g' twice).[13]

However, since 1979 four original signatures have been found. Two are located on draft letters in the Leiden archive (*fig. 2a*).[14] These letters were missing for decades, but luckily they were retrieved in 2017. A third signature can be found on the title page of book in the Universitäts- und Landesbibliothek of Bonn,[15] and a fourth is shown in an *Album Amicorum* preserved in the Universitäts- und Landesbibliothek of Sachsen–Anhalt in Halle.[16] All these documents show the same spelling 'MARGGRAFE'. This way of writing is confirmed by the spelling on MARGGRAFE's Wittenberg disputation (1634 – *fig. 2b*) as well as below three Latin poems by MARGGRAFE, two made in Stettin for the astronomer LORENZ EICHSTÄDT (1634; 1636), and one made in Rostock for a fellow student (*fig. 2c*). Finally, MARGGRAFE's only surviving manuscript map of the coast of Brazil (1643) is signed 'Georgius Marggrafius', the Latin way of writing his name, also spelled with the letter 'g' twice. This is the only time MARGGRAFE himself used a Latin spelling for his name, indicating an increased self-awareness. After all, the Latinization of a name stands for the identification of the person concerned with the milieu of the literate, if not the scholars. On his academic Wittenberg thesis he did not yet dare to take that step.

In conclusion, we therefore restore in our book the correct German spelling 'MARGGRAFE', despite the fact that most modern literature has followed WHITEHEAD in his incorrect assumption of writing 'Marcgraf'.[17]

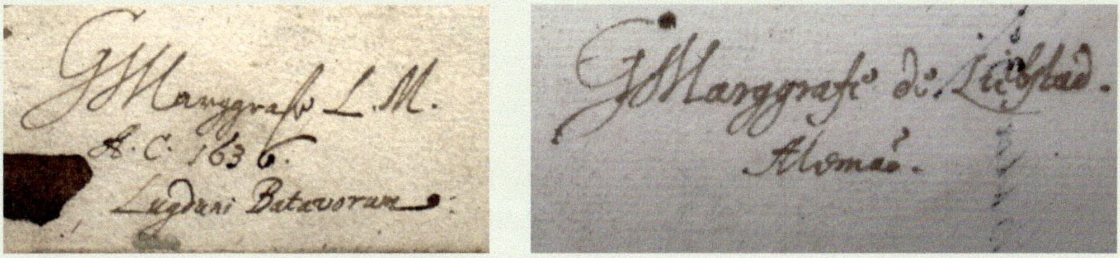

FIG. 2a. Two examples of MARGGRAFE's signature (Erfgoed Leiden en Omstreken (ELO)

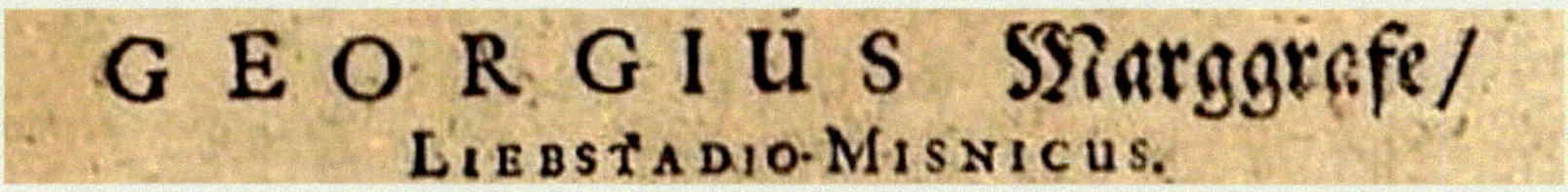

FIG. 2b. The spelling of the name on his 1634 Wittenberg disputation (Universitäts- und Landesbibliothek Sachsen-Anhalt, Halle)

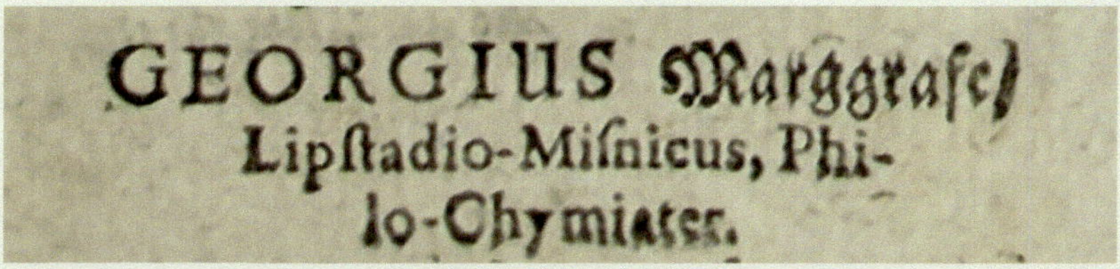

FIG. 2c. MARGGRAFE proudly presenting himself with the academic degree of *Philo-Chymiater* ('a practitioner of alchemical medicine', or Medical Candidate) obtained at Wittenberg University. Signature below a poem in a Rostock collection of verses for the graduation of a fellow student (1635; Sächsische Landesbibliothek und Universitätsbibliothek Dresden).

CHAPTER 2

CIRCULATION OF KNOWLEDGE IN EARLY MODERN DUTCH BRAZIL?

The Dutch trading companies and the emergence of a pre-modern knowledge society

Dutch expansion in the seventeenth century was primarily driven by the Dutch trading companies. The two largest were the *Verenigde Oostindische Compagnie* (VOC), established in 1602 for the trade in the Far East, and the *Geoctroyeerde West–Indische Compagnie* (at the time abbreviated as GWC, but nowadays usually referred to as WIC), founded in 1621, aiming at the Atlantic area. Both companies were governed by a group of merchants called *bewindhebbers* ('directors'). The VOC had seventeen of these officials and the WIC nineteen, which governing bodies were called respectively the *Heeren XVII* ('Lords Seventeen') for the VOC and *Heeren XIX* ('Lords Nineteen') for the WIC. Officially both trading companies were private ventures, but both the VOC and WIC received an official charter granted by the States General of the Dutch Republic. This charter made these private corporations hybrid entities because through this charter their aim was both economical and political. On the one hand, they were companies engaged in trade and transport, but on the other, as semi-governmental entities, they acted abroad as colonial governments, ruling foreign territories as sovereigns, employing not only merchants and sailors, but also soldiers and engineers, who used guns, built fortifications and staffed warships.

Both companies were organized into so-called chambers, representing shareholders from different merchant centres in the Dutch Republic. In this way the *Geoctroyeerde Westindische Compagnie* (WIC) had five chambers: (1) the *Amsterdam* chamber represented, besides Amsterdam, also shareholders from Dutch cities such as Leiden, Haarlem, and Deventer; (2) the *Zeeland* chamber consisted of Middelburg, Vlissingen (Flushing), Veere, Tholen, and Zierkzee; (3) the *De Maze* chamber included Rotterdam, Delft and Dordrecht; (4) the *Noorder-*

kwartier chamber combined the merchant cities in West Friesland, namely Hoorn, Enkhuizen, Alkmaar, Edam, Medemblik and Monnickendam, and finally (5) the *Stad en Lande* chamber represented the city of Groningen with the so called 'Ommelanden' (the surrounding countryside). Each chamber brought its own ships, equipment and soldiers into the collective enterprise.

The WIC was established with multiple goals.[1] Firstly, its founding was prompted by the commercial success of the VOC, which company traded with various places in Asia. The WIC was looking for similar benefits in the Atlantic, but last but not least, the WIC was also established as an instrument of war. In 1621 the *Twelve Years' Truce* had come to an end, during which period the Dutch war of independence with Spain (which had started in 1568) was interrupted between 1609 and 1621. Now that the war revived, it became an important strategy of the Dutch Republic to undermine the economic base of the Spanish war efforts as much as possible. Because at that time Portugal was still part of the Iberian empire, this strategy implied that all Portuguese settlements in the Atlantic were also regarded as enemy territory. The WIC therefore mainly focused on the conquest of Spanish-Portuguese colonies in South America.

To fulfil their wide range of tasks, the two trading companies were permanently busy gathering information: intelligence useful for commerce, but also for war and the consolidation of their colonial power. With this commercial and military information went a broader range of knowledge, which was transported from all parts of the globe to the homeland. This knowledge was very diverse. Sometimes it was very tangible, when it concerned goods from the exploited countries, but often the knowledge was more elusive, when it comprised intelligence about their geography, their nature, their people, their habits, etcetera. Maintaining large-scale trade flows and establishing colonial footholds in foreign countries which were essentially different in nature from the homeland was not an easy task. It required large scale organisation and cooperation. Or, to quote Arndt Brendecke and Susanne Friedrich, 'the prerequisite for this was that in the wake of the trading ships, a communication network was built up which connected places on opposite sides of the world with one another'.[2] It is not without reason that these Dutch trading companies have been presented as knowledge networks, as well as the world's first multinational corporations.[3]

This gathering of knowledge within an organized commercial setting prompted Harald Cook to state in his *Matters of Exchange*, a book with the subtitle *Commerce, Medicine, and Science in the Dutch Golden Age* (2007), that seventeenth-century Dutch commerce served as a chief impetus for the rise of modern science. According to Cook, the merchant class of the Dutch Republic was very pragmatic in the way in which they did business. They were primarily interested in merchandise that could be gathered, transported, used and sold, and this no-nonsense approach was also applied to the corresponding facts. In that respect, Cook made a distinction between '*kennen*' ('to know how), as opposed to '*weten*' ('to understand why'). If one knew how things presented themselves to the senses, one did not need to bother to understand the inner nature of things. It was important to know the facts, not so much the causes. According to Cook this resulted in a notion of 'objective' knowledge (in the sense that knowledge could be related to objects). By definition, this kind of commercially driven knowledge was not speculative, which way of looking to things eventually became an important ingredient of modern science. As the VOC and WIC were the most important Dutch commercial organisations of the period, these two large trading companies played a major role in the creation of this 'way of seeing and knowing'. Although Cook discussed only the fields of botany and medicine, he argued that this 'way of seeing and knowing' could also be applied to other 'scientific' fields of knowledge.[4]

Critics, however, argued that Cook 'perhaps overplayed practicality as the driving force behind the desire to know and perhaps underestimated the role of wonder or curiosity'.[5] To apply Cook's hypothesis to other fields of scholarship, such as mathematics, physics, optics, alchemy and astronomy was considered too easy. 'Changes in mathematics and cosmology, and even experimental method, could be considered to have far more significant motors for change than trade', Steven Paul Matthews argued.[6] And what about the role of religion? Wasn't it already long ago that the 'Book of Nature' was identified as one of the – theologically inspired – driving forces behind seventeenth-century scientific curiosity?[7] Neglected too in Cook's approach seemed the theme of 'princely patronage and the rules of gentlemanly discourse', discussed in earlier literature as being important for the emergence of modern scientific practices.[8] And what about the role of an academic education? Klaas van Berkel, for instance, noted that Cook 'almost completely ignores the importance of university scholarship'. Wasn't it precisely in the seventeenth-century Dutch Republic that academic culture flourished? Many merchants received a university education; wouldn't this training have shaped at least partly their way of looking at the world?[9]

Given Cook's hypothesis and the counter-arguments and questions of his critics, it is interesting to consider the case of Georg Marggrafe, a scholar working in medicine, botany and zoology (the fields discussed by Cook), but also excelling in subjects noted by Cook's critics, such as alchemy, mathematics, astronomy, meteorology, geography and cartography. Moreover, Marggrafe was educated at several universities, and was employed by a princely patron in the service of the *Geoctroyeerde West-Indische Compagnie.* So, all the elements mentioned above come together here in one and the same person. This triggers a number of questions. Do we recognize in Marggrafe's work a 'way of seeing and knowing' as described by Cook? Can we distinguish which elements in Marggrafe's methods and working conditions contributed the most to his scholarly achievements? What do we know about the networks he was involved in? And, as he died prematurely, how were his notes collected, processed and distributed? And finally, how was the knowledge acquired by him disseminated (or not) in the contemporary scholarly communities of Europe? In short, what were the parameters that led to the success and/or failure of Marggrafe's efforts to increase knowledge in the fields in which he was engaged?

Dutch Brazil and its first and only governor general

The question 'How did Marggrafe end up in Dutch Brazil?' can only be answered if first it is explained why the Dutch established a colony there in the first place. At the time Brazil was seen as a profitable country, mostly because of its very lucrative sugar industry. Conquering Spanish-Portuguese territory in Brazil could be a double-edged sword for the Dutch. On the one hand, this conquest would harm the Iberian war effort, but at the same time it could be lucrative for the WIC to take over the revenues of the Brazilian sugar business. After some skirmishes and unsuccessful attempts to establish a bridgehead in Brazil in the years after 1624, a fleet of 67 Dutch ships, carrying 7,000 enlisted men and 1,170 cannons, managed to capture the villages of Recife and Olinda in Pernamuco in Northern Brazil in 1630 (*fig. 3 & fig. 4*).

The *moradores,* the Portuguese inhabitants of the region, did not welcome this Dutch occupation and in the first years resistance from

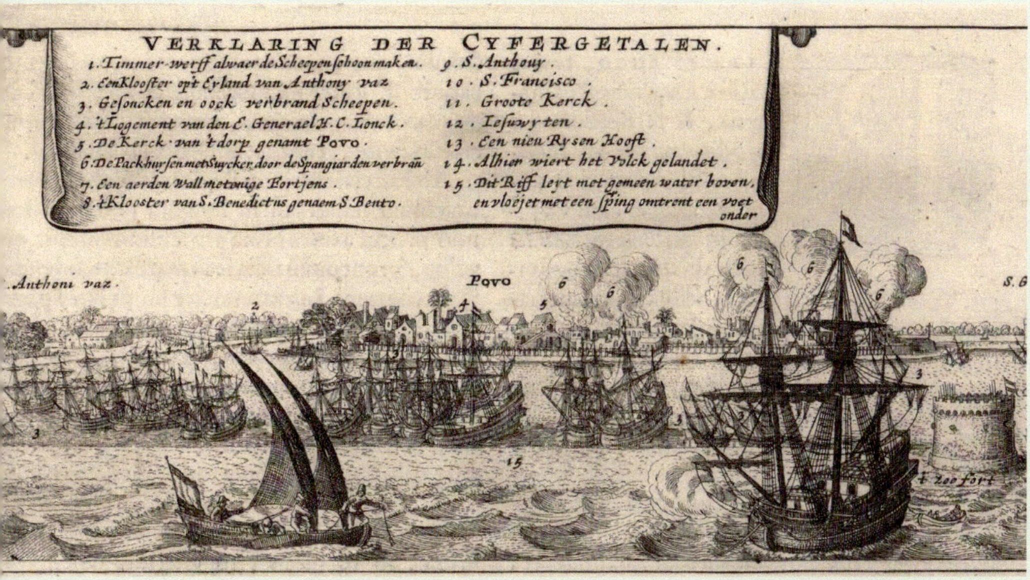

FIG. 3. The island of Antonio Vaz and the village of Recife (here called Povo, after Povoação, village) in 1630, after the conquest by admiral Hendrick Corneliszoon Lonck. Fragment of an engraving titled 'Olinda De Phernambuco', from *Civitates orbis terrarum. Illustriorum Hispaniae urbium tabulae* (Amsterdam: J.Janssonius, 1657). (Rijksmuseum, Amsterdam).

FIG. 4. Cannon of the *Geoctroyeerde West-Indische Compagnie* with its monogram GWC, with the letter 'Z' for the chamber Zeeland. Cast by MICHAEL BURGERHUYS at Middelburg in 1629. This is one of the 1,170 cannons used in the conquest of Pernambuco in 1630. Museu Historico Nacional, Rio de Janeiro. (Photos by the author, 2014).

guerrilla groups was very fierce. But in the mid-1630s it became clear that the Dutch occupation of Pernambuco and its surrounding area would become permanent. At that time the WIC had set up a system of colonial government in Recife, led by a *Politieke Raad* (Political Council) with representatives from the five chambers of the WIC. The voting power in this council was the same as with the *Heeren XIX*, the central administrative body of the WIC in the Netherlands.[10] But the local administrators in Recife fell short in establishing a stable and safe society. Stocks were mismanaged, corruption was widespread, and the productivity of the sugar mills dropped dramatically, with the result that by 1636 the debts of the WIC amounted to eighteen million guilders. The need for a more vigorous approach became essential. In July 1636, in a meeting of the *Heeren XIX*, a new governance model for the colony was proposed. The idea came forward that a strong governor would be desirable, not only as commander of the military, but also with a powerful say in the civilian administration of the colony. A man was needed with a social standing and background that would be acceptable for the two rivalling military commanders in Brazil, the Polish CHRISTOFFEL ARCISZEWKSI and the German SIGISMUND VON SCHOPPE.

After consultation with the States General of the Dutch Republic, the choice for this position fell to the German Count JOHANN MORITZ VON NASSAU-SIEGEN (*fig. 5*), better known by his Dutch name JOHAN MAURITS VAN NASSAU-SIEGEN (1604-1679). He was the eldest son from the second marriage of JOHANN VII VON NASSAU-SIEGEN (1561-1623), who from an earlier marriage already had twelve children. (His second marriage would produce thirteen more!) At first glance, it seemed a remarkable choice. Although JOHAN MAURITS was a distant relative of both Dutch stadtholders,[11] at that moment he was just one of the many younger sons of a German noble family, serving as an officer in the army of the States General. However, the count, who was in his mid-thirties, was very ambitious. As co-commander alongside his older half-brother WILHELM VON NASSAU-SIEGEN, he had received a lot of attention earlier that year for his prominent role in recapturing the fortress *Schenkenschans* (near Cleve) from a recent Spanish occupation. But more importantly, JOHAN MAURITS had invested in building a good relationship with the most important stakeholders within the stadtholder's court in The Hague since the late 1620s. One of these was CONSTANTIJN HUYGENS, the very influential secretary to the stadtholder, who would soon become his neighbour.

Already in 1633 both HUYGENS and JOHAN MAURITS had bought a building plot at the *Plein*, ('Square') in The Hague, next to the stadtholder's quarters in The Hague. In the next years they would both build a classicist mansion here, using the same architect: JACOB VAN CAMPEN, assisted by PIETER POST. On one side of the *Plein* the *Huygenshuis* was erected (finished in 1637) and on the other, the much more luxurious *Mauritshuis* (finished in 1644; *fig. 6*). ERNST BRINCK, a magistrate from Harderwijk, heard in 1638, during a walk in The Hague, the rumour that the construction of the *Mauritshuis* would cost at least two hundred thousand guilders, twice as much as the Huygens mansion.[12] It was only thanks to his enormous income from Brazil that the German count was able to finance its construction. No wonder that people mockingly called the building *Het Suikerpaleis* ('Sugar Palace'), referring to the count's lucrative trade in Brazilian sugar.

So, JOHAN MAURITS obtained his position as governor-general of Dutch Brazil, mostly thanks to his close relations to the stadtholder's court. After all, the WIC was to a large extent part of the Dutch war effort in its struggle to maintain its independence from Spain. So, a close relationship with the commander-in-chief of the Dutch army and navy, *viz* stadtholder FREDERIK HENDRIK, prince of Orange,

FIG. 5. JOHAN MAURITS VAN NASSAU-SIEGEN (1604-1679), as depicted in 1658. Engraving by CORNELIS II VAN DALEN after a painting by GOVERT FLINCK. At the bottom of the oval his motto *Qua patet orbis* ('Wherever the world will bring me') (Rijksmuseum, Amsterdam).

FIG. 6. The *Huygenshuis* (left) and the *Mauritshuis* (right) in The Hague, as depicted in 1690. Detail of a watercolour by JAN VAN CALL (Gemeentearchief, Den Haag, sign. kl. A 666).

was seen as very useful. It was the Amsterdam WIC director ALBERT COENRAADS BURGH (1593-1647) who proposed JOHAN MAURITS for the job, and who, on behalf of the *Heeren XIX*, led the negotiations of his contract which was signed at the beginning of August 1636. The agreement gave the count powers in Brazil, even greater than those of the stadtholders in the Dutch Republic. As governor general of Dutch Brazil (or 'New Holland' as it was called at the time), JOHAN MAURITS would not only be the commander-in-chief of the army and the navy, but he would also be president of the newly established *Hooghe en Secrete Raad* (High and Secret Council) in charge of the civil administration of the colony. In this council, he would be assisted by three *Raden* (members), all former *bewindhebbers* of the WIC. In the event of a tie vote, JOHAN MAURITS's decision would be decisive. Further, the *Politieke Raad* received a new role as the court of appeal for criminal and civil justice and as an organ to administer the finances of the colony.[13]

Within a month, a contract for five years was agreed. For his service to the West Indian Company, JOHAN MAURITS would receive a monthly payment of 1,500 guilders. In addition, he received a one-off amount of 6,000 guilders and a share of 2% of all goods that were confiscated as being captured from the enemy. A small retinue, including a personal pastor, secretary, and physician, was also granted by the WIC. Finally, in Brazil the count and his retinue would enjoy what was called 'a free table'.[14] This provision meant that the meals of his court would be paid for by the WIC. In practice, this provision would become extremely costly for the company. In 1643, when this part of the contract of the WIC was limited by order of *Heeren XIX*, the costs of 'a free table' had risen to 7,000 guilders a month, almost five times the governor's regular salary!

Johan Maurits's achievements in Brazil

The fleet of twelve vessels that would bring the new governor to Brazil, together with a contingent of 2,700 soldiers, left the Netherlands at the end of October 1636. Arriving in Recife, on 23 January 1637, Johan Maurits proceeded very energetically. Within a month, he had defeated the enemy at Porto Calvo (170 km south of Recife), with the result that from then on the Rio São Francisco formed a natural boundary between the Dutch and Portuguese territories (*fig.* 7). At the banks of this river a new fortress, 'humbly' named *Mauritius*, was established to guard this new border. With this result, the military achievements were not over yet. By the end of Johan Maurits's first year in office, Dutch troops had occupied the captaincies of Ceará to the north and Sergipe to the south. The Dutch thus controlled virtually the whole of north-east Brazil. Overseas, also a new conquest was gained. At the African coast the Portuguese fortress São Jorge da Mina (the centre for slave trade in present-day Ghana), was conquered in August 1637. With this, the essential supply of new slaves for the Brazilian sugar industry was assured.

With regard to the civil administration, the count also worked energetically. He tackled corruption and dismissed untrustworthy officials, replacing them by more reliable persons.[15] Crucial, too, was the revival of sugar production. Therefore, all *moradores* who were willing to cooperate with the Dutch government were restored in the ownership of their *engenhos* (sugar mills) as well as their civil rights. Sugar mills that were abandoned were sold to new owners, Dutch or otherwise. With success. The sugar industry recovered well during Johan Maurits's reign. The total production of sugar in that period has been valued at 28 million guilders.[16] Hermann Wätjen, in his analysis of the impact of Dutch Brazil on the economy of the Dutch

Republic, concluded that during JOHAN MAURITS's time the trade between the Republic and Brazil must have been substantial, including in areas other than sugar. In retrospect, in the 1640s Brazil was an important factor in the economic life of the seven Dutch provinces.[17]

THE BACKGROUND OF A GOVERNOR GENERAL

From a cultural perspective, JOHAN MAURITS's government in Brazil is considered very remarkable. What happened in Dutch Brazil has been qualified as an unparalled cultural highpoint, unheard of in any seventeenth-century colonial setting. During the seven years of the count's stay in Recife, JOHAN MAURITS enabled and stimulated the work of

FIG. 7. A modern map of Dutch Brazil around 1640. Graphic design: © Richard Bos, Leeuwarden, first used in: Van den Boogaart, *Het Land van de Suikermolen* (2021).

Dutch painters such as Frans Post, Albert Eckhout, Abraham Willaerts and perhaps Gillis Peeters (*fig. 8*). These painters documented visually there (or afterwards in the Netherlands) a world that Europeans had never seen before.[18] But cultural life extended beyond the presence of painters. The ivory furniture Johan Maurits brought back from Brazil, and the presence in Recife of the silversmiths Jeremias Sporen and Andries Ketelaer, as well as the skilled glass engraver Pieter van Coninxloo demonstrate that the activity of such artisans was not an accident.[19] Locally in Dutch Brazil, there must have been a respectable market for luxury products, of which unfortunately, as far as we know, nothing has survived. This, in addition to the support that Johan Maurits gave to scholarly projects, such as the construction of an astronomical observatory, the creation of a botanical garden and a zoo, combined with his support for research into Brazil's natural history, leads to the question, 'Why did he do this?'

To understand Johan Maurits's role – and his self-fashioning – as a Maecenas for the arts and sciences in Brazil, it is necessary to have some insight into his upbringing.[20] He was born in 1604 in the ancestry castle of the Nassau family, castle Dillenburg in Hessen. At the age of ten, he joined two sons of Moritz, Landgraf of Hessen–Kassel, who went to the universities of Basle and Geneva. In 1616 they returned to Kassel. For the next three years, Johan Maurits studied at the *Collegium Mauritianum*, a genuine *Ritterakademie* for young noblemen, founded there by the landgrave, who was also called *Moritz der Gelehrte* ('Maurice the Learned'). This nickname already shows how much emphasis the landgrave – who was married to a half-sister of Johan Maurits – placed on learning and scholarship. At the *Collegium Mauritianum*, Johan Maurits received an education from the best professors available, recruited from an internationally oriented Calvinist network. In this way, the academy trained an elite both willing and able to serve their country beyond its territorial borders.[21] Students of this academy were among the first professionally trained diplomats of their time. But at this court in Kassel, Johan Maurits also experienced a humanist culture in which the arts and sciences were highly regarded as cultural capital. He witnessed there the support given to artisans, the existence of a botanical garden and the presence of an astronomical observatory founded by the landgrave's father Wilhelm IV von Hessen–Kassel, nicknamed *der Weise* ('the Wise'). These youthly impressions will have stimulated Johan Maurits in pursuing a similar culture at his own court in Brazil. Or, to quote Whitehead and Boeseman:

FIG. 8. Panorama of Recife, the island of Antonio Vaz and mainland Pernambuco in 1637, painted by GILLIS PEETERS from Antwerp (1612-1653). (Sotheby's, New York City, auction 19 May 1995, no. 31; Photo RKD).

When he left Brazil in May 1644, he had built for himself two imposing palaces, *Vrijburg* and the country house *Boa Vista*; he had laid out a new city, *Mauritsstad*; he had built the first astronomical observatory in both the New World and the Southern Hemisphere; he was the first European to establish a zoo and a botanical garden in the New World, or to have a museum there; and all the while his splendid residence in The Hague, the *Mauritshuis*, was being constructed, ready for his return. Broadly educated, of an illustrious family, and with a tremendous zest for the job on hand, JOHAN MAURITS comes close to the archetypal Renaissance Prince.[22]

A COLONY BUILT ON SLAVERY

This very positive picture of JOHAN MAURITS in most historical literature is for a large part the consequence of the self-fashioning that he pursued after his return from Brazil, especially by commissioning – if not paying for – the publication of three lavishly illustrated folio books celebrating his performance in the colony. The first of these was CASPAR BARLAEUS's *Rerum per octennium in Brasilia*, published by JOHANNES BLAEU in 1647, in which BARLAEUS described the eight years of the count's governorship in a way that borders on hagiography.[23] Although it is far from certain that JOHAN MAURITS financed this publication, one of his preserved letters shows that without shame he tried to influence the content of BARLAEUS's book.[24] The second book was the *Mauritiados Libri XII*, a long laudatory poem, written by the count's personal minister FRANCISCUS PLANTE, published in the fall of 1647 in an exuberant fashion by the Leiden publisher JOHANNES LE MAIRE, re-using the illustrations from BARLAEUS's *Rerum per octennium*.[25] The last publication that lauded JOHAN MAURITS's role in Brazil was the *Historia Naturalis Brasiliae* (1648), edited by the WIC *bewindhebber* JOHANNES DE LAET, in which publication the count was portrayed as an enlightened Maecenas of the natural sciences.[26] This last book was jointly issued by the Leiden publisher FRANCOIS HACKIUS and his Amsterdam colleague LODEWIJK ELZEVIER.

Yet it is not alone the self-fashioning that constructed the positive tone with which JOHAN MAURITS's rule is discussed in much of the historical literature. For it is also true that the rule of JOHAN MAURITS was also very positively assessed by contemporaries in Brazil, certainly in comparison with those of other Dutch administrators.[27] It is an established fact that his departure in 1644 was sincerely mourned by the whole colony or, as CHARLES BOXER put it, 'by Calvinist

Netherlanders, Catholic Portuguese, and cannibal Tapuyas, alike'.[28] More recent literature, however, is more critical, also emphasizing that much of the praised achievements were made at the expense of many enslaved Africans.[29] WÄTJEN has calculated that in the timespan of ten years between 1636 and 1645, approximately 23,000 slaves were imported into Dutch Brazil.[30] In 1641 the count's 'table' fed some eighty enslaved *negros ende negerinnen*.[31] A list of the *domestique negros* in the service of high ranking WIC officials shows that of these, 35 men, seven women and six children were in the count's personal service.[32] Enslaved men and women were everywhere. This same list demonstrates that most high-ranking persons used several enslaved Africans in their households or businesses. Whether it was a member of the *Politieke Raad*, an admiral, or just the head of the WIC's bakery, the use of enslaved *negros* was seen as essential. Even most Calvinist ministers owned their own enslaved assistants. According to the list named above, WILLEM PISO used five slaves in his hospital and two in the WIC pharmacy. Although MARGGRAFE, in his notes, never explicitly refers to the involvement of enslaved men or women, we have to remember that it is nevertheless irrefutable that MARGGRAFE's achievements were only possible thanks to the functioning of the Dutch colony as a whole, which could not have existed without these enslaved Africans.

A princely court in Dutch Brazil: Vrijburg and Boa Vista

As described above, in Brazil JOHAN MAURITS built for himself two palaces, the country estate *Vrijburg* and the fortress-mansion *Boa Vista*, both on the island of Antonio Vaz. It is a huge misunderstanding (which unfortunately is repeated everywhere) that MARGGRAFE's astronomical observatory was housed in one of the two towers of *Vrijburg*. This is absolutely wrong, as will be explained elsewhere in this book. The observatory was built on top of the count's first house, elsewhere on the island. On *Vrijburg*, one tower was used as a lookout for ships, and the other served as a lighthouse.[33] *Vrijburg* was nevertheless an important location for MARGGRAFE, because it was here that the count had laid out a garden, or, perhaps better said, a *hortus botanicus*, in which various species of Brazilian plants were growing. Specimens of these plants were described, drawn or dried, both by MARGGRAFE and his superior, the physician WILLEM PISO.

Their findings were later included in the *Historia Naturalis Brasiliae*, edited and published by De Laet in 1648. A zoo was also located here, or perhaps at the fortress-mansion *Boa Vista*, containing a wide range of animals. These, too, were studied, depicted, and illustrated. Woodcuts of these illustrations eventually found their way into the *Historia Naturalis Brasiliae.*

As locations do matter for an assessment of the knowledge gathered in these places, it is important to consider this garden and its role more in detail.[34] What exactly was created on the spot? Very informative in this respect is the ground plan of the estate, published in Barlaeus (*fig. 9*). It shows a garden with a strictly geometric layout around the main building of *Vrijburg*. The plants and trees are accommodated in geometric patterns. The legend accompanying the map mentions, apart from a pleasure garden and a bowling green, several beds for vegetables, herbs, fruit trees, shrubs, hedges with citrus or lemon trees on espaliers, pergolas with grapevines or pomegranates, a banana plantation, and a field with several larger trees, everything framed by avenues with coconut trees. Further, the map shows a large fish pond, a poultry yard, a swan's island, a rabbit hill, and some outbuildings such as slave quarters, stables for horses, and in the farthest corner of the garden a small house for the *hortulanus* (the unknown main gardener). So, the character of the *hortus* was, as Wilhelm Diedenhofen put it, 'a pleasure garden with a practical purpose'.[35]

According to the eyewitness testimony of the Portuguese friar Manoel Calado do Salvador (1584-1654), one of the most striking features in the garden were the avenues with numerous impressive tall coconut trees.[36] Although it was the common opinion that such full grown trees could not be transplanted, Johan Maurits still ordered that it would be tried. Many of these palm trees were dug out in mainland Pernambuco. They were transported on a four-wheeled vehicle to the river Capibaribe. There, these trees were loaded on a pontoon boat and brought to Antonio Vaz. To the amazement of many, the transplant of these trees – some of which were as high as 50 feet – went well. Even during the first year, they bore fruit. Marggrafe witnessed this re-usage of fully grown coconut trees. In his 'Historiae Rerum Naturalium Brasiliae' (the second part of the *Historia Naturalis Brasiliae*), he wrote, 'In the year 1640, I saw large ones being transplanted, [...] with good results, with about 300 men (*HJZ: undoubtedly enslaved Africans*) being employed in this work'.[37] It is even possible that a sketch between Marggrafe's

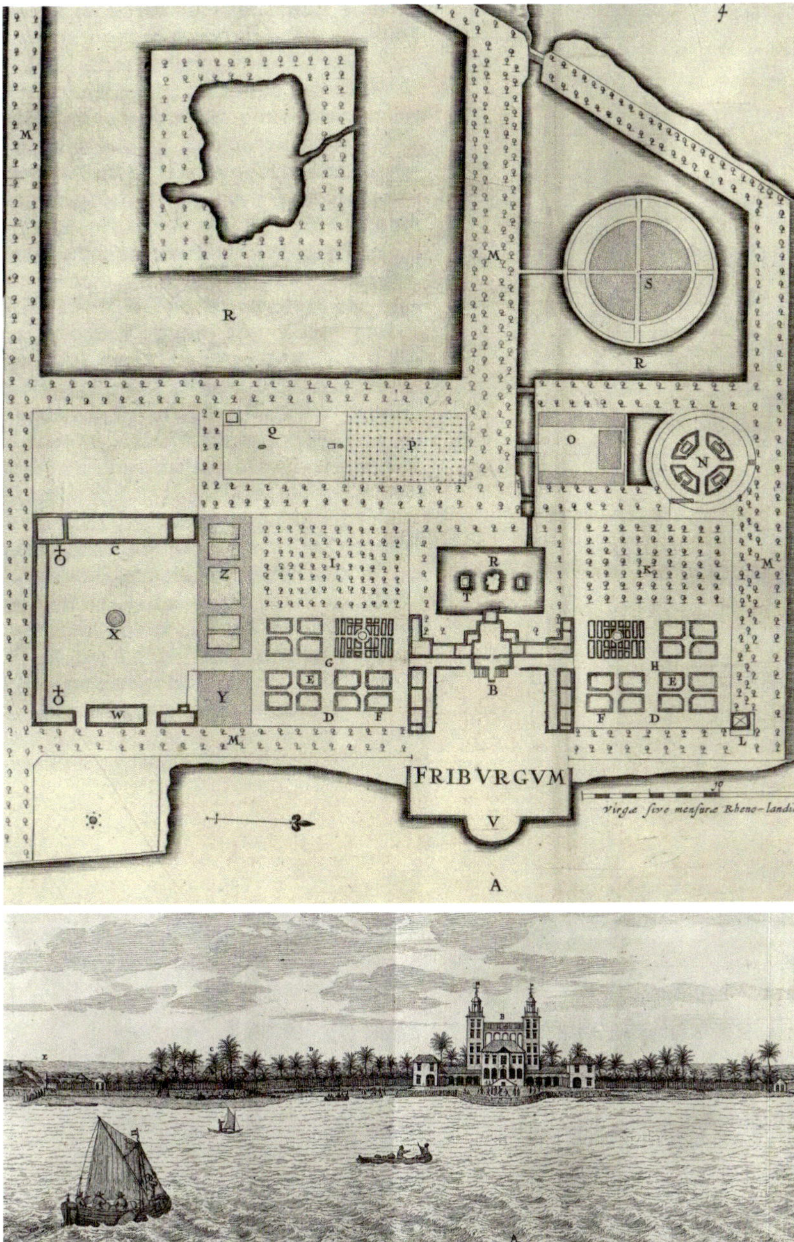

FIG. 9. Plan and side view of castle *Vrijburg* and its garden. From Barlaeus, *Rerum per octennium in Brasilia* (1647), between pages 144 and 145. (Rijksmuseum, Amsterdam).

Leiden manuscripts concerns the determination of the height of these coconut trees (*fig. 10*).

It's true that CALADO exaggerates when he talks about 2,000 coconut trees (BARLAEUS speaks of seventy!), but he probably wanted to indicate that the garden contained many different species of trees. JOHAN MAURITS indeed created at *Vrijburg* a real *arboretum*, a botanical collection of trees. BARLAEUS mentions hundreds orange and lemon trees, and other local fruit trees, including a banana. PIERRE MOREAU, a *geheimschrijver* ('cryptographer') of the government in Recife, who, after a huge uprising of the Portuguese settlers in late 1645, witnessed the destruction of the garden with pain in his heart, speaks of 'delicious trees of Brazilian wood, palm wood, ebony, cedar wood, wood as white as snow, violet wood, marble wood and other nicely smelling wood', which 'enchanted the spacious and long hiking trails further than one could see' around the 'wonderful and beautiful country house' that the count had made there.[38]

It is clear from the sources that JOHAN MAURITS was highly interested in gardening.[39] According to his treasurer JOHANNES GREVINGH, the count came every day to his garden, 'trying with his plants to do something new'. It was a pleasure to see what 'his excellency' had achieved in this field with growing, sowing and planting trees, herbs and all kinds of crops. Likewise, the count was fond of the breeding of all kinds of rare beasts and poultry, and he also stimulated the

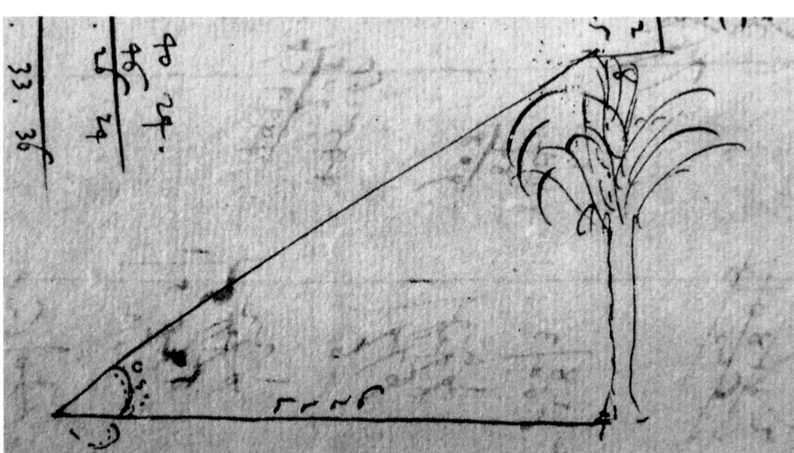

FIG. 10. Sketch by GEORG MARGGRAFE explaining how the height of a coconut tree can be determined (ELO).

craft of fishing.[40] BARLAEUS indeed mentions three fish ponds on the estate 'stocked with all kinds of fish that could serve as a relief for the people when other food was scarce'.[41]

But neither BARLAEUS, nor CALADO or other witnesses mention the presence at *Vrijburg* of wild animals, and – as noted by ARGUS VASCONCELOS DE ALMEIDA – the zoo is also *not* present on *Vrijburg*'s map. So, either the map represents a design that was adapted over time to locate the various wild animals, or the zoo was not housed at *Vrijburg* at all. The fortress-mansion *Boa Vista,* where the count 'spent his leisure time, when he was free to do so', was perhaps a far better location (*fig. 11*). There, too, were a garden and a fishpond, 'which displayed great beauty'.[42]

The fact that it was in the context of *Boa Vista* that BARLEAUS spoke of the count's museum as 'his treasure house', to which 'ships coming from the Indies, Africa, and other regions brought *foreign animals* [our italics], plants, household goods, clothing, and weapons of barbarian tribes that formed a pleasing and unusual spectacle', suggest indeed that the zoo was located there. Perhaps not such a strange idea, as one imagines how the smell of a zoo would have influenced life in *Vrijburg*'s garden. After all, according to CALADO, it was there where 'the ladies and their friends would go to pass the summer festivals, and to have their treats, and make their picnics and drinks as they do in Holland, with their musical instruments'.[43]

DUTCH BRAZIL AS A SEVENTEENTH CENTURY TOWER OF BABEL?

One other aspect of Dutch Brazil deserves our attention before we concentrate on GEORG MARGGRAFE's life and work. This last aspect concerns the question of communication. It is a real miracle that the diversity of the colony's population did not lead to complete chaos. As the Dutch Republic was only a small country with not that many inhabitants, the WIC was forced to hire its workforce from all over Europe. In Recife one could find soldiers and *vrijluiden* ('free people', i.e. persons not employed by the WIC), not only from Holland, but also from Germany, Poland, Bohemia, Denmark, Sweden, Norway, France, England and Italy. Further, there were the Portuguese *moradores* (residents), whose cooperation was essential to the Brazilian sugar industry, the cork on which the Brazilian economy floated. To facilitate the communication between the Dutch and the Portuguese

FIG. 11. View of Fortress/castle *Boa Vista* or *Schoonzicht*, finished in 1643, guarding the entrance of the Island Antonio Vaz at the Rio Capibaribe. JOHAN MAURITS's zoo may have been next to this location. From Barlaeus, *Rerum* (1647). (Rijksmuseum, Amsterdam).

speaking communities, many Sephardic Jews moved from Amsterdam to Recife, because this group mastered both languages as their mother tongue. But between themselves the Jews also used Hebrew. Then, of course, there were the native tribes of Brazil, mainly the Tupis and the Tapuyas (also known as Tarairiu), both with their own language. And finally, there were the African slaves, who also spoke their own languages. So, the so-called Dutch colony was in fact a real 'Tower of Babel'.[44] It was in this setting that MARGGRAFE, too, had to function, and had to gather his information about the local flora and fauna. With regard to the history of science, this makes the case of colonial Dutch Brazil a perfect example of a so-called 'Trading Zone'. This is a concept, developed by the historian of science PETER GALISON for the case of modern physics, where scholars, engineers and technicians had to cooperate in order to achieve results, although their 'languages' and mental frameworks were rather different. A decade ago, PAMELA LONG adapted this idea of the 'Trading Zone' to the artisanal world of Early Modern Science, where interchanges between scholars and artisans led to the creation of new knowledge.[45] This throws up another interesting question, namely can we notice something about of this aspect of language barriers in MARGGRAFE's work? How did he operate in these literal 'Trading Zones'? Was it necessary for him to cooperate with others, or could he achieve his results in spite of this enormous diversity in the society in which he lived and worked?

PART II

BIOGRAPHY

FIG. 12. The church of Liebstadt, where GEORG MARGGRAFE was baptized, with in front of the church (*left*) the former school of Liebstadt where his parents lived, and (*right*) the rectory, the home of his grandfather *pfarrer* JOHANN SIMON. On the facade of the former school, which today serves as the town hall, is a small plaque commemorating the birthplace of GEORG MARGGRAFE, placed there in 1907. Photo: Norbert Kaiser (2006).

CHAPTER 3

YOUTH IN GERMANY

Childhood in Liebstadt

Georg Marggrafe was born on 20/30 September 1610 (old style/ new style) in Liebstadt, a small German village in the region of Meissen, located 23 km southeast of the centre of Dresden near the border with Bohemia (the western part of the present-day Czech Republic). At that time the village was populated by some 350 people, living in about 70 houses and one castle overseeing the valley. Georg was the oldest child of his eponymous father Georg Marggrafe (1582-1634) and Elisabeth Simon.[1] The Marggrafe family lived more than a century in and around Liebstadt. Father Georg was a schoolmaster, who in 1609 had married the daughter of the local vicar, Johann Simon (1557-1618).[2] As the Liebstadt school was situated next to the church and its vicarage, the bride Elisabeth Simon was literary 'the girl next door' (*fig. 12*).[3]

Father Marggrafe was a learned man. He studied at the Lutheran Gymnasium in Freiburg until 1593, after which he continued his education at the University of Wittenberg, also known as the 'Leucorea'.[4] He probably was related to another Georg Marggrafe (1533-1575), who in 1553 had studied in Leipzig and worked since 1566 as a vicar in the village of Pappendorf, 100 km west of Liebstadt.[5] His son, Gottfried Marggrafe (1570-1632), studied in Leipzig (1587) and Wittenberg (1593), before becoming a vicar in Langhennersdorf, near Freiberg, some 60 km from Liebstadt.[6] The thorough education of both Gottfried and father Georg Marggrafe is so similar that a family relation seems likely.[7]

Education was also very solid on mother's side. Father Georg Marggrafe studied in Wittenberg together with his future father-in-law, Johann Simon (1557-1618), who originally came from the village of Jüchsen, north-west of Coburg.[8] Simon would become the vicar of the Liebstadt community in 1594. He must have been quite a character. When Simon died in 1618, in his memory a huge epitaph with his

FIG. 13. Full-size portrait of Georg's grandfather, JOHANN SIMON (1557-1618), *pfarrer* of Liebstadt from 1594 until his death. Epitaph in the church of Liebstadt. As no portrait of GEORG MARGGRAFE has survived, the figure of his grandfather gives an impression of what his physiognomy may have been like. Photo: Norbert Kaiser (2006).

life-size portrait was placed in the church, near the pulpit where he had preached for more than twenty years (*fig. 13*).[9] Two years later, father GEORG MARGGRAFE left his profession as schoolmaster to become a Lutheran vicar, too, preaching in the nearby village of Dobra (an hour walk south of Liebstadt) until his own death in 1634. According to the *Vitae*, a short biography published in 1685 by GEORG MARGGRAFE's sixteen years younger brother CHRISTIAN MARGGRAFE (1626-1687), both father GEORG and grandfather JOHANN SIMON noticed the eagerness towards learning of the young GEORG MARGGRAFE, and as either men were very knowledgeable in the Classical languages, they taught the boy Greek and Latin. When they saw that GEORG also had a talent for both music and painting, they took care that he was educated in these areas too, so that GEORG became an excellent musician, both vocally and instrumentally, and 'a painter not to be despised'.[10]

WANDERING YEARS IN GERMANY

On 6 May 1627, at the age of 17, MARGGRAFE left Liebstadt together with a classmate, for what would become a period of eleven years of learning. Four days later he followed his father's footsteps when both boys matriculated as students of Wittenberg University.[11] GEORG must have had excellent credentials, for next to his name in the Wittenberg register is noted *gratus receptus*; he evidently did not have to pay any fee. But as a result of the Thirty Years War, Wittenberg was not safe at the time. So, in the next years MARGGRAFE started to drift from university to university, being educated along the way in 'mathematics' (including astronomy), 'botany, physics and chemistry'. According to the biography cited above, GEORG visited first a number of cities in the south of Germany, successively (1) Strasbourg, (2) Basel, (3) Ingolstadt, (4) Altdorf (near Nürnberg), (5) Erfurt, and (6) Leipzig.[12] He then returned once 'to his father's house', probably in 1633 to visit his then sick parent (his father died in 1634), went thereafter back to (7) Wittenberg to obtain a candidate's degree in medicine, after which he set course to the north of the country, visiting (8) Greifswald, (9) Rostock and (10) Stettin. In 1636 he sailed from there around Denmark to the Netherlands, with the intention to go to (11) Leiden University.[13]

The registers of Strasbourg and Basel University indeed confirm that he was there. To be exact, on 22 November 1627 (o.s) GEORG enrolled in Strasbourg.[14] The fact that later in life he possessed a book by ISAAC II HABRECHT (1589-1633), the Strasbourg professor of mathematics and astronomy, published during his stay in this city, suggests that GEORG took his first courses on astronomy courses with this man.[15] After a winter in Strasbourg, GEORG exchanged this city for the University of Basel, where he registered in April 1628.[16] A high-profile scholar there was the old PETRUS RYFF (1552-1629), professor of mathematics since 1586. RYFF was the author of several books of mathematics, astronomy, and astrology, but he was also regarded as an expert in medicine. In that field was also active THOMAS PLATTER (1574–Dec 1628), professor of anatomy, botany and practical medicine at Basel's university. But we can only guess whether or not GEORG MARGGRAFE attended their lectures. The only thing known about his stay is that in Basel, too, the fee required for entering the university was free. When he registered it was again noted next to his name: *nihil propter paupertatem* ('no fee, because of poverty'). This lack of money is probably the reason that later on we don't find his name listed in the university registers of the other places listed in the *Vitae.* In these unregistered sites GEORG most likely only followed *privatissimo,* private courses given by professors at their home. GEORG's biography, for instance, informs us that in Rostock he attended lectures in botany given by the young professor SIMON PAULLI (1603-1680).[17] Indeed, in one of his later works, PAULLI confirms that MARGGRAFE had been among his first students at the Rostock Academy.[18]

BECOMING A MEDICAL CANDIDATE AT WITTENBERG UNIVERSITY (1633-1634)

Until now it is always stated that GEORG MARGGRAFE never obtained any formal academic degree, in spite of his wandering along various universities. But at some moment in time GEORG returned to Wittenberg University, the *Leucorea.* In 1633-1634 MARGGRAFE followed courses there given by the famous professor of medicine DANIËL SENNERT (1572-1637). In March 1634 he defended a disputation under SENNERT's supervision, entitled *De Causa continente morborum* (about the cause of diseases). A printed copy of this 17-page disputation has survived in a library in Halle (*fig. 14*).[19]

With this thesis, written in Latin and Greek, MARGGRAFE became officially a Candidate in Medicine. In it, MARGGRAFE followed SENNERT's teaching of combining the classic Greek medicine of HIPPOCRATES and GALEN with the alchemical experiences of PARACELSUS. But he also cited more modern writers, for instance the German botanist

TRIN-UNO DUCE,
Præside,
Amplissimo, Clarissimo, & Excellentissimo Viro,
DN. D. DANIELE SENNERTO,
Phil. & Medico celeberrimo, Fac. Med. P. P. Seniore, p. t. Decano, & Ser. Elect. Sax. Archiatro, Dn. Præceptore & Patrono meo optumo maxumo,
THESES HASCE
De
Caussa continente
morborum,
Candidis Philiatris ingenii lance trutinandas,
IN ACADEMIÆ WITTENBERGENSIS
theatro Apollinis publico,
Præsento
GEORGIUS Marggrafe/
LIEBSTADIO-MISNICUS.
Idibus Martl,
A. C. ∞ Iↄ CXXXIV.
Typis heredum GEORGII MULLERI.

FIG. 14. Title page of MARGGRAFE's disputation in Wittenberg. (Universitäts- und Landesbibliothek Sachsen-Anhalt, Halle, sign. 003 779 483).

and physician LEONART FUCHS (1501-1566), the anti-Galenic Italian scholar GIOVANNI ARGENTERIO (1513-1572),and of course 'the excellent' SENNERT, 'our patron, who chairs this meeting'.[20] According to MARGGRAFE's thesis nothing happened without a cause. Practising the art of healing meant searching for the cause of a disease.[21] GEORG addressed this quest in 58 propositions of varying length. We leave MARGGRAFE's reasoning here as it is, as being beyond the scope of this book. Important though, is the fact that SENNERT's approach as a practicing scientist, employing in his reasoning the evidence of experience, observation and experiment, left a crucial mark on how GEORG would explore both nature and the heavens. Interesting also is the expression in ancient Greek Σύν θεω (Sýn Theo) that MARGGRAFE puts on the top of the first page. In English translation, this means 'With God', a short form for 'With God's help'.[22] This was evidently his personal motto, as it returns several times in the manuscripts in Leiden, as well as in those preserved in Paris and Lisbon. In line with this motto was the dedication to the Creator at the very end of MARGGRAFE's thesis.[23] Both gestures unmistakably refer to MARGGRAFE's Lutheran education, in which servitude to the Creator was always the pivotal attitude.

Noteworthy too, are the last sentences of this Wittenberg thesis, in which GEORG showed awareness of the fact that, in the course of time, various learned physicians pointed to different causes of illnesses. 'I know the diversity of judgments', he wrote, an insight he summarized in the sentence: 'In the same meadow, the ox seeks the grass, the dog the hare, the stork the frog, and the pig the clay for the sweet-smelling bitterness'.[24] In this way MARGGRAFE also placed his own statements into perspective. So, he continued, 'It will, however, be yours, kind reader, to interpret all these things for the better'.[25] MARGGRAFE then ended his tract with the appeal: 'If you know something better, say it candidly and impartially; if not, don't chew on this'. This last expression – *haec rodere noli* – was printed in strong capital letters. It referred to a then well-known saying, meaning 'Don't say anything bad about this'.[26]

As customary in printed academic disputations, MARGGRAFE's thesis concluded with a small poem devoted to the academic candidate. The poet concerned was FRIDERICUS GOEBELIUS (or in German

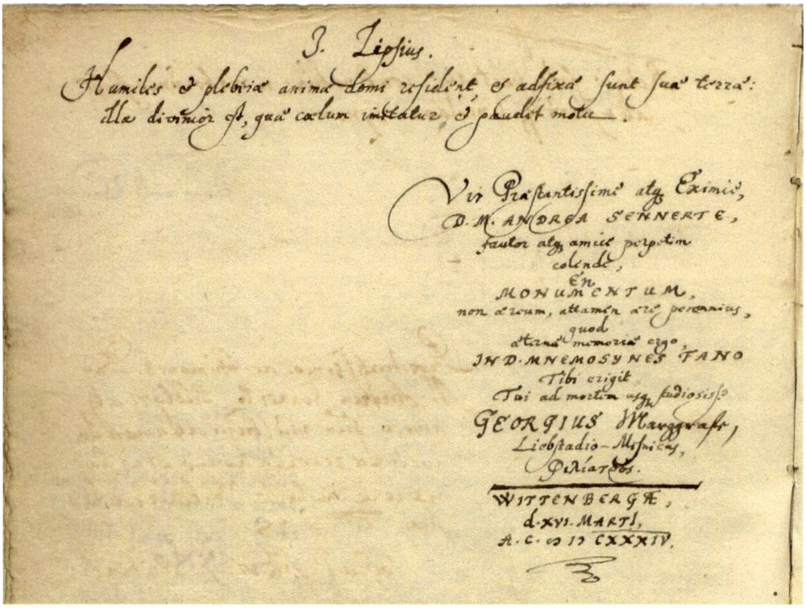

FIG. 15. MARGGRAFE's inscription in the *Album Amicorum* of ANDREAS SENNERT (1606-1689), with the quote from Justus Lipsius: '*Humiles et plebeiae animae domi resident et affixa sunt suae terrae: illa divinior est, qua coelum imitatur et gaudet motu*' ['The humble and ordinary people sit at home and are attached to their country; But divine is he, who imitates the air which loves movement']. (Universitäts- und Landesbibliothek Sachsen-Anhalt, Halle, sign. Misc. 55 oct., fol. 561vs).

'FRIEDRICH GÖBEL'), a fellow student of medicine from Nürnberg.[27] In it GÖBEL compared MARGGRAFE with the Greek god PHOEBUS APOLLO, the representative of rational beauty and order, as well as with HERMES, both the herald of these Greek gods, and the protector of travellers, thieves, merchants and orators.[28]

Another fellow student at Wittenberg was ANDREAS SENNERT (1606-1689), the son of MARGGRAFE's *Doktorvater*, whom he labeled his 'eternal friend' in the inscription he wrote in SENNERT's *Album Amicorum* (*fig. 15*). Later in life SENNERT Jr would become librarian of the Wittenberg *Leucorea* and professor of philosophy. MARGGRAFE attributed to him a phrase from an often cited letter by the famous Louvain professor JUSTUS LIPSIUS, written in 1578, in which LIPSIUS advocated the cultural travel (the *Bildungsreise* or *Grand Tour*) as a valuable means of education for a civilized student.[29] Special also is the Greek suffix that MARGGRAFE placed after his name, in which he called himself a *philiatrist* (Θιλίατρος), meaning 'a doctor-believer, who sees illness approach with tranquillity'.[30]

A horoscope for a Wittenberg professor of astronomy (1634)

Among the manuscripts in the Leiden archive is a horoscope in MARGGRAFE's own handwriting concerning CHRISTOPH NOTHNAGEL (1607-1666), professor of astronomy at the Wittenberg *Leucorea* since the spring of 1634, having taught there already since 1632 (*fig. 16*).[31] This horoscope proves not only that these two young men must have met during GEORG's stay at the university, but also that GEORG's passion for astronomy already emerged in his Wittenberg years. It is plausible to suppose that it was NOTHNAGEL who taught MARGGRAFE the first principles of contemporary astronomy and the art of constructing horoscopes. The surviving horoscope that MARGGRAFE made for himself may be seen as an exercise in this craft.[32] This skill would often come in very handy for MARGGRAFE. The Leiden manuscripts show that he must have regularly earned a living by making horoscopes for others.

After his graduation in Wittenberg, GEORG first went to Rostock, for a short stay with the newly appointed professor of botany SIMON PAULLI, who, like himself, was an alumnus of DANIËL SENNERT in Wittenberg. Being in Rostock, he probably was as interested in PAULLI's future father-in-law, the Rostock professor of medicine and higher mathematics JACOB FABRICIUS (1576-1652), to whom PAULLI had dedicated his Wittenberg thesis in 1630. In his younger years Fabricius had worked for four years as an assistant to Tycho Brahe, in his famous observatory on the then Danish island of Hven.[33] Given Marggrafe's fascination for Tycho Brahe in the years to come, it is tempting to assume that Fabricius shared with him some recollections from his time with the great astronomer. Anyway, it seems that Jacob Fabricius and Marggrafe indeed have become acquainted. When in 1635 a small booklet was printed to celebrate the doctorate of a fellow student of medicine from Rostock, we find among the poems printed in this collection, not only contributions by Fabricius and Paulli, but also one from our Georg Marggrafe. His poem is in typical Neo-Latin, in which allusions are made to the name of the laurate (a certain Salomon Fischer from Lubz in Mecklenburg), in which this *Piscator* was compared with Greek gods such as Jupiter and Neptune.[34]

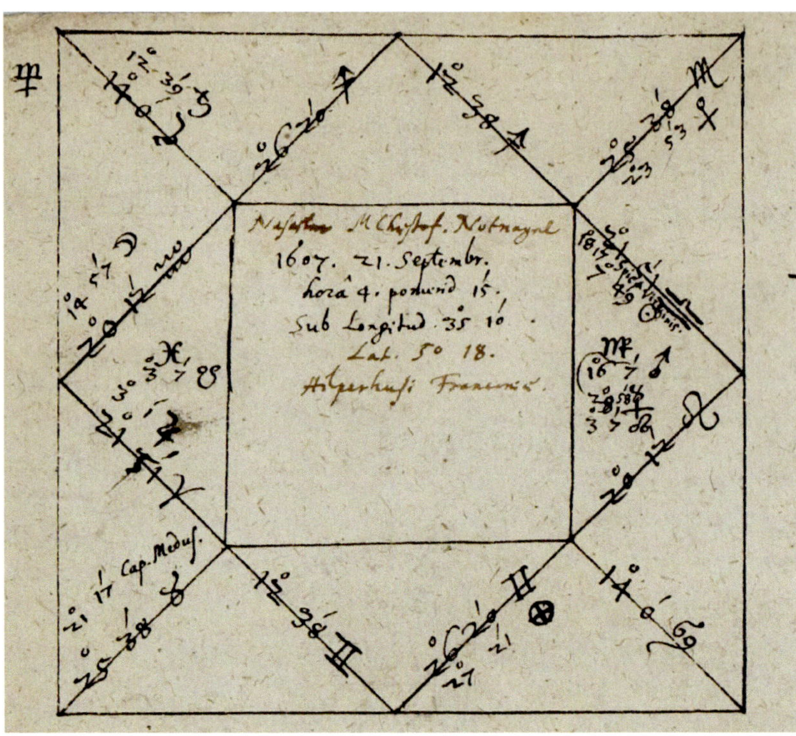

FIG. 16. Horoscope for Christoph Nothnagel (1607-1666), professor of astronomy at Wittenberg University. (ELO)

Constructing astronomical tables in Stettin (1634-1635)

From Rostock, Marggrafe travelled on to Stettin, another port on the Baltic Sea under Swedish rule, where for almost two years he became an assistant to Lorenz Eichstädt (1596-1660). Eichstädt was the local town physician, but he was foremost known as a compiler of astronomical almanacs, published since 1615. Eichstädt was also the author of a *Prognosticon* on a conjunction between the planets Saturn and Jupiter, in which booklet he identified himself as both a physician and an astronomer.[35] It is quite likely that Marggrafe had heard about Eichstadt from his Wittenberg professor Sennert, as it was in Wittenberg, under Sennert's guidance, that Eichstädt had obtained his medical degree in 1621.[36]

In the 1630s Eichstäd surely could use some assistance. In 1631 he had started to calculate some planetary ephemerides, following methods introduced by Johannes Kepler, which tables he published in the German language.[37] (An ephemeris is a table of values that gives the positions of astronomical objects in the heavens at a given time). These astronomical tables were much sought after, and in order to achieve a wider reach, Eichstädt decided to publish his next set of ephemerides in Latin. The first edition for the years 1636-1640 appeared in April 1634, but already in September of that year a second printing (with a slightly altered preface) could be issued. In this September edition Marggrafe was allowed to contribute a small verse in Eichstädt's honour. He proudly signed this small poem as *Philo–Chymiater*, meaning 'a practitioner of alchemical medicine', a clear reference to his recently acquired academic status (*fig. 17*).[38]

Marggrafe stayed in Stettin for almost two years. It must have been here where he acquired the skills to make complicated astronomical calculations. His remarkable skills were acknowledged in the next edition of Eichstädt's planetary ephemerides. In this version for the years 1641-1650 Eichstädt credited Marggrafe as a valuable co-worker, especially for his solution of a problem with regard to the angular distance of Jupiter in a configuration with two stars, which configuration they had observed on 25 February 1635 (old style).[39] The record of another observation made in Stettin on 21 December 1635 is preserved among Marggrafe's manuscripts in the Leiden archive.[40] This time the sighting involved the planet Mars.

Among the Leiden manuscripts are also a few horoscopes that most likely date from his time in Stettin.[41] One of these is constructed for the lawyer Johann Fabricius (1594-1652), *Obersekretär* of the city of Stettin (no relative of the aforementioned Rostock professor).[42] Coincidentally his *Stambuch* or *Album Amicorum* has been preserved, in which Eichstädt has written a scholarly contribution, but unfortunately Marggrafe has not.[43] However, Georg Marggrafe *did* write an entry in the *Album* of the young Christoph Loth, a student of medicine from Königsberg, when he paid a visit to Eichstädt and requested a contribution to his *Stammbuch* from both the master-astronomer and his mate.[44] But unfortunately this German *Album Amicorum* has been destroyed during World War II, as part of the *Stadtbibliothek* of Königsberg (today Kalinigrad). So, we will never know which message was passed. But, it may be clear that producing distinguished Neo-Latin poems was at the time very fashionable in intellectual circles. Poems were applicable on many occasions, at births, marriages, academic promotions, and – as we see in the following example – also at funerals. For Marggrafe left Eichstädt's service at a very sad moment. On Leap Day in February 1636, Eichstädt lost his four-year-old son and namesake, a loss that caused much anguish to the astronomer. His friends therefore offered him a collection of consolation poems, to which Marggrafe, too, contributed a small verse: in elegant Latin, of course.[45]

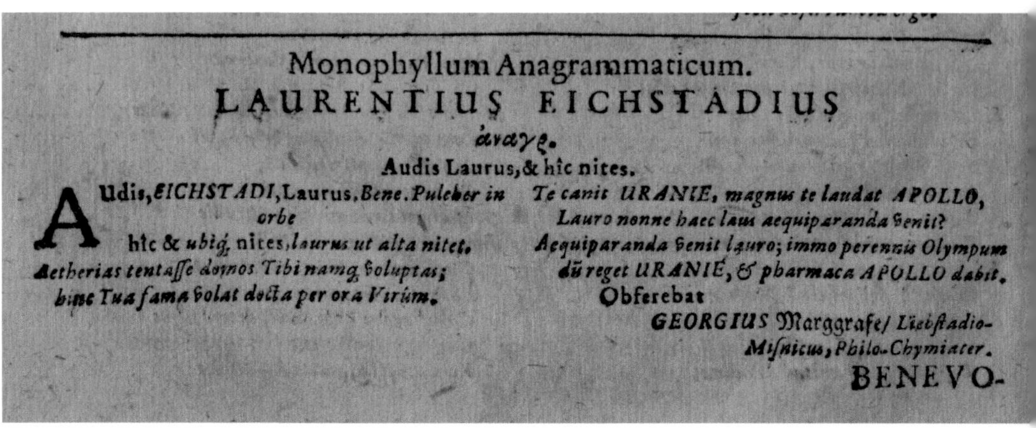
Monophyllum Anagrammaticum.
LAURENTIUS EICHSTADIUS
ἀναγρ.
Audis Laurus, & hîc nites.

Audis, EICHSTADI, Laurus. Bene. Pulcher in orbe
hîc & ubiq; nites, laurus ut alta nitet.
Aetherias tentasse domos Tibi namq; voluptas;
hinc Tua fama volat docta per ora Virûm.
Te canit URANIE, magnus te laudat APOLLO,
Lauro nonne haec laus aequiparanda venit?
Aequiparanda venit lauro; immo perennis Olympum
dû reget URANIE, & pharmaca APOLLO dabit.
Obferebat
GEORGIUS Marggrafe/ Liebstadio-Misnicus, Philo-Chymiater.
BENEVO-

FIG. 17. Marggrafe's poem in Lorenz Eichstädt's honour, signed 'Philo-Chymiater', printed in Eichstaedt, *Ephemerides novarum et motuum coelestium* (Stettin: D.Rethius, September 1634). (Det Kongelige Bibliotek, Copenhagen).

CHAPTER 4

A CRUCIAL YEAR AT LEIDEN UNIVERSITY

Arrival in Leiden

In the summer of 1636, Marggrafe left Stettin and embarked on a ship to Holland with the intention to enter Leiden University. He must have had good reasons to come, for the year before the city of Leiden experienced one of the worst plague epidemics in its entire history, killing a quarter of its population. In 1635 and early 1636 many students had fled the town to escape the disease.[1] Nevertheless, on 11 September Georg Marggrafe matriculated at the University, officially as a student of medicine (*fig. 18*).[2] In Leiden, Marggrafe first moved in with a certain Dirck Eduwaer, a boarding house owner at the Pieterskerkkoorsteeg (no. 1 on *fig. 19*).[3] But rather soon Marggrafe appears to have moved to another, probably cheaper, accommodation. His *Vitae* states that a mathematician, Samuel Carolus Kechel ab Hollesteyn, was his 'chamber-fellow'.[4] Other documents show that they rented a room at the house of Griettie Bailly on the Rapenburg, just across the observatory, which at the time was called *Theatrum Astronomicum* (nos. 2 and 3 on *fig. 19*).[5] Kechel, who had entered Leiden University in 1632, was

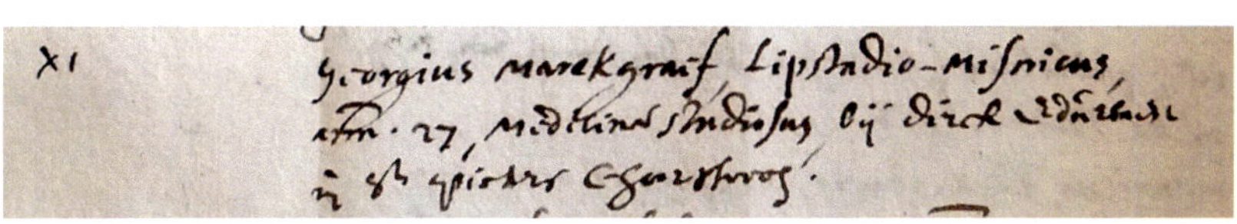

FIG. 18. Marggrafe's matriculation at Leiden University on 11 September 1636: '*Georgius Marckgraef Lipstadio-Misnicus, ann. 27, medicinae studiosus bij* Dirck Eduwaer *in [de] St. Pieters Choorsteeg*' (Georg Marggrafe, from Liebstadt near Meissen, 27 years old, student in medicine at Dirck Eduwaer in the alley leading to the St Pieters-church) (Leiden University Library, ASF 9).

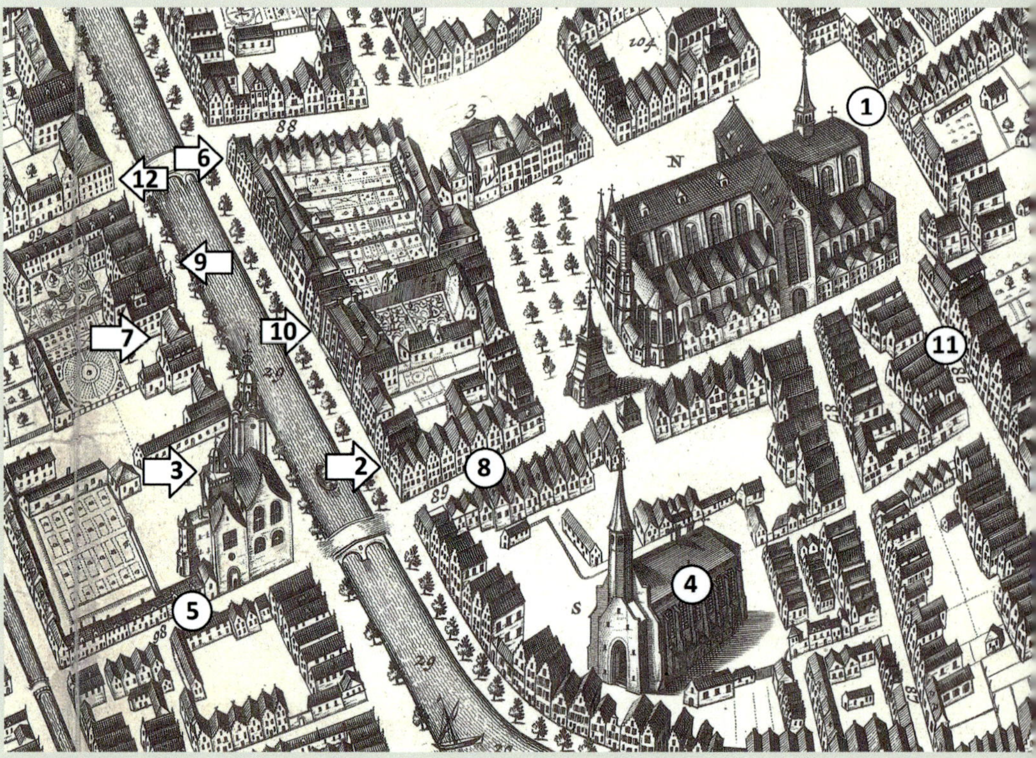

FIG. 19. Fragment of the Map of Leiden, 'Measured and drawn by the surveyor Johan Douw de jonge' in 1667-1670, engraved by Christiaen van Hagen in 1675. (ELO, PV362B).

Legend:

GEORG MARGGRAFE:

1. Pieterskerkkoorsteeg, the alley where Dirck Eduwaer lived, the man from whom Georg Marggrafe and Otto Von Uexküll-Güldenband rented their first lodgings in 1636. This is also the alley where in 1647 the bookseller and publisher Frans Hacke (or Franciscus Hackius) had his two printing presses, on which the *Historia Naturalis Brasiliae* is printed (or at least part of it).
2. Rapenburg 54, house rented by Grietgen Bailly, where Georg Marggrafe and Samuel Kechel shared a room in 1637.

THE UNIVERSITY

3. Rapenburg 73, the Academy Building, next to the *Hortus Botanicus*, on top of which the *Theatrum Astronomicum* (the Astronomical Observatory) was housed. (The tower is built in 1670). In the narrow gallery north of the Academy Building lived the 'opsiener' (supervisor) of the observatory, Pieter Vincentz van Heemskerck, also 'letterzetter' (typographer) in de adjacent printing workshop of the brothers Elzevier.
4. Falide Begijnhof: former church, where the *Theatrum Anatomicum*, the University Library, the Fencing School and the *Duytsche Mathematiek* was located.
5. Nonnensteeg: the alley where, next to *Hortus Botanicus*, the office of professor Adolph Vorstius was located.
6. Rapenburg 12, The house on the corner, bought by professor Jacob Golius in 1638, which he probably rented before.

OTHERS:

7. Rapenburg 69a, the house of the widow Anna de Nerée-Wijnhoff, where Jørgen From had his lodgings in 1636-1638.
8. Klocksteeg, the alley where the widow Anna Lus-Baudius, Kechel's future mother-in-law, had her boarding house, renting out rooms to the Bartholin brothers in 1637-1639. In June 1652 Georg's brother Christian Marggrafe also chose his lodgings in this street.
9. Rapenburg 29-31, double house of the *mercator sapiens* Johannes de Laet.
10. Rapenburg 44, house of the parents of Willem Piso until 1634, when they moved to the Nieuwsteeg. (Piso lived in Amsterdam after 1633).
11. Nieuwsteeg, alley on the west side of which the mathematical instrument makers Jan and Jacob David had their workshop from 1613 until at least 1639. They produced brass surveyor's instruments, such as the circle of Dou.
12. Rapenburg 23, on the corner of the Groenhazengracht: *Het kasteel van Antwerpen*, the house rented by Georg's brother Christian Marggrafe in 1665, where he established his chemical laboratory. In 1674, at the time of his marriage, he moved to the Pieterskerkhof (the area around the church at no. 1), where he wrote his brother's biography, the *Vitae*.

born in Prague in 1611, so he was almost the same age as MARGGRAFE and came likewise from central Europe.[6] They shared an interest in astronomy, so it seems plausible that they quickly became friends.

At first sight it seems that MARGGRAFE intended to finish his medical studies in Leiden, by getting a formal degree as a Medical Doctor, just as his Wittenberg study friend FRIEDRICH GÖBEL had done in Frankfurt at the Oder, the year before. However, if we look more closely how MARGGRAFE's career progressed until then, having worked as a computational astronomer for two years, it seems more likely that MARGGRAFE chose this Dutch university mainly for astronomical reasons.

LEIDEN UNIVERSITY IN THE SEVENTEENTH CENTURY

MARGGRAFE came to Leiden, because at that time the university's observatory was the only operational astronomical observatory in Europe. Its predecessors, the observatories of TYCHO BRAHE and JOHANNES KEPLER in Prague, as well as the 'Landgraf's Observatory' of WILHELM IV VON HESSEN in Kassel had ceased their activities, whereas the Leiden University Observatory was brand new.[7]

With the foundation, in 1633, of an astronomical observatory as a new tool for the instruction of practical mathematics, Leiden University followed a policy introduced almost at its establishment in 1575. This policy aimed to mix classical humanist knowledge with new empirically oriented branches of knowledge, rooted in observation and experiments. Under the same roof as the library (the familiar source of trustworthy knowledge), the university had installed a *Theatrum Anatomicum* (1590, no. 4 on *fig. 19*), where new knowledge was gathered by means of anatomical dissections of human and animal bodies. For similar reasons a *Hortus Botanicus* (1595, below no. 3 on *fig. 19*) had been laid out in the backyard of the academy building. In both places curiosities and natural history specimens were also collected, gathered from all over the world. This policy to combine at Leiden University classic bookish scholarship with practical research involving the observation of nature continued during the seventeenth century.[8] Shortly after the installation of the new observatory a new initiative was taken. In November 1636, at the very moment at which MARGGRAFE became a student of medicine, the medical professor of the university introduced a new system of clinical teaching. At one of the Leiden hospitals, the 'Caecilia Gasthuis', a

Collegium Medico Practicum was founded, where medical students could be trained in the practical 'bedside' aspects of their profession.[9] Later in the century other sites were added for performing operations with 'a mindful hand', such as a *Laboratorium Chymicum* (1669), a site for the try-out in the field of the art of fortification – then a branch of mathematics – (1672) and a *Theatrum Physicum* (1675).[10]

Especially in the 1630's and 1640's Leiden was the place to be. This period may be considered as probably the most successful period in the university's history.[11] As a 'dynamic and decidedly progressive spot for early modern learning' the city attracted hundreds of students coming from all over Europe.[12] It was here that during MARGGRAFE's stay GALILEO GALILEI's *Discorsi* as well as RENÉ DESCARTES's *Discours de la Méthode* were printed: the first at the Elsevier press, in the house next to the observatory; the latter at JEAN LE MAIRE's printing press in the Pieterskerkkoorsteeg, the same alley where MARGGRAFE found his first lodgings.[13] As DESCARTES stayed in Leiden from December 1636 until May 1637, meeting GOLIUS on a regular basis during the same months in which MARGGRAFE followed GOLIUS's courses and worked at the observatory, it even seems possible that DESCARTES and MARGGRAFE encountered each other. Perhaps the two even discussed telescopes and astronomy.

THE UNIVERSITY'S BRAND NEW OBSERVATORY

The foundation in the years 1632-1634 of an astronomical observatory at Leiden University was a logical step in a development which was started by Leiden's second mathematics professor, WILLEBRORD SNEL VAN ROYEN (1580-1626), latinized as SNELLIUS, the re-discoverer of what now is known as Snel's law of refraction.[14] SNELLIUS had been a university professor since 1613, but long before his appointment, in the years 1600-1601, he had spent some time in Prague with the astronomer TYCHO BRAHE, assisting him with astronomical observations.[15] With his own eyes SNELLIUS had witnessed the large quadrants and sextants that had made BRAHE famous. With such large instruments one could measure the coordinates of an astronomical object with the accuracy of a fraction of a degree, using the (naked-eye) Tychonic sights designed by BRAHE. After his return to the Netherlands, SNELLIUS ordered two similar Tychonic quadrants from the Amsterdam instrument maker WILLEM JANSZ (who would later take on the surname BLAEU). In 1595-1596 this artisan had lived for a few months with BRAHE on the island Hven, on which occasion he had learned the craft of making astronomical instruments.[16] After SNELLIUS had been installed in the Leiden chair of mathematics, he used these privately owned large iron quadrants, introducing practical 'Tychonic' astronomy at Leiden University. In fact he placed the practice of making measurements at the heart of his pedagogy. Thanks to his trigonometric measurements SNELLIUS was able to determine the length of the Earth's circumference, which result was published in 1617 in the book *Eratosthenes Batavus*, a title that referred to the old Greek scholar who had made a similar attempt in Ptolemaic Egypt.[17] SNELLIUS came up with a result that differed from the current value by 3.5 %. In his ambition to obtain a higher degree of accuracy, SNELLIUS continued his measurements and in the mid-1620s he ordered another, probably more accurate quadrant, made of wood and brass from WILLEM JANSZ (BLAEU).[18] It was this new quadrant that would initiate the establishment of the Leiden astronomical observatory.

In 1629, almost three years after SNELLIUS's death, his colleague JACOB GOLIUS, professor of oriental languages since 1625, was asked to take over his chair in mathematics (*fig. 20*). GOLIUS had been a student at Leiden University already at the age of 14, when he matriculated as a student in law.[19] However, with particular preference he focused on subjects that belong to the mathematical sciences. With great interest he followed courses at the so-called *Duytsche Mathematique*, a branch of the university where cartographers, surveyors, and engineers were trained in practical mathematics, not in the usual Latin, but in the Dutch language. His name can be found among the signatories who requested the appointment of FRANS VAN SCHOOTEN THE ELDER, as professor at the *Duytsche Mathematique*.[20] GOLIUS finished his basic education, the *trivium*, in 1616, but rejoined the university again in 1618, now with the ambition to study the oriental languages under the guidance of professor THOMAS ERPENIUS (1584-1624).

Provided with knowledge in both areas, GOLIUS succeeded in 1622 to get an appointment as an engineer to participate in a Dutch diplomatic mission to Morocco. He was assigned the task of mapping the bays on the west coast of that country. During that time GOLIUS not only enhanced his knowledge of the Arabic language, but he also collected many old manuscripts for his Leiden teacher professor ERPENIUS, who, however, died suddenly in 1624. Returned home that same year, GOLIUS was the right man to succeed him.[21] In the

FIG. 20. Portrait of JACOB GOLIUS, professor of oriental languages and mathematics at Leiden University. Painting by HIERONYMUS VAN DER MY copied after a canvas by JAN DE VOS, dated 1651. (Leiden University).

next years, GOLIUS returned once more to the Middle East to collect in total 300 Arabic manuscripts, many relating to the mathematical and astronomical sciences.

Because the chair of mathematics had been vacant since the death of SNELLIUS in 1626, GOLIUS now convinced the custodians of Leiden University that, according to him, the unification of the professorships for mathematics and the oriental languages had a great advantage. In that case they would have to pay the salary of only one professor instead of two, with the result that some money would be available for the purchase of study material, such as mathematical instruments.[22] So, at the end of October 1629 GOLIUS quickly was promoted to *Magister Artium*, in order that he could be appointed *ordinarius professor matheseos* the next month.[23]

In order to fulfil his new mathematical task, GOLIUS bought two large quadrants from SNELLIUS's heirs.[24] In 1632 he sold the newest one to the university, arguing that the instrument would be necessary to demonstrate to the students the course of the celestial bodies.[25] As the use of such a large astronomical quadrant required a suitable platform, GOLIUS asked the trustees of Leiden University for 'a proper place on top of the academy building, to place the instrument'.[26] The city architect of Leiden, JAN JACOBS VAN BANCHEM, was instructed to design such a platform. In 1633 this rectangular observation platform, surrounded by a balustrade, was constructed on the roof of the main university building. A room below the platform was made for instruction purposes and to shelter two globes. In the summer of 1634 a small octagonal turret was added to the structure, to house the quadrant.[27] With these acts the first astronomical observatory owned by a European university had been founded. This *Observatorium* stands out proudly on a prospect of the city, drawn in 1640 and printed in 1647 (*fig. 21*).[28]

MARGGRAFE'S ACTIVITIES AT LEIDEN UNIVERSITY

According to the short biography left by his much younger brother, during his stay in Leiden, GEORG was 'passing over the nights in the turret of the academy and spending the day in the [botanical] garden and fields', while being instructed by professor JACOB GOLIUS in mathematics as well as in the Arabic language, and by professor ADOLPH VORST (VORSTIUS) in botany (*fig. 22*).[29] Between MARGGRAFE's papers a short note is preserved, dated in both the

FIG. 21. Depiction of Leiden Observatory constructed in 1633-1634 on the attic of the Academy Building on the Leiden Rapenburg. Detail of a prospect on Leiden, drawn by CLAUDE RIVET in 1640 and printed by JACOB SAVRY in 1647. (ELO, PV1722.6a1).

FIG. 22. The Leiden *Hortus Botanicus*, next to the Academy Building, founded in 1595, where MARGGRAFE was instructed in Botany under the supervision of professor ADOLPH VORSTIUS. Drawing for an engraving, dated 1675. (ELO, PV12513).

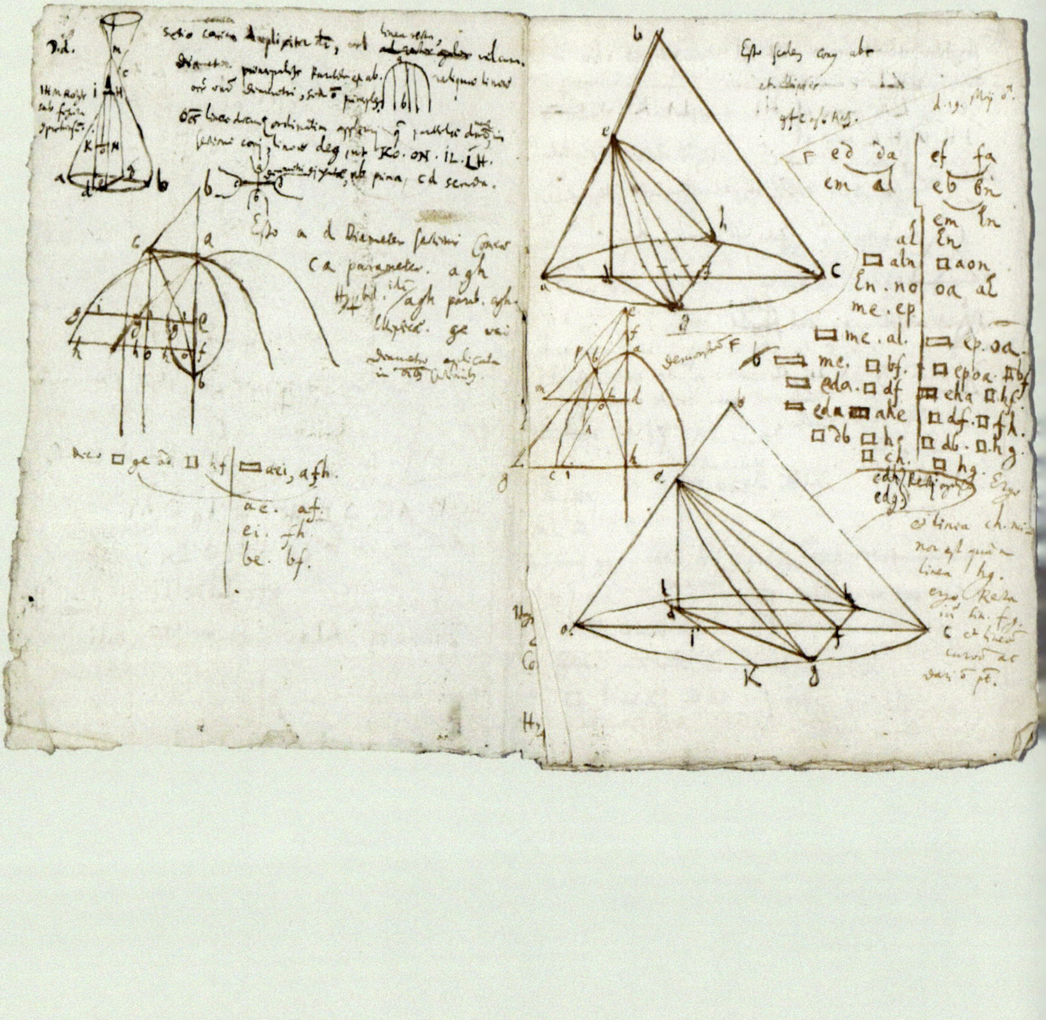

FIG. 23. MARGGRAFE's lecture notes about conic sections, noted at Leiden University in October 1636. (ELO, North doc. no. 66)

old (Julian) and new (Gregorian) calendar style: *29 September 1636 / 9 October 1636, Lugduni Batavorum* [= Leiden], this indeed bears witness to GOLIUS's mathematical training.[30] GOLIUS was renowned in Europe, among others things for his discovery of a precious ancient manuscript containing an Arabic adaptation by the famous THABIT IBN–KORRAH from Harran of APOLLONIUS PERGAEUS' book on conic sections.[31] GOLIUS often used this book in his lectures; he even had a series of conic sections made in boxwood.[32] So it is interesting to see that in MARGGRAFE's lecture notes conical sections were prominently discussed (*fig. 23*), next to other geometrical topics and logarithms.

Given MARGGRAFE's later career as a 'cartographer' in Brazil, it is possible that in Leiden he also witnessed practical mathematical instructions at the so-called *Duytsche Mathematique*, the vernacular branch of the university where professor FRANS VAN SCHOOTEN THE ELDER trained surveyors and engineers.[33] It is known that various other students followed this dual path of instruction.[34] If so, MARGGRAFE will have made his acquaintance here with the successful Dutch surveying instrument invented in 1610 by the surveyor JAN PIETERSZ DOU, known today as the 'Hollandsche cirkel' (Dutch circle). This 'astrolabe on a stick' was the standard surveying instrument in the Netherlands for almost two centuries.[35] Anyhow, the preserved documents testify only about the fact that MARGGRAFE made some preliminary astronomical calculations in November 1636 and further that he started regular astronomical observations at the observatory on 19 January 1637.[36] These observations, which were continued on a regular basis until 6 November 1637, will be dealt with later, in Chapter 9. However, a personal note from his Leiden observation register still deserves our attention here. MARGGRAFE observed at the observatory whenever he could, but at the end of June he evidently fell ill (or was he intoxicated?), as is shown in his observation register by two small sketches, in which he depicts himself lying down, as an explanation why he had not observed on two evenings (*fig. 24*).

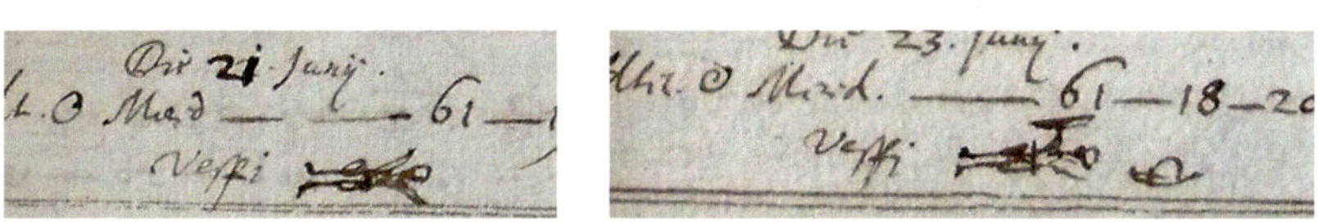

FIG. 24. MARGGRAFE's portrayal of himself, lying down because of illness, next to the word *vesperi* ('evening'), explaining why he could not observe on the evenings of 21 and 23 June 1637. On the right, he also pictured his hat. (ELO, North no. 52, fol. 8r).

Horoscopes for Leiden students

During his stay in Leiden Marggrafe made a living by constructing horoscopes for his fellow students. The drafts of some of these are still preserved among the Leiden manuscripts. These horoscopes were described extensively by John North in 1979, but as these astrological papers mostly provide only initials as well as a date and location of birth, North was not able to identify the recipients. A check, however, of these data with student names in the Leiden *Album Studiosorum* gives matches for at least four of these horoscopes. These names provide us with information about the people Marggrafe dealt with during his stay in Leiden. For instance Marggrafe's horoscope for 'G.H. à B.', born on 11 February 1609 [o.s.] at Altenburg [*today Altenberge, abour 12 km northwest of Münster*], matches exactly with the Leiden entry on 26 August 1636 of 'Georgius Henricus à Beerensteyn, Saxo. 26 years old [= born ~1610], student in law, *in de Molensteegh int Gecroonde Schaep*.[37] This was the Latinized form for the name of the German student Georg Heinrich von Bernstein. A similar match can be found for Adolfus Fredericus von Hagen from Mecklenburg, also a student in law, who rented lodgings in the same house *'t Gecroonde Schaep* in the Leiden Molensteeg.[38]

Another match concerned the medical student Georgius Kirstenius, or in German Georg Kirsten (1613-1660), from Stettin.[39] His horoscope is on the same paper as those of Bernstein and Hagen, so although he lived elsewhere in Leiden, Kirsten probably was one of their friends. In 1647 he became the first professor of medicine in his birth town Stettin, where he obtained some fame for his alchemical approach to medicine. So, just as Marggrafe, he was a kind of *Philo-Chymiater.*[40] Since Marggrafe lived in Stettin only shortly before, one can imagine that there were common grounds for conversation between the two, both in terms of residence and interest.

Interesting, too, is the identification of the Danish student Henrik Fuiren (1614-1659) as one of Marggrafe's astrological customers.[41] At Leiden University he was registered as a student of medicine, but like Marggrafe he devoted much of his time to the study of mathematics.[42] Fuiren was an exceptionally wealthy student, who could live from the fortune he had inherited from his grandfather, a successful merchant from Copenhagen, who had died in 1628, in the same year as Henrik's father. Fuiren is an extreme example of the eternal student. He began his studies at the University of Copenhagen in 1628, but finished his academic education only in 1646, when he went for the third time to Leiden University after having obtained his medical degree at Basel University.[43] After 19 years of study FUIREN finally returned to his home town Copenhagen, where he settled as a physician and soon became a professor of medicine, but foremost being renowned for his huge library and his impressive cabinet of natural curiosities.[44]

HENRIK FUIREN belonged to a small circle of Danish intellectuals. He was the son of the physician JØRGEN FUIREN and MARGRETHE FINCKE, daughter of the influential professor of mathematics and medicine THOMAS FINCKE (1561-1656) who served the University of Copenhagen for more than 60 years. His family dominated Danish university life in the seventeenth century. About one half of the professors at Copenhagen University between 1620 and 1720 were members of or related to the Fincke dynasty. Because HENRIK FUIREN's father had died in 1628, his uncle ØLE WORM (who had married another daughter of THOMAS FINCKE), served as his paternal tutor, a role WORM also performed for the sons of his sister-in-law ANNA FINCKE, widow of CASPAR BARTHOLIN (1585-1629), the late professor of medicine at Copenhagen University.

During MARGGRAFE's Leiden years, WORM corresponded not only with HENRIK FUIREN, but also with his brother THOMAS FUIREN and their full cousins BARTHOL and THOMAS BARTHOLIN, at the time all students at Leiden University.[45] These letters provide insight into what Leiden students experienced at that time. In 1635, for instance, the FUIREN brothers had to flee Leiden to avoid being affected by the plague, so they went to Franeker University instead. After their return in the spring of 1636, we learn that the *Hortus Botanicus* was badly affected by the drought, so due to lack of the proper herbs, the professor of botany VORSTIUS was not able to hold his botanical exams. FUIREN also reported about the start of the earlier mentioned *Collegium Medico Practicum,* in which WALAEUS, the professor of medicine, taught the basic rules of practical medicine. FUIREN was especially fond of the lectures given by the professor of eloquence, MARCUS BOXHORN, 'who with such great praise explains Ovid's metamorphoses that even professors who frequently attend his lectures always leave with increased knowledge'.[46] Interesting is that in June 1636 WORM requested HENRIK to buy for him the description of the Americas by JOHANNES DE LAET, the very person who a year later would mediate for MARGGRAFE to fulfil his desire to go to Brazil. FUIREN not only sent WORM the requested book volume, but added later also a sample of the skin of an American pig, which WORM recognized as coming from Brazil.[47]

Revealing, too, is the letter by FUIREN's cousin THOMAS BARTHOLIN about the high cost of living as a student in Leiden, a circumstance that also will have affected MARGGRAFE. To WORM he wrote:

> Just as my glorious cousins, I chose lodgings in Leyden. But where I found the grace and fame of this place only appealing, the more repulsive I found the amount of my money expenditure, which was way beyond my expectations. Because everything here comes at such a high price that you would think that the Dutch merchants weigh the goods with gold. We pay six florins a week for food, but also for our lodgings, which is certainly the cheapest in Leiden, because there are those who pay seven, some even nine guilders in other hostels. Some believe that we can easily live on our own diet and take care of it ourselves: but whoever we tell our plans, they are out to make our life miserable. [...] The gentlemen FUIREN and other fellow countrymen have given us a striking example; [...] They all claim that they sometimes fell short because of the enormous expenditure.
>
> In other words, we live at the lowest possible cost, without worries and without hunger. When our linen is to be washed, a florin goes with it every month; equally, the supply of necessary books is not obtained for nothing. Wine, and everything else that serves to keep the fragile body in good condition, must be paid for separately. For the care of the body it is quite necessary in these parts to swim in the sea. [...] Until now, we have avoided the fairly well-known physical exercises, such as dancing, boxing, ball games and the like, because 'our scrotum is full of cobwebs'.[48] Private lectures are held at a high price. And if we ever have to put on new clothes, when these are worn out, I fear that the two hundred rigsdalers that are donated to us annually are insufficient for this.
>
> During the winter, three dissections of a human body were held. OTTO HEURNIUS[49] dissected only a single individual, even one of the opposite sex. But FALCOBURG,[50] who in his care and accuracy in detection, teaching, and dissection far surpasses HEURNIUS, opened and dissected two male corpses. [...] The botanist ADOLF VORSTIUS has not yet opened the public garden, but without a doubt he will soon start his lectures. If this garden nourishes something in its womb, either seeds or plants or roots, which your well-cultivated garden may miss, then I would send it without hesitation.[51]

Friends of Marggrafe's roommate Kechel

Henrik Fuiren is also an interesting person, because it appears that he became a friend of Marggrafe's roommate Samuel Kechel ab Hollensteyn, of whom is known that he became an assiduous observer at Leiden Observatory from 1637 onwards (*fig. 25*).[52] Years later, when Fuiren returned to Leiden in April 1646, he directly took his lodgings at Kechel's house, together with his cousin Thomas Bartholin, whose younger brother, the mathematician Erasmus Bartholin would join them the next month.[53] That this choice for the Bohemian mathematician indeed referred to a friendship becomes evident in October 1646, when Kechel's daughter Anna Elisabeth was baptized and both Henrik Fuiren and Erasmus Bartholin stood witness at this event.[54] That this friendship was an old one from the late 1630s becomes clear from the mere fact that in the years 1637-1639 three brothers Bartholin (Barthol, Thomas and Caspar) rented a room from the lodging-house keeper Anna Baudius.[55] She was not only the daughter of the late professor of eloquence and history Dominicus Baudius (1561-1613), but more importantly, she was foremost the mother of Kechel's fiancée Aeltgen (or Alida) Lus. So, Anna Baudius's house in the Klocksteeg was surely a place where Kechel could be found (no. 8 on *fig. 19*). Having seen all these connections, it seems rather plausible to conclude that when Marggrafe constructed Henrik Fuiren's horoscope, he was also witnessing the emergence of Kechel's friendship with these Danish students.

A special friendship with Jørgen From

In Leiden Marggrafe made a friendship of his own with another Danish student, a man who came from a more common background. This concerned Jørgen From (1605-1651), also known as Georgius Frommius, who earned his living as a *preceptor*. In that capacity he had to assist wealthy Danish students as a tutor, and to act as their butler and guide during their *Grand Tour* throughout Europe. In this way, From had already accompanied students to universities in Angers, Oxford, Basel, Geneva, Padua, Rome and Naples. In the fall of 1636 he undertook a new journey with two rich Danish students that would bring him to the Netherlands, France and England.[56] But From was also a scholar in his own right. He had studied at the University of

FIG. 25. Samuel Carolus Kechel ab Hollensteyn (1611-1668). Drawing by Dirk Druyf made in 1655 for the lecture notes of his pupil Joos Crommeling Jacobz. (Kupferstichkabinett Staatliche Museen zu Berlin, KdZ-nrs. 71.167).

Helmstedt and followed *privatissimo* in Rostock, Leipzig, Wittenberg and Frankfurt an der Oder.[57] In Denmark he had been a student of the Danish astronomer CHRISTEN SØRENSEN LONGOMONTANUS (1562-1647), whose chair in astronomy at Copenhagen University JØRGEN FROM would take over after LONGOMONTANUS's death.

It was more or less by accident that FROM arrived in Leiden in September 1636. Their Danish ship had set sail to England, but violent storms prevented his small company from going there. Instead they went to Leiden, despite concerns about the plague and rumours of war in the Southern part of the country.[58] The three Danes matriculated at Leiden University on 30 September 1636 – just three weeks after MARGGRAFE.[59] The university's register shows that FROM and his two rich pupils had their lodgings in one of the most elegant pensions in Leiden, the boarding-house of ANNA WIJNHOFF (1588-1647), widow of the Calvinist minister RICHARD JOHANNES DE NERÉE on the Rapenburg canal (no. 7 on *fig. 19*).[60] Her garden bordered the *Hortus Botanicus* and was next to the well known printing house of IZAAK and ABRAHAM ELZEVIER. No wonder FROM could send home the news of a controversial book that was printed there at that very moment. To his acquaintance, the king's Chancellor CHRISTEN FRIIS (1581-1639), whose nephew JØRGEN FRIIS also just had arrived in Leiden,[61] FROM wrote:

> ELZEVIR published recently GALILEO's little book, which I dare to send to your Highness, since you asked me to keep my eyes open [...] GALILEO is still alive, but his health is poor. For more than a year he has been blind in one of his eyes, and recently he has also lost the sight of the other. A mathematician from Amsterdam has been in Italy to speak with him about the determination of longitude by means of Jupiter's moons. GALILEO did not dare to write letters on this matter, and he is too old to travel abroad. It is inconceivable that many great men are persecuted, with the result that their most beautiful thoughts are thereby lost or remain unfinished. [...] And now GALILEO is not allowed to publish his ideas in his own country. For this reason this treatise is published here without his knowledge, although in accordance with his wish, and it is distributed secretly, from one hand to another, by ELZEVIR JUNIOR.[62]

He brought this news also to JESPER RASMUSSEN BROCHMAND (1585-1652), professor of theology and the father of one of his pupils, writing 'the ELZEVIRS have printed a book by GALILEO on mechanics [which] I will send to you'.[63] He also informed him that it was his intention that BROCHMAND's son and his friend would become acquainted 'with the learned men' of the city.[64] Among these was the linguist professor MARKUS BOXHORN, who was asked to privately instruct the two boys for a few days, which result 'was to the satisfaction of both him and the other professors'.[65]

As a student of the astronomer LONGOMONTANUS, JØRGEN FROM was naturally very interested in the Leiden *Theatrum Astronomicum*. So it is not surprising he soon became acquainted with JACOB GOLIUS, 'who for long has lived in the Orient and has brought with him many treatises by the most distinguished Arab astronomers'.[66] In May 1637 he reported to his patron LONGOMONTANUS what he had seen and experienced at the Leiden Observatory:

> In Leiden we have a large Tychonic quadrant (which the Academy bought last year from the heirs of SNELLIUS) and a sextant of just one foot *which a certain student* has equipped himself with. [Note: *our underscoring*] We have even practiced these instruments on most clear nights this winter. [...] We have made observations of Mercury, which agreed almost precisely with the Rudolphine tables, deviated slightly from yours and much from LANSBERGEN's.[67]

As can be established from the MARGGRAFE documents in the Leiden archive, that very month, April 1637, GEORG MARGGRAFE used the observatory every single day with a clear sky, so this 'certain student' can be identified with reasonable certainty as being MARGGRAFE. Indeed he used a sextant, and, as we shall see, one of his special interests was certainly a study of planet Mercury's orbit. That the two astronomers became friends appears also from the fact that among MARGGRAFE's legacy in the Leiden archive a draft letter can be found, addressed in 1638 to a certain 'FRAEM'.[68] This name – as appears from the Leiden *Album Studiosorum* – was the Latin expression with which FROM introduced himself at the time.[69]

According to the historian of science HELGE KRAGH, it was at the Leiden Observatory that JØRGEN FROM most likely became the first Danish scholar to use a telescope for astronomical observations. In a letter to LONGOMONTANUS, dated 1638, he wrote about this, emphasizing, however, that his effort was merely an exercise:

> Last night we observed with a telescope how the Moon covered the western of the three bright stars as they appear in the sketch, beautifully drawn by *my Bohemian friend*, with whom I usually observe.

> [*our underscoring*] [...] When he was told that I would write to you about our observations, he urged me to send you his sketch. [...] If you have time to look at them, I will send all our observations from the past year, even if they are for private practice use only. In the next month, February [1638], we will make careful observations provided God gives us good health and clear weather.[70]

This 'Bohemian friend' can be identified with certainty as SAMUEL KECHEL, of whom it is known that he was a skilled draughtsman (see *fig. 26*). In MARGGRAFE's letter to JØRGEN FROM, cited above, he also refers to KECHEL as FROM's 'loyal colleague when observing'.[71] Their friendly relation is also demonstrated by FROM's gift to KECHEL of a book with his handwritten dedication, sold in auction in 2005.[72]

In Leiden, JØRGEN FROM also established contact with MARTIN VAN DEN HOVE, better known as MARTINUS HORTENSIUS, who was a professor of mathematics at the Amsterdam *Athenaeum Illustre* since 1634.[73] He had been educated at the Latin school of Rotterdam by its rector, the natural philosopher and telescope investigator ISAAC BEECKMAN. Under his guidance HORTENSIUS made astronomical observations from 1625 onwards, sometimes with the naked eye and on other occasions with a telescope.[74] He enrolled at Leiden University in March 1628, where he became a student of WILLEBRORD SNELLIUS, but a year later he went to Middelburg in the southern part of the country, where for one year he was the assistant of the astronomer PHILIPPUS LANSBERGEN. Returned to Holland in 1630, HORTENSIUS performed many astronomical observations, probably also with the quadrant from SNELLIUS's heritage before its installation at the Leiden Observatory.[75] In short, HORTENSIUS was the most skilled Dutch astronomer of the day, because SNELLIUS, BEECKMAN, as well as LANSBERGEN died before FROM's arrival in 1636. No wonder that JØRGEN FROM wrote to HORTENSIUS in the autumn of 1636:

> There is no one in the entire Netherlands I would rather meet than you, especially after [the Danish mathematician ERIK OLUFSEN] TORM has told me about your learning and formation. I'm sorry I could not visit you when I was recently on a quick visit to Amsterdam.[76]

This missed encounter was compensated for a year later. In the late summer of 1637 FROM succeeded in meeting the Dutch astronomer.

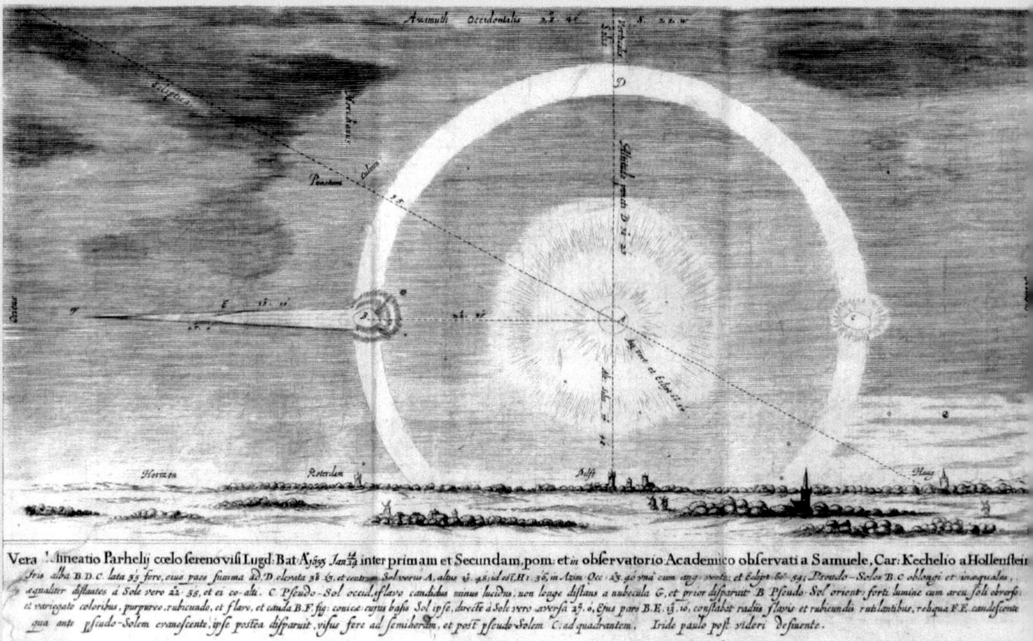

FIG. 26. Parhelium observed and drawn by SAMUEL KECHEL AB HOLLENSTEYN at the platform of Leiden Observatory in 1653. At the horizon, toward the town of The Hague (*Haag*) can be seen; in the middle the two churches of Delft; to the left the city of Rotterdam. The church in the foreground is those of Voorschoten. (Leiden University Library).

'Occasionally HORTENSIUS has come to us from Amsterdam', he wrote to BROCHMAND.[77] To LONGOMONTANUS he wrote that 'HORTENSIUS now looks somewhat more gently to [the world system of] TYCHO [BRAHE],' adding that, 'he told me that our king seriously considers building an observatory based on TYCHO's principles', referring to plans for the Round Tower observatory in Copenhagen, which facility indeed was established in 1642.[78]

Whether or not MARGGRAFE and KECHEL were included in these encounters with HORTENSIUS remains a mystery. Our guess is that they were not, although HORTENSIUS was aware of MARGGRAFE's presence in Leiden at least since March 1637, when he wrote to one of his correspondents that 'some German' made regular observations there.[79] But if MARGGRAFE had met the Amsterdam professor, he probably would have written to him from Brazil. But HORTENSIUS is not included in MARGGRAFE's short list of correspondents, compiled in June 1638, which suggests that he never met this Dutch astronomer.[80]

MARGGRAFE AS PRECEPTOR FOR ESTONIAN STUDENTS?

The Leiden *Album Studiosorum* reveals two other remarkable things. First it appears that when MARGGRAFE matriculated at Leiden University on 11 September 1636, he was not alone. On the same day the Estonian nobleman OTTO VON UEXKÜLL from Tallinn also registered as a student, but more importantly, he went to the same boarding house as MARGGRAFE.[81] That this was no coincidence becomes evident on 13 July 1637, when OTTO's younger brother KONRAD VON UEXKÜLL also entered the university, together with two noble relatives.[82] Where did they take up their lodgings? *Nota bene* in the house of GRIETTIE BAILLY on the Rapenburg canal, the same boarding house to which MARGGRAFE had moved, to become KECHEL's roommate! Evidently the Estonian OTTO VON UEXKÜLL had moved with him. This raises the question whether MARGGRAFE was perhaps also employed as a *praeceptor* or private tutor for these high ranking and rich Estonian noblemen, whose fathers belonged to the top of the Estonian nobility.[83] For less fortunate students, such a job as a tutor for wealthy students was quite common in those days. The Leiden *Album Studiosorum* mentions such a position as *famulus* (servant) often separately, but unfortunately not consistently. We know, for example, that JØRGEN FROM taught his pupils the basic principles of mathematics in Leiden and arranged meetings for them with the university professors.[84] FROM's Danish friend HANS JØRGENSEN MULE, who during a short-lived illness took over FROM's tasks for a while, performed a similar job for his own noble pupil, who also had a second servant with him. MULE even came to Leiden three times as a *preceptor* for different wealthy students.[85] This also happened to students from Germany. The servant CHRISTOPHORUS URSINUS, for example, a fellow townsman from MARGGRAFE's birthplace Liebstadt, twice accompanied a group of German students to Leiden, first in 1635 and later in 1639.[86]

So, did GEORGE MARGGRAFE play a similar role as tutor for these Estonian nobles? If so, it was only until the summer of 1637. By coincidence it is known that starting on 13 August 1637 two of the newly arrived Estonians followed extramural lessons in fortification from the East-Prussian mathematician CHRISTIAN OTTER (1598-1660). He was one of several mathematical teachers who in the late 1630s worked in the margin of Leiden University. OTTER started this course during the Siege of Breda (21 July - 11 October 1637) and ended shortly after the city's surrender.[87]

In the context of these Estonian students, it is worth mentioning that among the horoscopes made up by MARGGRAFE, there is one marked with the birthplace 'Rewalia' (today's Tallinn). However, the date on the horoscope (15 June 1611, old style]) does not seem to fit one of the above named Estonian students, as deduced from the Leiden *Album Studiosorum*. However, sometimes the age mentioned during registration does not fit reality.[88] So, in theory it may be possible that this unidentified horoscope was made for OTTO VON UEXKÜLL, the oldest of this Estonian group, who gave in 1636 as his age 21 [= born c. 1615], but whose real birth date is unknown.[89]

THE LEIDEN LUTHERAN CHURCH

One aspect of MARGGRAFE's Leiden sojourn still deserves attention. As the son and grandson of a Lutheran pastor, it is to be expected that GEORG MARGGRAFE also attended the local Lutheran church in Leiden, which congregation flourished enormously in those years. Leiden had a Lutheran congregation since 1588. A newly built church – the first in Leiden to be built for a Protestant community – was erected in 1618, after a fundraising campaign in several German cities. The modest building was hidden behind a row of houses at the Hooglandse Kerkgracht. The building was not allowed to stand out, because in the Dutch Republic denominations other than the dominant Calvinist church were not allowed to have church buildings visible from the street. To fund a new west wing for the Lutheran church (to be built in 1640), a new fundraising campaign was set up in 1634, this time among Lutheran students and other visitors from abroad, who for some time came to Leiden. The register in which these gifts are recorded is in a way a precious addition to the *Album Studiosorum* of the University.[90] Indeed, financial contributions by most of the German and Danish students mentioned above can be found in this gift book. For example, from the students for whom MARGGRAFE made horoscopes, there are donations by ADOLFUS FREDERICUS VON HAGEN, GEORGIUS KIRSTENIUS and HENRIK FUIREN, but gifts are also recorded from *preceptors* such as JØRGEN FROM, HANS JØRGENSEN MULE and CHRISTOPHORUS URSINUS and most of their pupils. Finally, all Estonian noblemen mentioned in the last paragraph, without exception, generously donated to the Leiden Lutheran Church. However, no donation from our GEORG MARGGRAFE was found in this gift book, which omission may be seen as a testimony of the oft-repeated claim that he was rather poor and always short on cash.[91] Nevertheless, thanks to the gifts of his fellow students, the Leiden Lutheran Church could be enlarged with a gallery in 1640.

PREPARATIONS FOR BRAZIL

In the fall of 1637 MARGGRAFE probably tried to make arrangements that enabled him to travel to South America. In the light of his later involvement in publishing MARGGRAFE's Brazilian notes, and the fact that they corresponded during MARGGRAFE's time in Brazil, it is plausible to suppose that the Leiden-based *Mercator Sapiens* JOHANNES DE LAET (1581-1649) was crucial in making these arrangements (*fig. 27*).[92]

DE LAET was not only a respectable gentleman-scholar, but he was also one of the founders and directors of the Dutch West Indian Company (WIC). He is the perfect embodiment of the entanglement of the rising interest for new philosophy with the concerns of commerce. Although he never crossed the Atlantic, DE LAET had a reputation for being the most knowledgeable person with regard to the two America's.[93] According to MARGGRAFE's younger brother, 'going there' had been MARGGRAFE's aim right from the start:

FIG. 27. JOHANNES DE LAET (1581-1649), portrayed at the age of 60 by JAN GERRITSZ. VAN BRONCKHORST. (Rijksmuseum, Amsterdam).

> He burnt with a great desire of contemplating the southern stars. [...] He knew [...] that there was in America a harvest of no small praise: therefore he turned over every stone, he grabbed at every opportunity that could lead him to America.[94]

Making contact with DE LAET was rather easy. He lived nearby, on the Rapenburg canal, in a house just a stone's throw away from the academy building, the centre of Leiden University (no. 9 on *fig. 19*).[95] In his capacity as *bewindhebber* of the West Indian Company, DE LAET was a fierce supporter of emigration to the small Dutch colony in Brazil. The minutes of the meetings of the board of the WIC (which are only preserved for the years 1635-1636) show that he made several calls to encourage the transport of *vrije luyden* ('free people' – not employed by the company) to Brazil.[96] But of course, a crossing of the Atlantic would always require a certain amount of money, and for MARGGRAFE that really was the problem. But when in July 1636 the board of the WIC received an urgent letter from the government in Recife, an opportunity arose that could take away this obstacle. The letter reported that a certain WILHELM VAN MILANEN had died during the last military campaign. In itself this message was not so special; it happened all the time, only VAN MILANEN was not just anyone. He was the court physician of JOHAN MAURITS, the governor of Dutch Brazil. *The Heeren XIX* were requested to search for 'another capable person, worthy of supplementing this place'.[97] In a next letter, written at the end of August 1637, but received in Holland at the end of September, the urgency of the matter was underlined, because it appeared that JOHAN MAURITS also had fallen ill. The request was very clear:

> Please look for another competent and experienced person and send us him soon, because at the moment there is no one here able to cure any serious or dangerous illness, so that His Excellency [JOHAN MAURITS], in the absence of [his own medical doctor], now consults a Portuguese physician.[98] We have here two good surgeons, but they should not go beyond their capabilities, for they are experiencing at the moment more than they can cope with.[99]

The message left nothing to be desired in terms of clarity: a new court physician to replace the late VAN MILANEN had to be found very quickly.

Hired as assistant to the court physician Willem Piso

Whether it was De Laet, or the Amsterdam magistrate Albert Coenraads Burgh, who contacted the young medical doctor Willem Pies (1611-1678) is not known. Both *bewindhebbers* of the WIC are credited in the 1648 preface of the *Historia Naturalis Brasiliae*.[100] De Laet certainly must have known him since childhood, as one of the young dwellers on the Rapenburg canal (no. 10 on *fig. 19*). Piso – as he preferred to call himself – was born in Leiden as the son of a German church organist, who originally also embarked on a study of medicine.[101] Willem Piso was educated at the Leiden Latin School and probably also at the Leiden surgeon's guild.[102] After obtaining a medical degree in Caen (France) in 1633, he practiced medicine in Amsterdam, where he became involved in the city's cultural circles. After some hesitation, Piso agreed to become the new Brazilian court physician. But he was advised to hire two assistants with medical backgrounds, given the circumstance – underlined in the letter quoted above – that medical expertise in Brazil was really scarce.

This vacancy gave Marggrafe the opportunity to travel to Brazil without paying a transport fee. On the contrary, he would even receive a salary for this passage. So, on the evening of 7 October 1637 he noted in his Leiden observation register: *vesperi impeditus negotiis fui*, or 'this evening I was prevented [from observing] by business'. It is an educated guess that the nature of this business consisted of a consultation with Piso (and perhaps De Laet) about a position as *domestique* or assistant of Piso for medical matters. That Marggrafe accepted this job offer becomes evident from the remark he wrote under his final Leiden observation of 6 November 1637: 'This is my last observation in Leiden, because [...] I am leaving for South America on 18 November' (*fig. 28*). In fact, Marggrafe would only set sail on 1 January 1638. In the meantime he will have stayed, first in Amsterdam and later on the roadstead of Texel, where the fleet of the West Indian Company always prepared itself for a crossing of the Atlantic. From that period the Leiden manuscripts only preserve some notes, made between 22 december 1637 and 1 januari 1638, containing preparatory calculations for solar heights and moon positions, evidently made when he waited on board the ship that would take him to Brazil.[103]

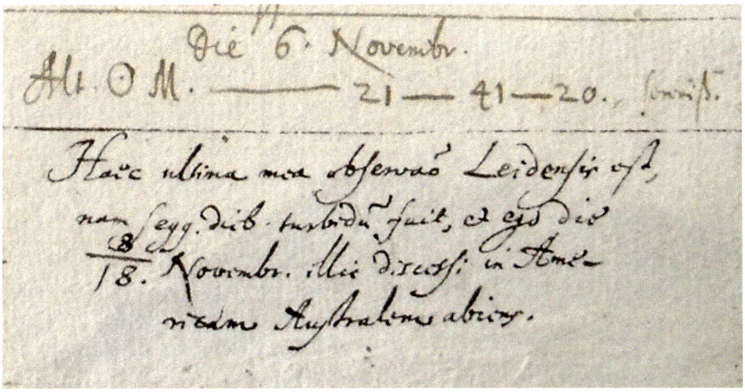

FIG. 28. MARGGRAFE's remark, written below his observation of 6 November 1637: '*Haec ultima mea observatio Leidensis est nam sequentibus diebus turbidem fuit, et ego die 8/18 November illic discessi: in Americam Australem abiens*' (This is my last observation in Leiden, because the following days the weather was turbulent, and I left for South America on 18 November) (ELO).

THE JOB DESCRIPTION OF THE COURT PHYSICIAN AND HIS ASSISTANTS

When WILLEM PISO was hired as the count's personal physician, he entered the service of the WIC as an employee of the Chamber Amsterdam. As the archive of this branch has been almost completely lost (being sold as scrap paper in the early nineteenth century), it is impossible to track down more about the conditions of PISO's contract. However, by a lucky coincidence, the minutes of the meetings of the directors of the Amsterdam Chamber for the years 1635-1636 have been preserved. These were exactly the years in which JOHAN MAURITS was appointed. Therefore, these minutes also reveal the circumstances and conditions on which PISO's predecessor, the late WILHELM VAN MILANEN, was hired. We can reasonably assume that the agreements made in that procedure were comparable with those concerning PISO's appointment. So, what can we retrieve from these minutes about the task and conditions of a Brazilian court physician and his assistants?

Paragraph 10 of JOHAN MAURITS's contract with the WIC stipulated that a personal minister, a physician and a personal secretary would be assigned to him at the expense of the West Indian Company.[104] But before anyone could suggest a candidate, it was JOHAN MAURITS personally who in August 1636 suggested VAN MILANEN as his court physician.[105] Why the count suggested this man, or who had recommended this physician to him, is not known. About WILHELM VAN MILANEN, a city physician from Utrecht, only very little is known. Two notary deeds attest to his competence, which suggest that he was very experienced.[106] At the time of his marriage, in 1618, he is already noted as a medical doctor, which implies that he practised medicine for almost thirty years.[107] It is unknown where he obtained his degree. In Utrecht he lived in the house *Cranenborch* at the Oude Gracht, next to the *St. Barbara gasthuis*, which indicates that most probably he was the main physician of this hospital.[108] In the meeting of the Amsterdam Chamber of the WIC of 1 September 1636, it was reported that the directors VAN CEULEN, DE VRIES and BURGH had negotiated with 'doctor MILANEN', with the result that, 'after many words back and forth', it was agreed that VAN MILANEN, for his salary of 150 guilders a month, would treat all officers and soldiers of the WIC, if they were injured in battle. But others would have to pay him. Further, a servant was granted with the salary of an *adelborst* ('cadet'), as well as a *distillateur* ('distiller'), evidently for his pharmacy. The first assistant would receive ten guilders a month, the second eight.[109] This arrangement was slightly amended in the meeting of 15 September, where after a request of JOHAN MAURITS, it was permitted that VAN MILANEN would charge a reasonable sum for his consultation for 'natural diseases' to those officers and others who were mediated enough to pay a customary fee. This with the restriction that 'those who cannot afford to pay' were treated for free.[110]

From these conditions, agreed with the WIC's first court physician, it appears that the fact that PISO was allowed to hire two assistants had already been agreed with his predecessor. These assistants probably would receive a salary of ten guilders a month. This was almost half of the payment to an ordinary soldier in Dutch Brazil, who received eighteenth guilders a month, which often was supplemented by 'regular plunder'.[111]

MARGGRAFE was probably the second *domestique* who was appointed. The first certainly was HEINRICH KRALITZ (1607-1638), who came from the same region in Saxony as MARGGRAFE.[112] Just as MARGGRAFE, he possessed a degree as a 'candidate of medicine'. KRALITZ had already signed up for a trip to Brazil in October 1636, then as a *vrijman* (a freely established entrepreneur, not employed by the WIC).[113] But he must have changed his mind when he learned that he could go for free when he went in the service of the court physician. KRALITZ was born in Potewitz, a small village, near the town of Zeitz in Saxony. In 1622 he went to the (still existing) Landesschule Pforta in Naumburg, founded in 1543 as one of the first Lutheran schools in a former Cistercian monastery.[114] After his graduation in April 1628, KRALITZ entered the University of Leipzig as a student of law, a subject he soon changed for medicine.[115] The fact that in 1631 KRALITZ published a poem in memory of the wife of the Jena professor of medicine JOHANN FRIEDRICH SCHRÖTER (1559-1625) suggests that he was encouraged to make this new choice as a result of their influence.[116] KRALITZ came to Leiden in 1632, where he worked as assistant to the professor of botany ADOLPH VORSTIUS. Under his supervision, KRALITZ made a new inventory of all the plants growing in the Leiden *hortus botanicus*. His final work, the *Catalogus plantarum horti academici Lugduno-Batavi*, listed over 1,100 cultivated plants and over 280 species found wild near the Leyden *hortus*. The small booklet was finished in May 1636 and was published by the academy printer ELZEVIER (*fig. 29*).[117] Through this work, KRALITZ had become an expert in botany. Being thoroughly acquainted with all the herbs used in medicine, he promised to be a great asset for the Brazilian community. However, KRALITZ was not that fortunate. He drowned, probably even before his arrival in Brazil.[118] One of his last actions in the Dutch Republic may have been his donation of one 'Reichsthaler' to the Leiden Lutheran Church.[119]

LEIDEN OBSERVATORY AFTER MARGGRAFE'S DEPARTURE

MARGGRAFE's regular observations at Leiden Observatory were atypical. This observatory was *not* a research institute *avant la lettre*, such as TYCHO BRAHE's observatories on the island Hven. In line with Leiden University's mission to provide to its students a merger of classical scholarship and practical instruction, the observatory was intended only as a training site. It was a place for demonstrating the heavens, not for charting them. That's why, from JACOB GOLIUS, the observatory's founder, we only know three random observations.[120] Observing the heavens on a regular basis simply was not GOLIUS's mission as university professor. As one of his students observed, GOLIUS was more inclined to the Arabic languages than he was engaged in mathematics and practical astronomy.[121] Although GOLIUS would remain in office until his death in 1667, from the last 25 years no involvement

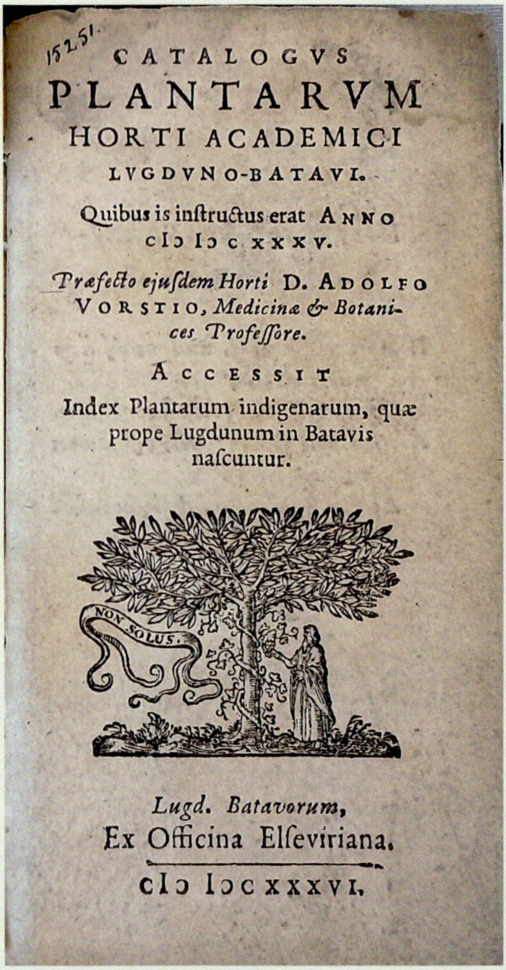

CATALOGVS
PLANTARVM
HORTI ACADEMICI
LVGDVNO-BATAVI.
Quibus is instructus erat ANNO
CIƆ IƆ C XXXV.
Præfecto ejusdem Horti D. ADOLFO
VORSTIO, *Medicinæ & Botanices Professore.*
ACCESSIT
Index Plantarum indigenarum, quæ
prope Lugdunum in Batavis
nascuntur.

Lugd. Batavorum,
Ex Officina Elseviriana.
CIƆ IƆ C XXXVI.

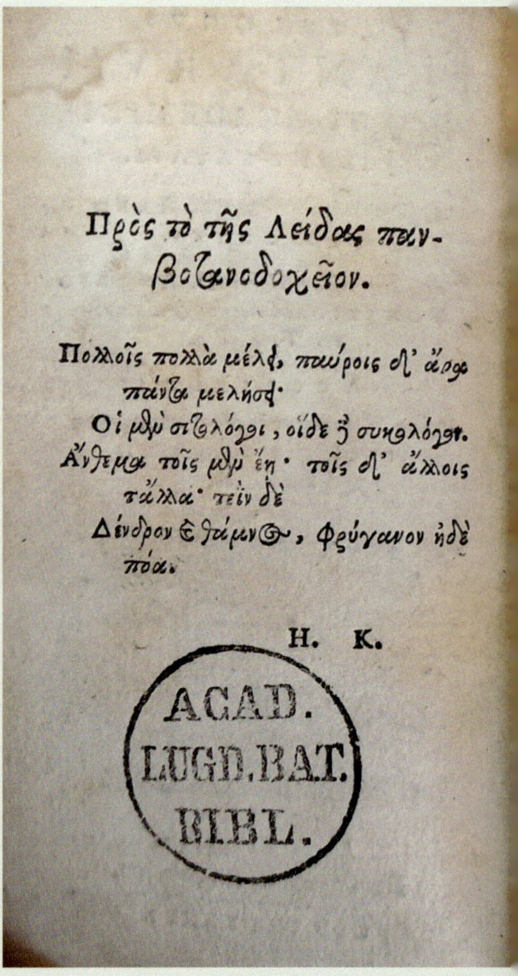

Πρὸς τὸ τῆς Λείδας παν-
βοτανοδοχεῖον.

Πολλοῖς πολλὰ μέλει, παύροις δ' ἄρα
πάντα μελήσει·
Οἱ μὲν σιτολόγοι, οἱ δὲ γε συκολόγοι.
Ἄνθεμα τοῖς μὲν ἔη· τοῖς δ' ἄλλοις
τἄλλα· τεῖν δὲ
Δένδρον ἢ θάμνος, φρύγανον ἠδὲ
πόα.

H. K.

FIG. 29. Title page of the *Catalogus plantarum Horti Academici Lugduno-Batavi* (1636). With an accompanying poem in Greek, signed 'H.K' [= Henricus Kralitz], titled 'Πρός τό τής Λεδαχ παν Βοτανοδοχέον' (= [dedicated] to the Botanical Garden of Leiden). Although the name of the professor of botany, ADOLPH VORSTIUS, is on the title page, the real compiling work was done by KRALITZ, as is recorded in 1637 by the Amsterdam physician, and later Franeker and Leiden professor of medicine and botany, JOHANNES ANTONIDES VAN DER LINDEN. (Leiden University Library. Photo by the authors).

at all with the observatory has been recorded. GOLIUS simply left the supervision of the observatory to an *opsiener* ('overseer'), appointed by the university.[122] This also is the reason why in 1646 one of Leiden's students wondered whether the observatory had fallen into disuse.[123]

However, this was not completely true. MARGGRAFE had created some followers in practical astronomy; firstly he had inspired his roommate and friend SAMUEL KECHEL, who after a few incidental observations in 1638-1639, would make long-term regular observations at the observatory after the death of his wife and children in 1647. He even discovered independently the comets of 1661 and 1664.[124] A second follower was the mathematician ADRIANUS HOVIUS (c. 1609-1655), who had entered Leiden University as a student of philosophy in June 1633, and who started working as a private mathematician in January 1637, the very month in which MARGGRAFE started his observations at the observatory. In November 1640, the university's governors granted HOVIUS permission to use the observatory with three of his private pupils. But rather soon HOVIUS changed his career. In 1643 he obtained a degree in law, after which he became rector of the Latin School in Gorinchem.[125] In the mean time another student in practical mathematics had followed in MARGGRAFE's footsteps. Between July 1642 and April 1643, BENGT HEDRAEUS (1608-1659) from Uppsala (Sweden) used the observatory for regular observations. In his book *Nova et accurata [...] quadrantis astronomici azimuthalis* (1643) he described his adjustments to the quadrant, made with the assistance of the Utrecht instrument maker JOHANNES SNEEWINS.[126]

All in all, Leiden Observatory served as a model, not only for MARGGRAFE's astronomical observatory in far away Recife (1638), but also for similar observatories in nearby Utrecht (1642) and in Sweden's Uppsala (1648).[127]

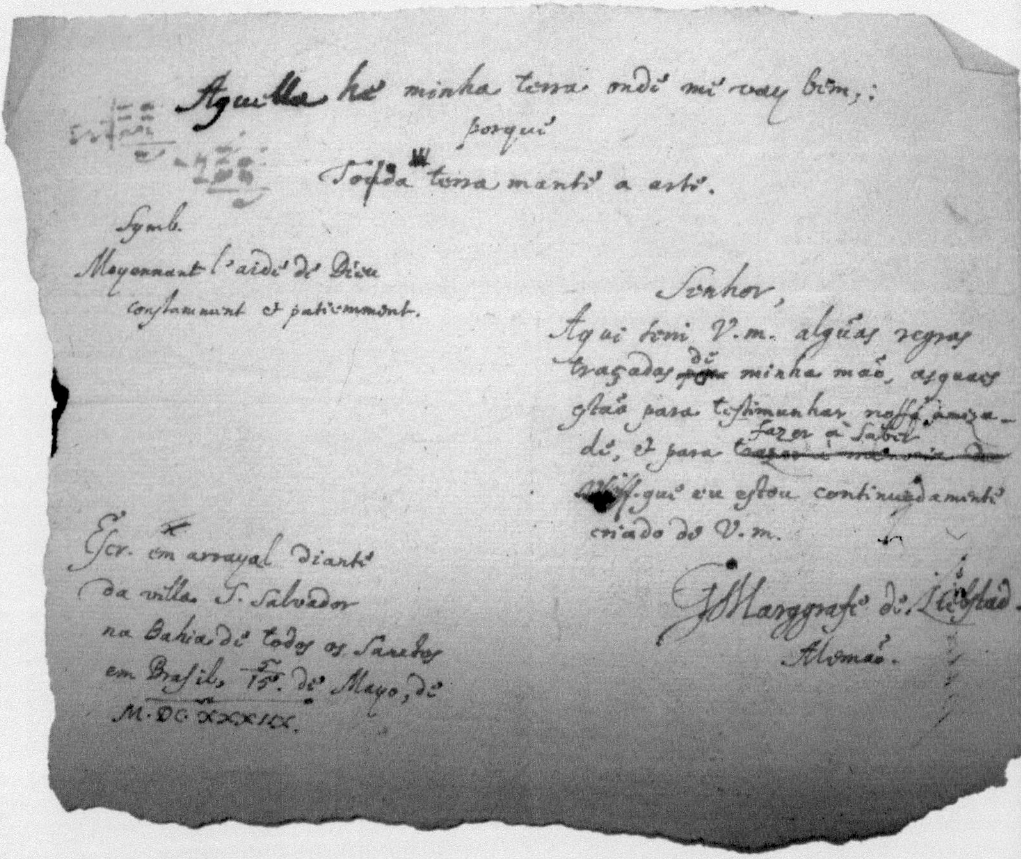
Aguilla hé minha terra ondé mí vay bem; :
porqui
Toda terra mantê a artê.
Symb.
Moyennant l'aidé de Dieu
constamment et patiemment.
Senhor,
Aqui temi V.m. algũas regras
traçadas de minha maõ, asquais
estaõ para testimunhar nossa amiza-
de, et para fazer a saber
V.M. que eu estou continuadamente
criado de V.m.
G Marggrafe de Liebstad.
Alemaõ.
Escr. em arrayal diante
da villa S. Salvador
na Bahia de todos os Santos
em Brasil, 15. de Mayo, de
M.DC.XXXIIX.

FIG. 30. MARGGRAFE's draft of a letter to JOHAN MAURITS VAN NASSAU, dated 15 May 1638. (ELO) *Transcription*: Aguilla hé minha terra ondé mí vay bem: | porqui | Toda terra mantém a arte. | Symb. | Moyennant l'aidé de Dieu | constamment et patiemment. | Senhor, | Aqui tem V.M algũas regras | traçadas de minha maŏ, as quais | estaŏ para testimunhar nostra amita- | de, et para fater a saber | V.M. que eu estou continuada mente | criada de V.M. | G MARGGRAFE de Liebstad | Alemaŏ | Escr. em arrayal diante | da villa S. Salvador | na Bahia de todos os Santos | em Brazil, 5/15 de Mayo, de | MDCXXXIIX.
Free translation: For this eagle [= 'Aquila' is also Latin for 'boy'] his land is where it goes well: because all the land is kept according to the arts. [Symb.] | [*in French*] With God's help, constant and patiently. | [*in Portuguese*] Sir, Here you have some drawings outlined by me, which are to testify to our friendship, and to know Your Highness that I am continuously at Your Highness's service. G MARGGRAFE from Liebstad, Germany. Written on the day of arriva at the city of São Salvador, near Bahia, in the Bay of All Saints in Brazil, 5/15 May 1638.

CHAPTER 5

LIFE IN BRAZIL

Travel to and arriving in Brazil

In the beginning of January 1638, Marggrafe set sail for Brazil, for a crossing of the Atlantic that would last some two months.[1] During the crossing he prepared for what was to come. The Leiden manuscripts show that on board he made numerous calculations about, among others, the height of the Sun during the entire year 1638; various aspects of the planets during that year; the position of the vernal equinox; and in a small pocket book he made daily notes about the Moon and the planet Mercury.[2] Before he left Leiden, he had been so clever as to ask De Laet for a letter of recommendation for Count Johan Maurits van Nassau, in which his capacities as a mathematician and astronomer were elucidated.[3] So, already before he had started his job, Marggrafe made sure that his employment in the service of Piso would not last forever.[4] Indeed, soon after his arrival Marggrafe contacted Johan Maurits. By means of a short letter, he fashioned himself as a person with various capacities (*fig. 30*).[5] In this letter, Marggrafe showed his courtesy, by praising his new home country guided under the count's care; he demonstrated his drawing skills by attaching to his letter some drawings he made during his voyage to Brazil. By writing in Portuguese, he also showed his command of the Portuguese language, a tongue he seems to have acquired during his passage to Brazil.

The siege of São Salvador in Bahia

Whether the count paid any attention Marggrafe's letter is questionable. Johan Maurits really had something else on his mind, being in the midst of a siege of the Portuguese-controlled city of São Salvador, the centre of the area around the Bay of All Saints (Bahia).

On 5 April 1638 the count had set sail for Bahia with a force of 4,300 men (including MARGGRAFE), distributed over thirty ships.[6] The aim was to conquer São Salvador to prevent any further attacks on the Dutch Brazilian territories. If this pivotal city would fall into Dutch hands, it would surely result in the collapse of Portuguese Brazil. There would be no stronghold for the Portuguese from where such hostile operations could be undertaken. So, there was much at stake. However, as it turned out, the artillery brought along was not heavy enough to do much damage to the city, and the siege itself proved insufficient to starve its population.

Therefore, on the night of 17-18 May 1638, only two days after MARGGRAFE had written his letter, JOHAN MAURITS decided on a storm attack of the city. During this battle MARGGRAFE escaped death by only inches, when a cannon ball 'of 18 pounds' flew right past his head, crushing the head of the soldier next to him.[7] But the attack failed, and – as many soldiers had fallen ill (MARGGRAFE's itinerary mentions that he was hit by severe diarrhoea) – JOHAN MAURITS had no other choice than to break up the siege and return to Recife. In retrospect this failure to seize São Salvador would determine the fate of the Dutch colony. Soon a large Portuguese armada with reinforcements would leave Lisbon. A second chance to safeguard the Dutch colony in the long term would not arise again.

THE DECOR OF MARGGRAFE'S LIFE IN BRAZIL: THE FAST-CHANGING ISLAND OF ANTONIO VAZ

The Dutch armada from Bahia returned to Recife on 5 June 1638. MARGGRAFE had been away from Recife for two months. After his arrival from Holland in March 1638, he had been ordered to sail with the fleet to Bahia on 8 April; so only now, after his return, could he make the Dutch colony his new home. Being an assistant to the court physician, he was assigned to the headquarters of the count on the island of Antonio Vaz. This island was probably named after the sixteenth-century Portuguese explorer ANTÔNIO VAZ DE LLACERDA (fl. 1533).[8] The area was separated from the peninsula of Recife by the river Biberibi at one side and from Pernambuco's mainland by the river Capibaribe on the other (*fig. 31*). After the conquest of Recife by the Dutch in 1630, Vaz was completely abandoned by the Portuguese. It housed only one or two mansions and a Franciscan cloister, which soon was converted into the Dutch fortress 'Ernestus'.[9]

FIG. 31. Map of the island of Antonio Vaz and the peninsula of Recife with its small corridor to the city of Olinda at the mainland of Pernambuco. Engraving by WILHELMUS HONDIUS, after a design by the engineer ANDREAS DREWISCH. ['Grond-Teyckening van het Eylandt Anthony Vaaz het Reciff ende Vaste Landt aen de haven van Pernambuco in Brasil met alle de Schansen, Redouten en andere wercken aldaer gemaekt.1640']. (Leiden University Library, COLLBN Port 191 N 66).

Upon his arrival on Antonio Vaz, MARGGRAFE will have been received by an almost namesake. This was the *wachtmeester* of the island, CHRISTOFFEL MARCKGRAEFF from Huppin (probably a German town in Brandenburg), who was appointed as 'duty officer' not long before. Whether or not the two men had any contact afterwards is not revealed by the available sources. The only further mention of CHRISTOFFEL is when he stood witness at a baptism in 1638.[10]

That year most newcomers were located on Antonio Vaz, because Recife itself was 'literally packed to the rafters'.[11] Since the start of Dutch rule in 1630, the construction of new houses on Recife's peninsula had developed very rapidly, especially after the destruction of the nearby city of Olinda on the mainland. It therefore was also very expensive to buy or rent lodgings in Recife, which after all was the mercantile and administrative centre of the Dutch colony. It was here that all the cargo was shipped to and from the Netherlands. This city was always a very busy and crowded place.

The island of Antonio Vaz, on the other hand, still had a lot of space to offer. When MARGGRAFE arrived in 1638, the residential area on the island, called *Het Grootkwartier* ('The Great Quarter'), was walled like a bastion. These fortifications as well as the fortresses 'Ernestus' and 'Frederik Hendrick' were constructed in the early 1630s when the island was still frequently under attack from Portuguese guerrillas from the mainland of Pernambuco.[12] Inside the bastions of the *Grootkwartier* were only a few buildings.[13] As the architectural historian VAN OERS has observed, the *Grootkwartier* was *de facto* an army encampment, supplementary to the adjacent Fortress 'Ernestus'.[14] A detail of a painting by GILLIS PEETERS, dated 1637, illustrates this situation (*fig. 32*).[15]

But change happened quickly. In the years 1634-1640 the military function gradually gave away to civilian purposes. Incoming settlers and sailors needed space to live, so at first barracks were built inside the fortress 'Ernestus' to house some 200 men,[16] but soon also the *Grootkwartier* was filled with new houses. But when this was not enough, plans were made to construct a completely new city outside the *Grootkwartier*. This became a town designed according to the principles for urban design outlined in the early seventeenth century by the Dutch engineer SIMON STEVIN.[17] As a result, the location was renamed into 'Mauritsstad' (in Portuguese 'Mauricía'), not in December 1639 – as stated by NEDERVEEN – but already as early as May 1639, as is demonstrated from MARGGRAFE's letter to GEORG FROM, signed proudly from the *Nova civitate Mauritia in Insula Antonio*

FIG. 32a. The island of Antonio Vaz, as depicted by GILLIS PEETERS in 1637. Detail of the painting in fig. 8.

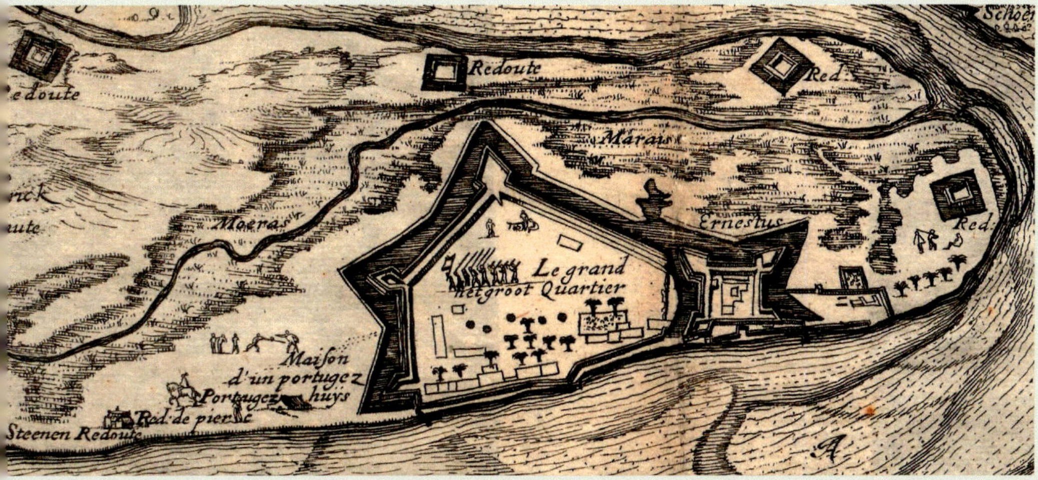

FIG. 32b. The island of Antonio Vaz, showing the fortress Ernestus and the fortified '*Groot kwartier*', to the left of which can be read '*Portugezen huys*' ('a house built by a Portuguese'). This refers to the mansion where Count JOHAN MAURITS established his headquarters, on top of which in 1639 the astronomical observatory would be built. Detail of fig. 31.

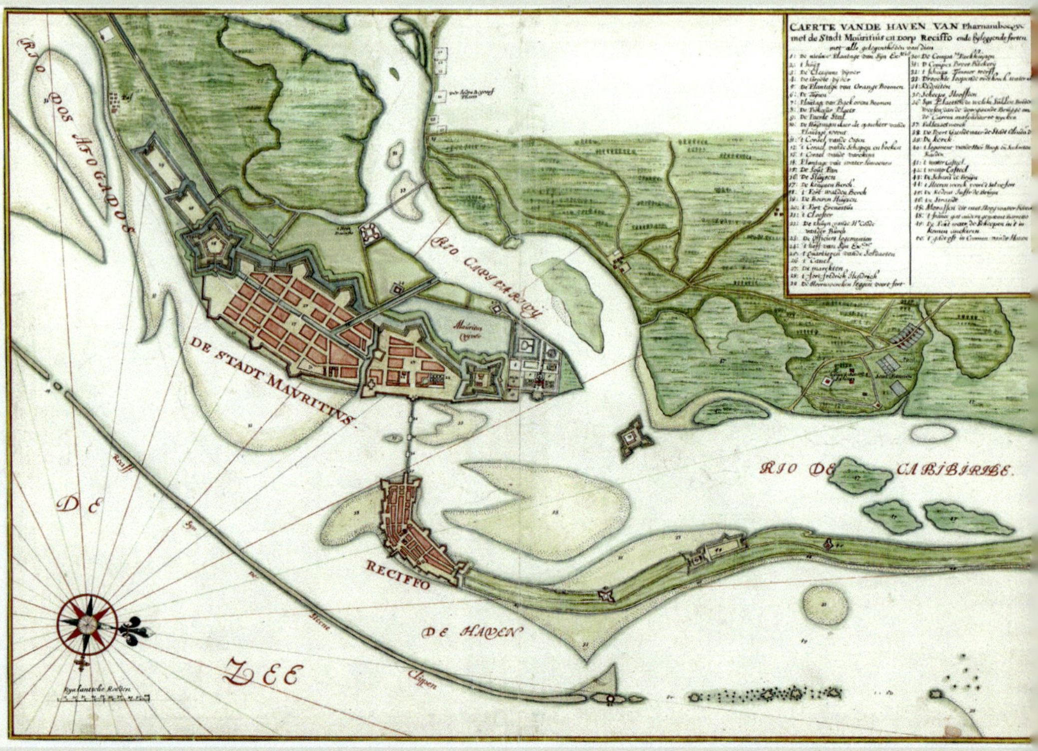

FIG. 33. 'Caerte van de haven van Pharnambocque met de stadt Mauritius en het dorp Reciffe met bijleggende forten en alle gelegentheden van dien' (situation ca. 1644). Map by JOHANNES VINGBOONS (1616-1670). (Nationaal Archief, Den Haag, Coll. Leupe, VEL 619-74).

Vaaz, Brasilia ('the new city of Mauritia on the island of Antonio Vaz in Brazil', see *fig. 33*).[18]

The count's first residence on Antonio Vaz

After his arrival in January 1637 Count Johan Maurits took up residence in a large, distinguished, but old Portuguese mansion inside the *Grootkwartier.* The house was located on the river bank, near the point where the ferry to Recife docked, which boat was connected to that peninsula by a cable. With this ferry a large number of people could be transported to and from Antonio Vaz. The mansion itself was a two-storey building, once built for the former owner of the surrounding lands.[19] Thanks to the efforts of Zacharias Wagener, the book-keeper of the count's kitchen,[20] we have a picture of this first residence (*fig. 34*).

FIG. 34. *Der Hoff seiner Excellenz* ('the residence of the governor'), depicted in 1639 in this watercolour by Zacharias Wagener. On top of the house stands the observatory constructed on behalf of Georg Marggrafe. (*Thierbuch*, Ca 226, fol. 107. Kupferstich-Kabinett, Staatliche Kunstsammlungen Dresden, Photo: Herbert Boswank).

This watercolour is made before April 1641 (when WAGENER left Brazil), but after September 1639 (when MARGGRAFE's astronomical observatory was finished). The picture shows a complex of buildings around a courtyard, limited on two sides by a row of small houses, possibly the quarters of the enslaved Africans. Such slaves were depicted on the courtyard, next to some sawing carpenters. In a corner goats are kept and in the middle a group of servants is lined up, smartly dressed in uniform liveries, probably ready to serve a meal in the dining room. This could be a hall on the first floor, but also on the second. From later sources it is known that there was also a stone staircase at the back of the house.[21] In the garden next door – where presumably vegetables were grown – more service buildings are visible, as well as a dovecote (in Holland a symbol of nobility).

In 2005 the Brazilian architect JOSÉ LUIZ MOTA MENEZES made a reconstruction of 'the House of His Excellency', using his elaborate knowledge of Portuguese building methods, abroad as well as at home (*fig. 35a & b*). He concluded that most likely the ground floor was used for domestic purposes: living rooms, offices, perhaps a larger hall and a kitchen. The upper floor, which could be reached by a staircase in the right front corner, contained rooms and bedrooms, which had their access from the balcony. This gallery was equipped with decorated arcades with stone pilasters, as becomes clear from the inventory of all the buildings on Antonio Vaz, compiled after the Dutch defeat in 1654.[22] It was probably in this gallery that WILLEM PISO lived. Being the count's personal physician he was required to stay in close proximity of the count. So, MARGGRAFE, being PISO's assistant, also lived in this mansion, or at least in one of the outbuildings nearby.

WAGENER undoubtedly also stayed in this complex. His *Thierbuch* demonstrates that he had contact with MARGGRAFE, as some pictures of animals in his *Thierbuch* are evidently copied from MARGGRAFE's watercolours.[23] As WAGENER's picture shows, the observatory on top of the house was an eye-catching element of the count's residence. This extension was constructed in 1639 according to MARGGRAFE's instructions and under his direction. Before the construction of *Vrijburg* palace, this platform was most likely also used as a lookout for incoming ships. The platform provided a nice view of Recife. No wonder that the painter FRANS POST used this high vantage point for the set-up of one of his best works, showing several newly built houses on the island, some with the stepped gables so characteristic of Dutch architecture of the seventeenth century (*fig. 36*).[24] POST's

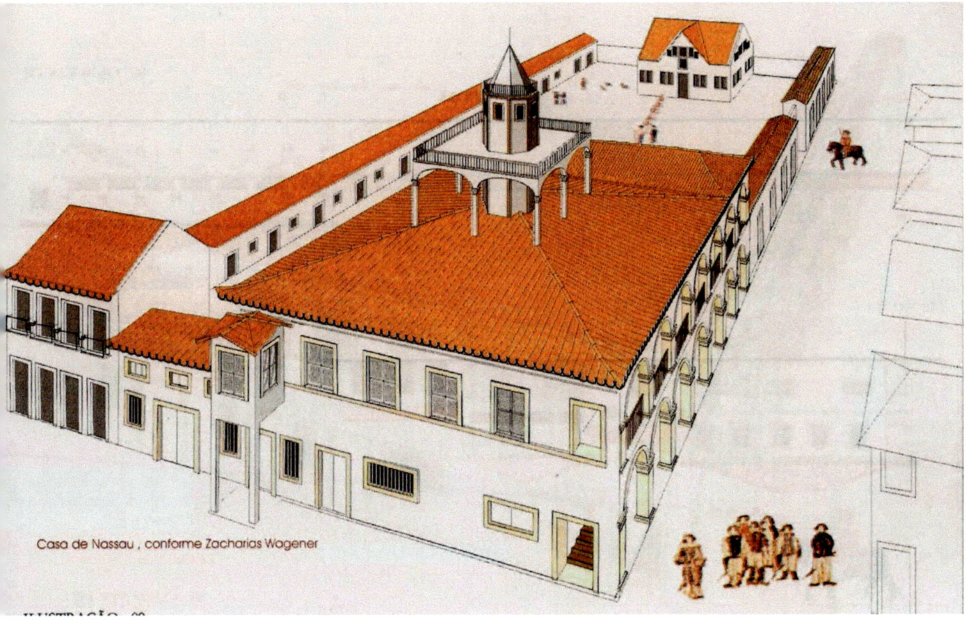

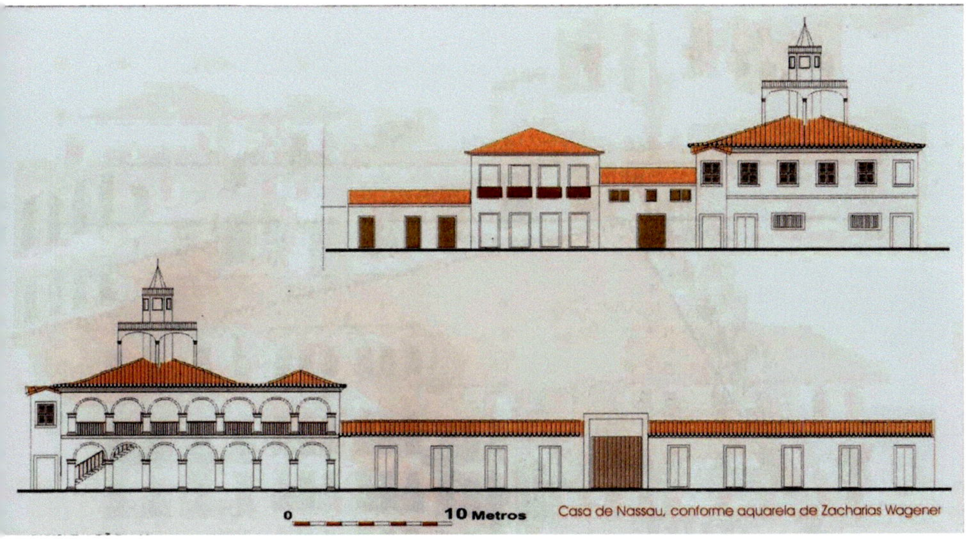

FIG. 35. *Der Hoff seiner Excellenz* ('the residence of the governor'), according to the reconstruction made by the architect José Luiz Mota Menezes in his 'A casa de sua excelencia' (2005).

FIG. 36. View from MARGGRAFE's observatory on the island of Antonio Vaz and the city of Recife on the horizon, painted by FRANS POST in 1653 after his return to Holland. According to the legend of the map VEL 619-74 (*fig.19*), the *Politieke Raad* DIRCK CODDE VAN DEN BURGH (d. 1644) lived in the house on the right (also to be seen on the right in WAGENER's watercolour). (Private collection, Brazil. From: Corrêa do Lago (2007), no. 22).

FIG. 37. Same view as in *fig. 36*. Drawing made by FRANS POST depicting the situation in 1644, the year in which the bridge to Recife was finished. On the horizon the two towers of castle *Vrijburg* can be seen. Former collection JOAQUIM DE SOUSA-LEAO, Rio de Janeiro, 1945. Drawing presumably destroyed by moisture around 1980. (From: De Sousa-Leao, *Frans Post* (1973), 7; See Corrêa do Lago (2007), D55).

painting is dated 1653; it was made after his return to Holland, but certainly after sketches he made *in situ.* One of these sketches, a pencil drawing depicting the same viewpoint, was present in the collection of a Brazilian art collector (*fig. 37*).[25] On it the castle *Vrijburg* can be seen, as well as the bridge between Recife and Mauritsstad, finished in February 1644, shortly before POST's return to the Netherlands.[26]

The painting shows an unknown event that took place between September 1639 (when the observatory was finished) and 1641 (when on the horizon the towers of palace *Vrijburg* – built that year – would be visible). As POST's biographers PEDRO and BIA CORREA DO LAGO concluded, the canvas was presumably commissioned by an official of the West India Company who wanted to rekindle memories of a particular occasion that had taken place during his years in Brazil.[27] It must have been quite an event, for the composition depicts more than 120 people in the fore- and middle-grounds, more human figures than any other painting by POST. Similar to WAGENER's depiction of the setting, all sorts of activity can be seen, officers on horseback, soldiers, civilians and of course ... the enslaved on whose labour the colonial economy depended.

A change of profession

In the first years in Brazil MARGGRAFE worked as an assistant of WILLEM PISO, just as he was supposed to do according the agreements made in Leiden in 1637. For that reason it was probably MARGGRAFE who made the list of contaminated drugs in the local pharmacy, which PISO requested and obtained permission to sell in July 1638.[28] According to the *Vitae*, MARGGRAFE indeed organised a pharmacy in Brazil, and sometimes he also worked as a medical practitioner, an activity he occasionally also had performed during his stay in Leiden.[29] Whether or not MARGGRAFE worked in that capacity in the hospital PISO bought in September 1638, is not known.[30] However, since Piso testified in 1655 that Marggrafe was no longer his *domestique* during the last three years of his stay in Brazil, because he had entered the personal service of Count JOHAN MAURITS, this implies that MARGGRAFE was employed by the count at least in the years 1641, 1642 and 1643. Therefore, his change in profession must be dated somewhere in 1640.[31] Indeed, in 1655, in a letter of his own, the count confirmed that MARGGRAFE had been 'my mathematician'.[32] Another statement, made by one of the count's officials, confirmed that MARGGRAFE for

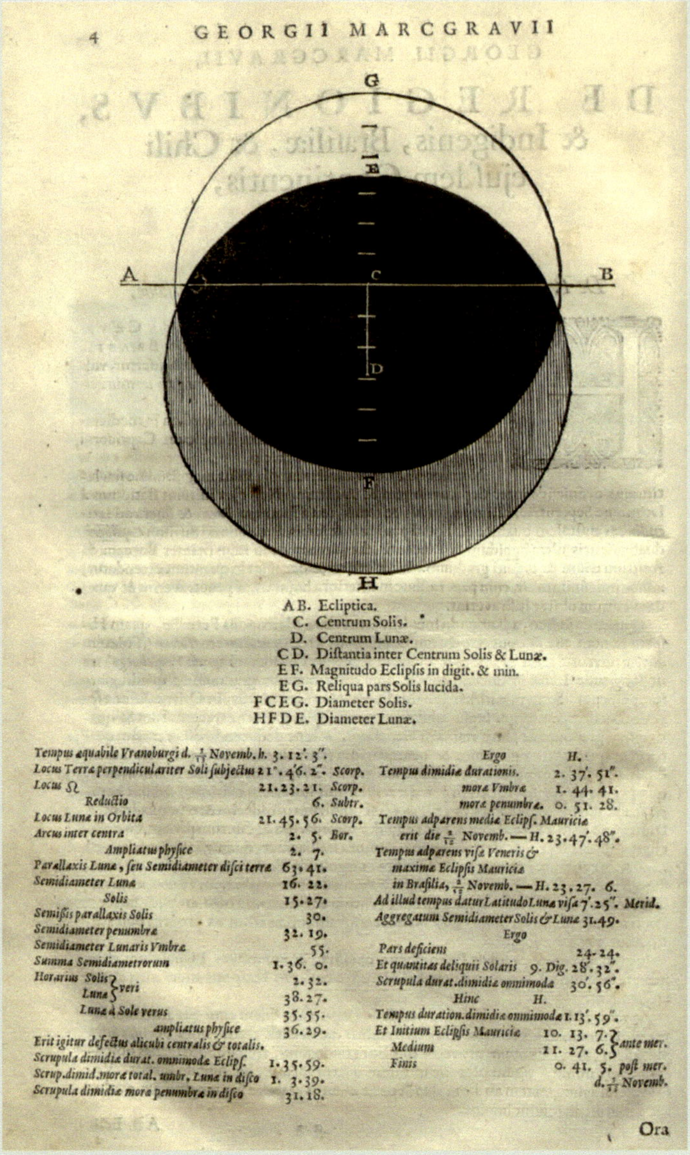

AB. Ecliptica.
C. Centrum Solis.
D. Centrum Lunæ.
CD. Diſtantia inter Centrum Solis & Lunæ.
EF. Magnitudo Eclipſis in digit. & min.
EG. Reliqua pars Solis lucida.
FCEG. Diameter Solis.
HFDE. Diameter Lunæ.

Tempus æquabile Vranoburgi d. $\frac{3}{13}$ *Novemb. h.*	3. 12′. 3″.	
Locus Terræ perpendiculariter Soli ſubjectus	21°. 46. 2″.	*Scorp.*
Locus ☊	21. 23. 21.	*Scorp.*
Reductio	6.	*Subtr.*
Locus Lunæ in Orbita	21. 45. 56.	*Scorp.*
Arcus inter centra	2. 5.	*Bor.*
Ampliatus phyſice	2. 7.	
Parallaxis Lunæ, ſeu Semidiameter diſci terræ	63. 41.	
Semidiameter Lunæ	16. 22.	
Solis	15. 27.	
Semiſsis parallaxis Solis	30.	
Semidiameter penumbræ	32. 19.	
Semidiameter Lunaris Vmbræ	55.	
Summa Semidiametrorum	1. 36. 0.	
Horarius Solis } *veri*	2. 32.	
Lunæ } *veri*	38. 27.	
Lunæ à Sole verus	35. 55.	
ampliatus phyſice	36. 29.	

Erit igitur defectus alicubi centralis & totalis.

Scrupula dimidiæ durat. omnimodæ Eclipſ.	1. 35. 59.
Scrup. dimid. moræ total. umbr. Lunæ in diſco	1. 3. 39.
Scrupula dimidiæ moræ penumbræ in diſco	31. 18.

Ergo H.

Tempus dimidiæ durationis.	2. 37′. 51″.
moræ Vmbræ	1. 44. 41.
moræ penumbræ.	0. 51. 28.

Tempus adparens mediæ Eclipſ. Mauriciæ erit die $\frac{2}{12}$ *Novemb.* — H. 23. 47′. 48″.

Tempus adparens viſæ Veneris & maximæ Eclipſis Mauriciæ in Braſilia, $\frac{2}{12}$ *Novemb.* — H. 23. 27. 6.

Ad illud tempus datur Latitudo Lunæ viſa 7′. 25″. *Merid.*

Aggregatum Semidiameter Solis & Lunæ 31. 49.

Ergo

Pars deficiens	24. 24.
Et quantitas deliquii Solaris	9. Dig. 28′. 32″.
Scrupula durat. dimidiæ omnimodæ	30′. 56″.

Hinc H.

Tempus duration. dimidiæ omnimodæ	1. 13′. 59″.
Et Initium Eclipſis Mauriciæ	10. 13. 7. } *ante mer.*
Medium	11. 27. 6. } *ante mer.*
Finis	0. 41. 5. *poſt mer.* d. $\frac{3}{13}$ *Novemb.*

Ora

FIG. 38. MARGGRAFE's elaborate prediction of the solar eclipse of 13 November 1640, as depicted in Piso's, *Indiae Utriusque re Naturali et Medica* (1658). The same picture was first published in Barlaeus, *Rerum* (1647), 197. (Leiden University Library).

his service received a salary of a *Vaendrig* (a military ensign-bearer).[33]

How and when MARGGRAFE entered the full service of the count is not exactly known. It seems to have been a gradual process. According to his brother CHRISTIAN, he already succeeded in grasping the count's attention shortly after the siege of São Salvador. That battle had 'destroyed an abundance of men', including a number of engineers. So, after his return in Recife, the governor had made a public request 'by the sound of a trumpet', requesting that 'all skilful masters of that art' should come forward, as the count was in desperate need of such artisans.[34] Indeed, a draft letter from MARGGRAFE to GOLIUS, written not long after the siege of São Salvador, already mentions that he had obtained an allowance (*stipendio*) from the governor, which he found rather exceptional.[35] In that letter he also informed GOLIUS that this illustrious count 'MAURICIA NASSOVIA' enjoyed his desire for astronomy and wanted to help him. This 'help' is confirmed in a next letter to GOLIUS, written on in October 1638.[36] In this epistle MARGGRAFE already mentions the efforts to construct a quadrant with a diameter of 10 Rhineland feet [= 3.14 meters], made of the best Brazilian wood, called by the locals *Pao Santo*. This instrument would be part of an astronomical observatory, to be placed 'on a building'. This all, thanks to the generous count.[37]

In short, it can be established that MARGGRAFE's career change was a gradual one. He already started with astronomical observations and setting up the astronomical observatory when he still was employed by PISO. But it is clear that at some moment in the year 1640 MARGGRAFE became fully employed by the count. This precedes WHITEHEAD's suggestion that MARGGRAFE obtained his position as the count's 'mathematician' or cartographer after the departure for Europe of the cartographer CORNELIS BASTIAENSZ GOLIATH in June 1641.[38]

A COURTIER FOR A RULER WITH IMPERIAL AMBITIONS

The most plausible event to mark the start of MARGGRAFE's new career is the upcoming partial solar eclipse in Brazil that was predicted for 13 November 1640. An elaborate advance calculation by MARGGRAFE can be found in CASPAR BARLAEUS's *Rerum per octennium in Brasilia* of 1647. But BARLAEUS puts this rare and impressive event particularly in a non-astronomical context. First of all he tells his readers that 'a number of good-hearted people' in Recife interpreted the eclipse as

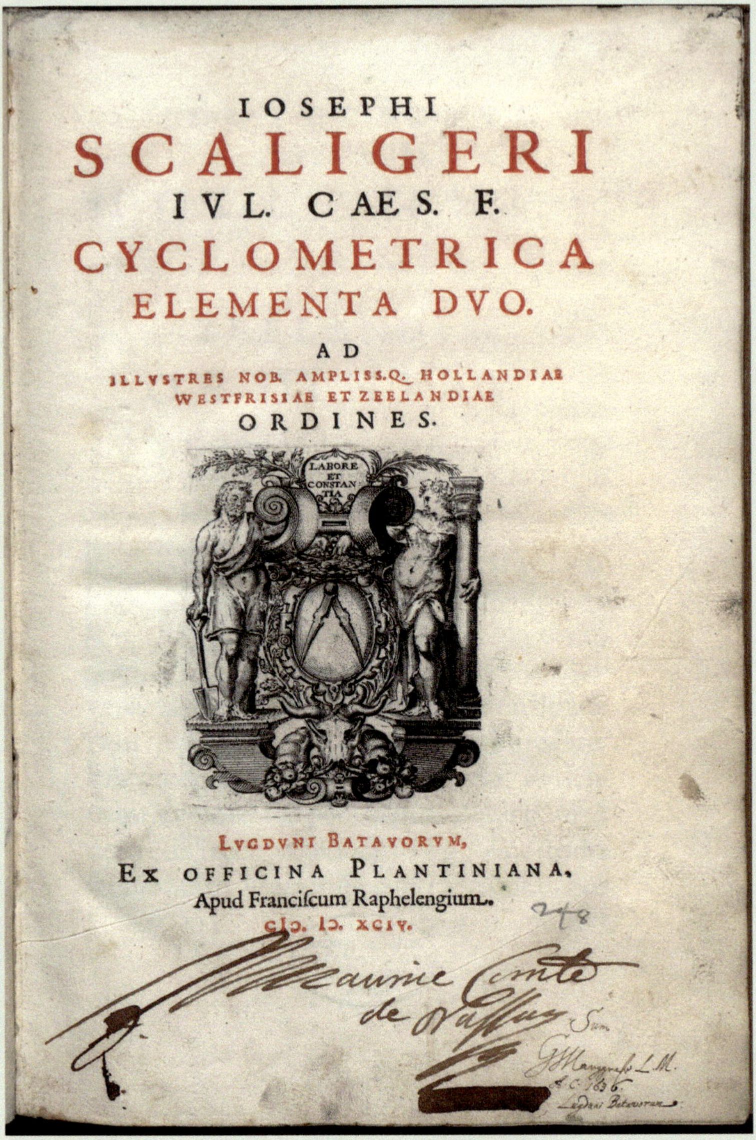

Fig. 39. Copy of Scaliger's *Cyclometrica Elementa Duo* (Leiden: Plantijn, 1594), with the autograph of both GEORGE MARGGRAFE [*GMarggrafe, L.M., Lugdunum Batavorum, A.C. 1636*] and JOHAN MAURITS VAN NASSAU [*Maurice comte de Nassau*] (Universitäts- und Landesbibliothek Bonn, sign. O 236).

a sign that, thanks to 'the prince of Nassau's bravery, care and effort', the 'Hispanic Dominance' in the region would lose its brilliance 'as the Sun also had lost its shine'.[39] Further he revealed that 'the Prince of Nassau' had been extremely fascinated by 'this darkness', in such a way that he had ordered all the ships' pilots, wherever they would be, to witness the event and report carefully to him what they had seen about the shape and duration of the eclipse.[40] BARLAEUS also states that the prince ordered a description and a drawing of 'all the Sun's properties' by 'his Mathematicum G. MARCKGRAVE' (*fig. 38*), similar to what the Roman emperor JULIUS CEASAR had done 'in the old days', when he – also 'in the midst of war' – had enjoyed the course of the heavens; a gesture applauded by the poet MARCUS ANNAEUS LUCANUS (39–65 A.D.) in his epic poem *Pharsalia*.

It cannot be said more clearly how JOHAN MAURITS used MARGGRAFE's astronomy – and certainly also his observatory (visible all over Recife and Antonio Vaz) – as an element to fashion his image as a God-given imperial ruler. As the historian ERIK ODEGARD has noted, building stately palaces, gardens, a zoo and an astronomical observatory 'underlined the elevated position of JOHAN MAURITS in Brazil. These were the symbols of his power and, though unproductive from the company's point of view, they were crucial for JOHAN MAURITS'.[41] Therefore, it is our conclusion that the upcoming event of the solar eclipse of 13 November 1640 marked the moment on which MARGGRAFE officially became the count's 'mathematician', which sets his appointment at some point in the fall of 1640. Was it to celebrate this career change that MARGGRAFE donated to the count a book on the squaring of the circle? A book that he had obtained in Leiden, in 1636? No one can tell, but their joint signatures on the title page certainly seem to reflect their mutual respect (*fig. 39*).[42]

HOROSCOPES FOR LOCAL OFFICIALS

The appointment as the count's personal 'mathematician' gave MARGGRAFE entry to the elite group of some fifty courtiers, who were invited every day to dine at the count's table. Or, to cite ODEGARD again: 'The creation of a true nobleman's court required the creation of social mechanisms of control, so that the COUNT OF NASSAU-SIEGEN could tie the colonial elite to himself, rather than to the official hierarchy of the WIC'.[43] The maintenance of a princely household had been part of the contract between JOHAN MAURITS and the *Heeren XIX* made at his original appointment in 1636.[44] But the increasing cost of this Brazilian court was the subject of several disputes between the *Heeren XIX* and the count.[45] In November 1641 this led to the compilation of a list of persons who were allowed to have a free dinner at the count's table.[46]

This list of 'guests' was compiled by Pierre Bonjour, the French steward of Count Nassau's household, who was also on the list as *Maitre d'[h]ostel.* How close he was with Georg Marggrafe cannot be asserted, but at least they knew each other rather well. It is again a horoscope that proves their contact. On 3 May 1643 Bonjour's son Henry was born, and it was Marggrafe who drew up the baby's horoscope.[47] It seems therefore no coincidence that on the list he is referred to as *Margrav d'astrologue,* or Marggrafe, the astrologer!

Thanks to a cross reference with Recife's baptismal register, more previously unidentified horoscopes can now be attributed to their 'owners'. In this way, just as from Marggrafe's time in Leiden, we obtain a glimpse of his Brazilian network. For instance, the horoscope for a *Filiola* (a Latin expression for 'little daughter'), born on the island of Antonio Vaz on 1 July 1636 (that is before Marggrafe's arrival in Brazil), is most likely made for Elisabeth Silnaer, daughter of a certain Hans Silnaer, of whom nothing else is known than that he lived on the same island where Marggrafe was stationed. The witnesses at her baptism – Pieter Janss, Elsjen Thielemanss and Maria Claess – are also completely without context, which seems to indicate that they belonged to the ordinary working class of *vrije*

Fig. 40. Inscriptions by Pierre Bonjour 'Maistre d'Hotel de Son Excellence en Pernambouc' and Andries Falloo, 'Commandeur van 't Fort De Bruyn', in the *Album Amicorum* of the ensign-bearer Johann Philipp Mulheiser. (Koninklijke Bibliotheek, KW 129 F 6).

luijden ('free people') in Dutch Brazil.[48] A later note by MARGGRAFE states that this little girl died in 1639, so the horoscope must predate that moment, probably long before MARGGRAFE's rise in social status.

More context is available with regard to *Freder. Valloy*, born on 12 March 1643 *int Fort De Bruyn, Mauritia*. This concerns FREDERICK FALLOO, baptised in Recife three days later. His parents were the army captain and military engineer ANDRIES FALLOO, commander of this fortress in the centre of Recife, and his wife ANNA GARIGES, a local Portugese lady.[49] One of the witnesses at this event was her sister MARIA GARIGES, the wife of ... the above named PIERRE BONJOUR![50] Another horoscope marked with only the birth date '22 May 1639' was evidently made for FALLOO's daughter MARIA, who was baptised in the church of Recife three days later.[51] Witness at this occasion was again the steward PIERRE BONJOUR, the baby's uncle through marriage. Both BONJOUR and FALLOO belonged to the upper class of Recife, as is evident from the fact that at the baptism of FALLOO's firstborn son HENRICUS, in September 1641, Count JOHAN MAURITS was a witness in person.[52] No wonder the inscriptions of both men can be found in the *Album Amicorum* of the former Leiden student of theology JOHANN PHILIPP MULHEISER (ca. 1603–ca. 1677), who as an army officer in Brazil continued the academic habit of collecting entries in a small booklet (*fig. 40*). In this way, between February 1640 and May 1644, he brought together the autographs of 93 members of the Dutch Brazilian elite.[53] In MULHEISER's *Album* there is no entry by MARGGRAFE probably because in the years 1640-1641 the latter was frequently on expedition in the interior of Dutch Brazil. Moreover, when in December 1642 MULHEISER was stationed for fourteen months at Fortress Ceulen, he just missed MARGGRAFE, who had surveyed there in April 1642.

One of the most colourful and high-ranking persons on whose request MARGGRAFE calculated a horoscope in Brazil was certainly JOHANNES BODECHER BENNING (1607-1642), at that time a member of the *Politieke Raad* in Recife.[54] Before his appointment in Brazil BODECHER had been a professor of ethics and physics at Leiden University. He was asked to join the *Politieke Raad* in January 1638. Why is unclear, as he had no managerial or judicial experience whatsoever. But BODECHER was well acquainted with quite a number of people from Holland's high society, including some with ties to Dutch Brazil. His collection of poems, published in 1637, has contributions devoted to, among others, CONSTANTIJN HUYGENS, the influential secretary of the Dutch stadtholder; ALBERT COENRAADS BURGH, WIC-

bewindhebber, LAURENS REAAL, VOC- and WIC-*bewindhebber*, and many academic colleagues, both from Leiden (e.g. JACOB GOLIUS and ADOLPH VORSTIUS) and Amsterdam (e.g. GERARD VOSSIUS, CASPAR BARLAEUS and MARTINUS HORTENSIUS).[55]

In February 1638 BODECHER requested – and obtained – permission from the university's trustees to leave his job for Brazil, while retaining his title as professor.[56] He arrived in Recife in June 1638. In a letter written shortly after BODECHER's arrival, MARGGRAFE displayed his great surprise that this former university professor now in a way had become a colleague.[57] However, BODECHER's career switch was no success. Soon reports began to circulate about a 'curious melancholy' with 'strange imaginations' that bothered him.[58] In April 1639 his dysfunction went so far that his colleagues in the *Politieke Raad* lost all confidence in him. 'He might have served better in the academy', the High Council wrote to the *Heeren XIX*, concluding: 'he should have stayed with his friends'.[59] Especially striking was the unhinged lifestyle BODECHER displayed in Recife. Or, to quote from another letter of his colleague BALTHASAR VAN DE VOORDE, 'throughout his life [here], our colleague BODECHIUS has continuously, through drunkenness, fornicating, and otherwise, established himself in such a way that our college was insulted in the eyes of all the inhabitants'.[60] BODECHER had been admonished without improvement, and even a denial of access to the Lord's Table had failed to produce any effect. 'By God, he sometimes has lost his senses', VAN DE VOORDE sighed. This was all the more painful, because it was precisely BODECHER who, in a controversial book published in 1631, had argued against the loose morals in academic circles.[61] Finally, the High Council, the daily Brazilian government, pushed for his departure 'with the nearest ship home'.[62] To save his honour, in August 1639, BODECHER was given the *pro forma* command of the return fleet that would bring him back to the motherland.[63]

Given this resume, it is noteworthy that BODECHER's horoscope is by far the most comprehensive of MARGGRAFE's surviving horoscopes. Twenty-four pages are devoted to numerous calculations with regard to 'I.B.B.'. MARGGRAFE calculated *directiones* (significant astrological events) for BODECHER for no less than seventeen birthdays, ranging from the age of 5 (in 1614) until the future age of 42 in 1649.[64] Had BODECHER requested this, as a way of testing the credibility of the astrological predictions? It seems so. Unfortunately, MARGGRAFE's handwriting in these drafts is so hard to decipher that one can hardly make any sense of what he claims. One document, however, is written in a neat hand. From this *Judicium Generale* we learn that 'according to the planets' BODECHER had 'a tendency for the liberal arts, poetry and eloquence', which was of course nothing more than MARGGRAFE's clever understatement.[65] Especially curious is a prediction for 27 June 1645, when BODECHER would have finished exactly 37 revolutions around the Sun.[66] Unfortunately the receiver of the horoscope would not live to see that day. BODECHER BENNING would die in Leiden in 1642, after having published his own impressions of Brazil in a rare *Epigrammata Americana* in 1639 and being officially declared insane in 1640.[67]

STRIVING FOR THE FULFILMENT OF AN ASTRONOMICAL DREAM

MARGGRAFE was aware that on the night of 25 to 26 June 1638 a total lunar eclipse would be visible from Brazil. But at the *moment suprême*, it was raining with strong winds, with the result that he could only catch a glimpse of the partially eclipsed Moon at rare moments.[68] Despite the fact that he must have been disappointed, the next day MARGGRAFE dutifully set himself to writing two letters to astronomers in Holland. It demonstrates his determination to make contributions to the field of astronomy. In one of the letters, to the Leiden professor JACOB GOLIUS, he solemnly promised to chart the southern sky and to improve the theory of Mercury's orbit. He would regularly submit his observations to him, 'noted down according to the faithful rules of mathematics'.[69] MARGGRAFE congratulated himself to be in this lucky position to work on such a challenging project. Before his leave he had purchased a *Tuba Optica* (telescope), similar to the one he had used at the Leiden Observatory. He also had bought an essential copy of the *Tabulae Rudolphinae*, as well a good quality glass for projecting the image of the Sun in an isolated dark room. But, and that was a kind of prophesy, MARGGRAFE also observed that his new country was very appealing for botanical enthusiasts, not knowing at the time that one day he would become the leading author for obtaining knowledge about the Brazilian *Flora* and *Fauna*. At that moment, being PISO's assistant, he was mainly interested in one specific botanical aspect related to his main task, i.e. the study of herbs for medical purposes. In that context he had brought with him a pair of scales, resembling the one GOLIUS certainly had seen in Leiden, in the house of PISO's mother.

MARGGRAFE expressed similar enthusiasm in the other letter to his Danish friend JØRGEN FROM, also written on the 26th of June 1638. From that epistle we learn that he already had made some *ad hoc* observations of some stars, when his was still on the Island of Texel in the Netherlands, the location where every Dutch East- and West-India vessel waited for the proper winds to sail.[70] With FROM he also exchanged the data of some observations, which went together with MARGGRAFE's salutations to SAMUEL KECHEL and their unnamed landlady (*hospeta*), evidently meaning GRIETTIE BAILLY.[71] Whether FROM ever received this letter is questionable. He had left Leiden in the spring of 1638 and arrived in London in August. In May 1639, when MARGGRAFE wrote a second letter to FROM, this uncertainty was even worse. He then had no idea where to send the letter. So, he asked a couple of merchants in the Netherlands if they could track him down and deliver in this way his letter to FROM.[72]

Performing astronomical observations was not a random thing to do for MARGGRAFE. He had the explicit aim to compose a book on the astronomy of the Southern Hemisphere, to be entitled *Progymnastica Mathematica Americana*, a name obviously inspired by TYCHO BRAHE's *Astronomiae Instauratae Progymnasmata* (1598/1602). In 1648, in the *Historia Naturalis Brasiliae*, its editor DE LAET remarked that, among MARGGRAFE's working papers, he had found the outline of this ambitious project. The scheme of the book would be divided into three parts. In a first chapter on 'Astronomy and Optics', MARGGRAFE intended to present observations of the southern stars, or more precisely those between the tropic of Cancer and the southern celestial pole. In this part, his findings relating to the planets would also be placed. Particularly he hoped to present a new theory for the orbits of the inner planets Mercury and Venus. Further he intended to expand the current theories on astronomical refraction and the solar parallax. This chapter would be finished by a new determination of the obliquity of the ecliptic, some observation of sunspots, and – if observed – other celestial rarities, such as comets and supernovae. A second chapter on 'Geography and Geodesy' would present theories on the determination of terrestrial longitude and the size of the Earth, complemented with a discussion of errors made by ancient and other geographers. Finally, the book would contain a series of astronomical tables, all based on new observations, called the *Tabulae Mauritii Astronomicae*, in honour of his benefactor, the governor of Dutch Brazil.

So, MARGGRAFE's desire to erect an astronomical observatory and to perform daily observations was no impulse. As will be outlined elsewhere in this book, this intention was put into practice on 19 September 1638, when for a short time, he started his first daily astronomical observations. But this happened in fits and starts. In May 1639 he informed his old roommate KECHEL that he had not been able to perform many observations because the construction of the observatory took so much of his time. But after its completion, it would be as useful as Tycho Brahe's *Stellenburgia*, he assured him.[73] That moment of completion occurred on 15 September 1639, more than a year after he had made his first astronomical observation. Finally, MARGGRAFE could make astronomical observations according to the standards he aspired.

EXPEDITIONS INTO THE INTERIOR OF NORTH-EASTERN BRAZIL

MARGGRAFE's ambition to become *the* astronomer of the Southern Hemisphere, in a similar way as TYCHO BRAHE had fulfilled that role for the Northern Hemisphere in the sixteenth century, was traversed by several setbacks. Of course, the weather had to cooperate, but his astronomical ambitions were also hindered by the assignment to participate in several voyages to mainland Brazil. Next to some random journeys to neighbouring places, such as a trip to the Island of Itamaracá in the beginning of October 1638,[74] MARGGRAFE took part in at least three such expeditions between 1639 and 1643.

The first one, a mainly military expedition into the interior of North-Eastern Brazil, started in June 1639. MARGGRAFE's brother CHRISTIAN, who possessed GEORG's Brazilian diary over the years 1637-1640, reports in his *Vitae* the following about this trip:

> The first departure was on the 21st of June 1639, which journey lasted for one month and nine days, until the 6th of August. He set out on that journey with 250 Brazilians, 150 Tapuyas, and 15 whites, waged war with the wild Tapuyas of the jungle, [...] and was engaged in skirmishes with those who ate whatever they killed.[75] During this trip, they hunted many wild beasts and caught many amazing fish. Amongst these fish, one so large that it was sufficient to satisfy the hunger of 15 men.[76]

In 2000, REBECCA PARKER BRIENEN discovered in the archive of the Dutch royal family (Koninklijk Huisarchief, in short KHA) a four-page account of this expedition which she identified – in our view on not such solid grounds – as being a copy of part of MARGGRAFE's itinerary. Indeed, some information about the expedition is almost the identical to the *Vitae*:

> The KHA document states that the party left on 28 June not 21 June, although the end date of 6 August is the same. Furthermore, it reproduces much of the same information, stating that the party left with 250 Brazilians and 150 Tapuyas and 15 whites in order to wage war against the wild Tapuyas. Like the biography, this document also mentions that they caught a fish large enough to feed fifteen men and still leave leftovers.[77]

Of course, MARGGRAFE was among the '15 whites' who probably ate the large fish, but in none of his documents does he write in Dutch, nor are there any specifics mentioned that would identify MARGGRAFE as the author. For example, no mention at all is made of his catch of a living female tortoise on the 20th of July 1639, which animal survived as his pet until April 1641.[78] BRIENEN overlooks further the fact that among the participants in an expedition were more educated persons able to write an account of the same journey. For instance, it was the habit to send with each expedition a Dutch minister who also could comfort the sick (*een predicant of sieckentrooster*). It was HENDRICK FREDERICK VAN MANSVELT – an officer mentioned in CHRISTIAN's *Vitae* as being involved in these expeditions – who had personally requested this assistance during a meeting of the Brazilian Church leaders in October 1638.[79]

Nevertheless, the document found by BRIENEN, and published together with VAN DEN BOOGAART, is the one of the few available sources that provide some insight into the conditions of such an expedition.[80] The account firstly reveals that the point of departure was Ceará, north-west of the Rio Grande, some 800 km from Recife. Ceará (in Dutch 'Siara') was a Brazilian captaincy, captured from the Portuguese in 1637. The area was found interesting for the Dutch because of the assumed presence of silver mines (*fig. 41*).

The document from the Koninklijk Huisarchief further shows that the expedition primarily served a military purpose. The aim was to weaken the support given by the 'wild Tapuyas' to the Portuguese still present in the environment. The battle on the 21st of July between those native Americans who fought on the Dutch side (the Tupís and the 'civilized' Tapuyas)[81] and others supporting the Portuguese side (the 'wild Tapuyas'), receives therefore prominent attention. In this

FIG. 41. The small Dutch stronghold in Maracanaú, Ceará (in Dutch 'Siara'). Detail from Barlaeus, *Rerum per octennium in Brasilia* (1647). (Rijksmuseum, Amsterdam).

battle 150 'enemies' were killed, while the losses on the Dutch side were 'only' nine Brazilians, three Brazilian women, and seven of the 'Dutch' Tapuyas. To the amazement of the Dutch, the Tapuyas 'chopped up their dead into pieces and ate them'.[82] After the battle fifty children of the 'wild Tapuyas' were taken hostage, which group was set free only after the promise (broken not long after that) that the 'wild Tapuyas' would support the Dutch colony in the future (see *fig. 42*).[83]

One may wonder why MARGGRAFE was present during such a military campaign? He definitely served not as a military man, or as a cartographer as he would do in 1641. He must have served the expedition in his role as PISO's assistant, searching for medical herbs and plants in the uncharted parts of the Brazilian inland. Needless to say, MARGGRAFE, driven by his eager curiosity, also had a keen eye for the other parts of nature. As the account about Ceará reveals, during their circa forty-day trek the company came across many animals and fish. They saw or hunted, for instance, wild boars, armadillos, coati, hedgehogs, rheas, ducks, aperea and tigers. We may easily assume that he drew some of these during this journey. It is certain at least that after his return, he planted some seeds from the *Ambaiba Brasiliensibus*, a high tree, as well as some other plants brought from this expedition in his garden on Antonio Vaz.[84]

FIG. 42. Dance of native Americans belonging to a Tapuya tribe. Painting made in the years 1641-1643 by ALBERT ECKHOUT (c.1610–c.1666) (Nationalmuseet, Copenhagen)

THE EMERGENCE OF AN ADDITIONAL GOAL

Mid-January 1640 MARGGRAFE was forced to stop his daily routine of performing astronomical observations because, as he wrote in his observation journal: 'In the remainder of the month as well as on February, for the most part we lacked fine weather and several times I went away'.[85] But from MARGGRAFE's letter to DE LAET, written on 6 February 1640 from the recently renamed city of Mauricía, we learn that there was another reason that caused a delay in the astronomical ambition. As this letter is one of the few drafts that more or less can be deciphered, we give the text here in full:

> To Mr. JOHANNES DE LAET, who lives in Leiden.
>
> Dear Sir,
>
> Your great favour towards me has been shown many times in the letters you have sent to the illustrious Count MAURITS OF NASSAU and

also to Mr. PISO. Please note that I send Your Honour this letter, as a testimony of my respect, and to let you know that I [always will be] Your Honour's most humble servant. Therefore, I urge the request, Your Honour, to take good notice of what I report to you briefly in my message. [Furthermore], I do not want to bother you, because I know all too well how busy you are every day.

Thanks to God, my health is still good, but my studies are not going as predicted, [but] everything is set up to track down the beauties of nature, which is very rich here, in heaven as well as on earth. At present, I have a little more than three hundred and fifty plants, diligently outlined in word and fine artwork. I have also [drawn] many animals, birds, fish and other bloodless animals (which we call insects). Each day the number increases, as a result of Mr. Piso's work and my own diligence. By God's grace, in a short while, I will present it all to the wise and grateful world, when I indeed will deliver it to the homeland of our Lord.

We have already sent some animals, fish, as well as various seeds, fruits, roots, teeth, and other natural things. In due time, we will send more. The forthcoming presents are to demonstrate our study and to gain the goodwill of our favorers and friends. I am running out of words. Most of all, regarding Your Honour to be my favorite friend, may I never relent to pay [back] – possibly everything – with a job well done.

Being forever Your Honour's most devoted servant, Marggrafe.
In the city of Mauricía, Brazil, 6 February 1640.[86]

From this letter several things can be concluded. Firstly, by that time WILLEM PISO already prepared a text for what later would be published by DE LAET as the first part of the *Historia Naturalis*. Secondly, it is clear that meanwhile, MARGGRAFE, too, had become fascinated by the beauty of – for Europeans – the very unfamiliar beauty of Brazilian nature. He had moved towards collecting the natural resources of which this country is so rich. As a skilled draughtsman, he already had made watercolours of 'a little more than three hundred and fifty plants', and worked further on drawings of animals, passerines, fish and insects. Thirdly, MARGGRAFE confirmed that he, too, worked on an accompanying text to his drawings and watercolours, as a way of returning to DE LAET all the favours that this *Mercator Sapiens* had done for him. Finally, this letter shows that several natural curiosities had already been sent home to the Netherlands.

Indeed, as early as June 1639 the Danish student THOMAS BARTHOLIN informed his uncle ØLE WORM in Copenhagen that 'the excellent doctor PISO, private physician to Prince MORITZ, [...], had donated to the public garden of this university some exotic plant seeds, a few large pods of cassia tree, a snakeskin and a piece of sugar cane'.[87] BARTHOLIN even succeeded to steal some seeds of the cassia tree from the Hortus's gardener.[88]

SETBACK IN MARCH 1640: THE COLLAPSE OF PART OF THE COUNT'S RESIDENCE

The next months, MARGGRAFE was mostly engaged with Natural History. In the beginning of March 1640 he was on a trip some 330 km south of Recife, in what he called the Aquapetida cove, where he discovered a fruit tree with orange seeds, described later in the 'Historiae Rerum Naturalium Brasiliae', MARGGRAFE's part of the *Historia Naturalis Brasiliae*.[89] These chapters mention also the flowering of a stone apple in his garden on Antonio Vaz in April, a dissection of an iguana in May, and the observation of butterflies in July.[90] He would not resume his astronomical observations until mid-June 1640. But the main reason for this lack of astronomical efforts was of a completely different nature: it appears that MARGGRAFE was temporarily handicapped in mid-March. About what happened he informs us in his astronomical notebook:

> In the night following 18 March [1640], our house where we live, spontaneously collapsed altogether while we all were sleeping. But thanks to the extraordinary power of God we all, seven in number, were spared alive, although wounded thence in the limbs, my furniture thence suffered damage, including the chest. Until the house was repaired, what was made after a time span of a trimester, I was constrained to abstain from mathematical works; having also occurred a dislocation of my left armpit, what made me useless for accomplishing anything and detained me more than two months.[91]

The collapse of the count's residence concerned only the part of the house where some of the count's *domistijken* (servants) slept, 'seven in number', according to MARGGRAFE. One of them was the *advocaat-fiscaal* HENDRICK DE MOUCHERON (1612-1670), whose head was severely hurt by the collapse.[92] He was married, so his – then pregnant – wife CORNELIA STRUYS and their young child COSMO were probably also accomodated in this section of the building. The count's physician, WILLEM PISO was a witness at the baptism of both DE MOUCHERON's children, so he, too, was close to this family.[93] Was PISO also among the seven affected by the collapse? Living in the count's proximity would indeed make sense, which would imply that MARGGRAFE shared his lodgings with his superior. Finally, it is known that the count's personal pastor and devotional poet FRANCISCUS PLANTE also lived in this part of the building.[94] But there are no known injuries from his side. This shared accommodation explains why in later years, CHRISTIAN MARGGRAFE named PLANTE – then a professor at the Illustrious School in Breda – as one of the persons who had known his brother well.[95]

Now that we know that HENDRICK DE MOUCHERON lived under the same roof as MARGGRAFE, it is interesting to learn more about this extraordinary man with special qualities. All the more, because, before Brazil, DE MOUCHERON had worked for a short time as a military engineer in Russia.[96] So, it is evident that he shared with MARGGRAFE an interest in mathematical topics. Therefore, it is conceivable that they have discussed astronomical or cartographic matters together. Born in Amsterdam, DE MOUCHERON matriculated at Leiden University threc times, (1) on 12 June 1629 as a student of philosophy, (2) on 8 November 1631 as a student in mathematics, and (3) on 17 August 1635 as a student of law.[97] In 1634, DE MOUCHERON worked for a year in Russia. Returned to Leiden, he evidently was a close friend of BODECHER BENNING, who not only devoted two of his poems to him (in his *Poemata* of 1637), but also shared with DE MOUCHERON one of Leiden's prostitutes, a certain JANNETJE JANSDR JUWEEL.[98] In 1639 DE MOUCHERON was recruited by the West Indian Company to become the new *Schout* (sheriff) of Recife. There he quickly rose to the rank of *advocaat-fiscaal* (public prosecutor) of the entire colony.[99] In 1648 he returned to the Netherlands, but joined the service of the Swedish king two years later. In Sweden DE MOUCHERON would become an influential magistrate. The text of an important Swedish Sea Law – passed in 1667 – was mainly his work. DE MOUCHERON was so much appreciated in Sweden that he was included in the Swedish nobility in 1666. He died in Stockholm four years later, leaving behind a large library with mostly books of Dutch origin.[100]

Fortunately for MARGGRAFE, the collapse of the front part of the count's residence seems to have had only little impact on the newly built observatory. Had the extra weight of the observatory been the reason for the collapse, this superstructure would certainly have been removed, which would have meant the end of MARGGRAFE's astronomical project. Nevertheless, it is telling for the overall state of the complex that the buildings adjacent to the count's residence were also in a bad shape. For instance, when in August 1642 one of these was sold for scrap, it was described as 'the dilapidated house, next to the old residence of his excellence in Mauritsstadt, being a house that had long fallen into disrepair'.[101] Anyway, MARGGRAFE's dislocated right shoulder healed, and on 11 June 1640 he resumed his astronomical observations in a series that continued until 17 October 1640.

Collector of naturalia for a German prince

In the fall of 1640, MARGGRAFE participated in two minor expeditions, one (with a duration of three weeks) starting in October and the other (with a duration of 11 days) in December.[102] Both expeditions were probably similar to the first one, combining military objectives with the exploration of Brazilian nature. No destination is known this time, but given the rather short duration of both journeys, it is plausible to assume that they took place in Pernambuco, the hinterland of Recife. About the first trip, the *Vitae* only speaks about 'a journey on horseback, being guarded by a company of Soldiers', with as the only peculiarity the observation of 400 birds' nests in a tree called 'Uti'.[103] The daily registers of the Brazilian High Council reveal that this first tour was guided by Major VAN MANSVELT, who led a company of 'twelve Brazilians, a sergeant and a batch of soldiers'.[104] The *Vitae* also mentions VAN MANSVELT in this context, stating that these expeditions were the result of a direct order of Count JOHAN MAURITS, and that their main objective was the gathering of information about the 'geography and natural things' of the country. Important, too, was to catch 'many birds and fishes', which were 'brought to ye Prince'.[105] Apparently these trips were also meant to fill the count's zoo with various Brazilian animals. This private character, especially of the second expedition, would also explain why almost nothing about these expeditions can be found in the records of the High Council, where in normal cases every expedition for the benefit of the West Indian Company would be recorded.

It is not such a bold supposition that MARGGRAFE, after his return from these trips, made one or two maps on its own initiative, just to demonstrate his capacities in this field. Being an astronomer, he could make better maps than an ordinary surveyor.

Gathering cartographic information

Cartography was an important business for the Dutch Brazilian authorities. Or to cite the cartographic historian Kees Zandvliet:

> After 1637 many Portuguese settlers left the area concurred by the Dutch, leaving behind their farms, many of which were equipped for growing and processing sugar cane. The government in Recife was therefore forced to sell or lease these concerns and the land that went with them to new owners. It was essential to survey and map this land both for administrative reasons and as a way of getting to know the hinterland where the Dutch hoped to find gold and silver. For this reason, they tried to penetrate deep into the interior along several rivers.[106]

The desire to have general map of the colony was first mentioned in January 1638.[107] And indeed such a map, primarily based on information gathered from Portuguese sources, was shipped to the Netherlands on 2 March 1640.[108] But as Zandvliet remarked: 'The geometry of the map was poor due to insufficient time for extensive field work'.[109] So, early in 1641, shortly before the earlier map maker of the count – Cornelis Goliath – left Brazil, Marggrafe received the assignment to collect better geographical coordinates. Whether or not he was instructed by Goliath is unknown. It is very likely that the two knew each other. In the small world of Dutch Brazil, a meeting between kindred spirits was inevitable. Besides, it probably was Goliath who had drawn the previous map of the Dutch territories, so an exchange of knowledge about Brazilian cartography was the obvious thing to do.[110] Anyway, based on the verified new information that Margrave would collect, a better map of the Dutch territories had to be produced. This new assignment, however, would give a severe blow to Marggrafe's astronomical project. He was forced to refrain from the daily observations required for his astronomical program; he had to leave his coveted observatory unoccupied for nearly 20 months! So, after his nightly observation on 7 February 1641, he recorded in his astronomical notebook:

On 9 February 1641, I departed and I was absent nine uninterrupted months for the sake of Geography. Coming back on November, I spent one month in Mauritia, but most part absent in neighbouring places also for the sake of Geography. On 10 December I departed to the southern part from here of Brazil for the sake of Chorography[111] and Natural History. Meanwhile I was forced to abstain from astronomical works. In the future however, God granting life and health and resuming uninterrupted work, I will finish all that still remains (to be done) in this part (of the world).[112]

The daily minutes of the Political Council leave no other conclusion than that Major Van Mansvelt was once again commander of this cartographic expedition. Already in December 1640 he had received a large amount in cash for a mission into the interior of the country. More money was added in January 1641.[113] From scattered evidence we can deduce that Van Mansvelt's troops crossed the country in all directions. At the end of February 1641 he probably was not far from the bay of Porto Calvo,[114] and we can determine where his company was in April 1641. That month, Van Mansvelt received an urgent order to chase some Portuguese guerrillas who had just burned down the premises of the sugar mill of a certain Nicasio Winnenvelt. This *ingenio Nicas* lay on the river Capibaribe in the interior of Pernambuco. Van Mansvelt came to the rescue from the north, so before that he was in the inland of Itamaraca.[115] But then Van Mansvelt reluctantly was forced to accept the appointment of military commander of the city of Olinda. The soldiers there indulged themselves in all kinds of excesses and needed a tough hand.[116] So, Van Mansvelt had to leave Marggrafe's expedition. He evidently did not appreciate his new job, for in the early summer of 1641 Van Mansvelt returned to the Netherlands. He bought a nice canal house in Amsterdam, where he and his wife continued their life as wine merchants.[117]

We have not been able to determine who became the new commander of Marggrafe's expedition. Anyway, in mid July 1641, Marggrafe received a special order 'to discover the area of the St Francisco River'. At that occasion his name is recorded for the first time in the minutes of the High Council in Recife:

The [council] has written to Mr. Gijsbert de With, director in the [region] Rio St Francisco,[118] now that this area has not been deserted, [with the request] that in case of hostilities by the Portuguese, he will provide assistance to the commander of the

> boat and to the Mathematician MARCKGRAEF, to ensure that they will have enough sailors and soldiers [with them] to be protected from the Tapuyas, and also [that they will have] sufficient tools and any other necessities that they may find useful to rise higher [up the river], beyond the falls, to explore the western regions of Brazil. This so that we can receive a relevant report on the minerals and other [valuables] that may be located there.[119]

Since this area is now known to be devoid of such resources, we can be sure that such 'relevant report on minerals' was never produced. The archives of the WIC further mention only the expenses of the skipper who took the group from Porto Calvo to Fort Mauritius on the Rio São Francisco in August 1641.[120] This fortress functioned as a base for MARGGRAFE's cartographic work. In his 'Historiae Rerum Naturalium Brasiliae' he mentions that he kept an ornate hawk eagle. called 'Urutaurana' (currently *Spizaetus ornatus* – see *fig. 43*), there for quite some time.[121]

Unfortunately the contents of the preliminary expedition report, sent to the High Council in September 1641, are not preserved. The daily minutes mention only its receipt and more expenses made:

FIG. 43. Woodcut of the ornate hawk-eagle named 'Urutaurana' (*Spizaetus ornatus*), a bird MARGGRAFE kept as a pet during his stay in Fortress Maurício next to the Rio São Francisco in the summer of 1641. From *Historia Naturalis Brasiliae* (1648), part 2, (1648), 204.

> From Rio St Francisco on the discovery of the western region: [received a report] from Rio St Francisco from Mr. Gijsbert de With about the state of affairs there, and particularly with regard to the discovery of the western parts beyond that river, ordered by him. About which he showed that 300 men were needed, which at the time caused new costs for that garrison; [expenses were] also [made] for more and other necessities, which are specified in a list made by him and the Mathematician Marckgraef, on which document one will further deliberate.[122]

This document, too, has failed to survive. But it evidently touched on the costs of the survey, for it is recorded on 30 September 1641 that Gijsbert de With returned to his base at the Rio São Francisco with a thousand guilders as payment for the expenses named above.[123] De With came back to Recife in January 1642, where he presented orally a more elaborate account of the expedition's findings in a meeting of the Political Council.

> Mr. Gijsbert de Wit, former director in Rio St Francisco, gave an oral presentation at our meeting concerning the situation of that district. His Honour was told that he would put it in writing and hand it over [to the council], so that one could be better informed about everything there.[124]

Not a word has been spent here on Marggrafe, who did all the work. Where every trace of De With's written report is missing, fortunately Marggrafe's map of the area of the Rio São Francisco (the *Praefectura De Cirîiî*) has been preserved (*fig. 44*).

From this expedition only one single observation in the field has been handed down to us. In his meteorological notes, on 2 July 1641, Marggrafe mentions a visit to a high mountain called *Itapuamuru*, where he experienced in midday that his hand froze and that droplets from his exhaled breath covered his hair and beard.[125] Henk den Heijer and Ben Teensma have recently identified this spot on Marggrafe's map of the *Praefectura De Cirîiî*, where in the region Itâ Puáma (a corrupted name for Itâ Baiana) the route of the expedition can be followed as a dotted line.[126] In this region Johan Maurits expected the presence of a silver mine, for which he had high hopes. Indeed, two mines are shown on the map, as well as the high mountain called on the map *Ita puámucû*, closely resembling *Ita puamuru* in the meteorological notes (*See fig. 45*).

FIG. 44. Map of the southern part of Dutch Brazil around the Rio São Francisco, an area surveyed by MARGGRAFE in the summer of 1641. The animals were added in 1647 by the Amsterdam cartographer JOHANNES BLAEU, after pictures delivered by FRANS POST. (Koninklijke Bibliotheek, Atlas Van der Hagen).

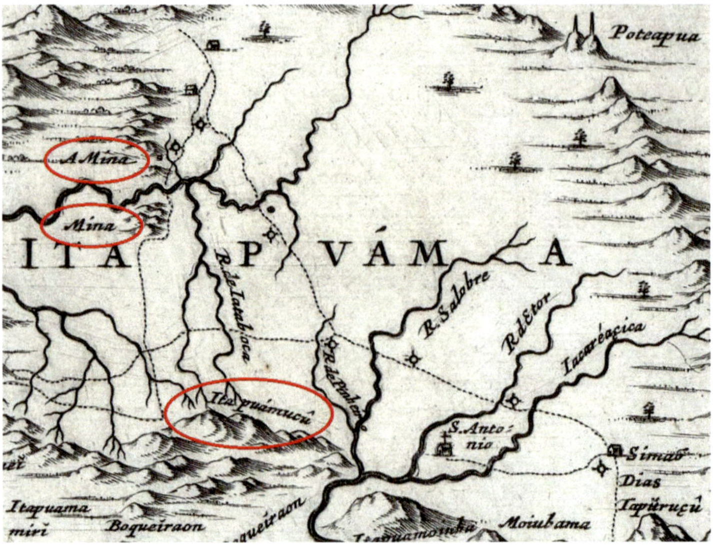

FIG. 45. Detail of the upper left part of *fig. 44*, showing the route of MARGGRAFE's expedition as a dotted line in the area of Itâ Puáma, where on the mountain *Ita puámucû*, MARGGRAFE's exhalation froze his beard. Also shown are the two mines inspected.

Interesting, too, is that TEENSMA also has identified an anonymous six-page manuscript in the Koninklijk Huisarchief, with the title *Wegwijser van de stad Mauritia tot aen 't fort Mauritius aen Ryo St. Francisco geleegen* ('Guidebook for travelling from the city of Mauricía to the fortress Mauritius, located on the Rio St. Francisco'), as belonging to the map drawn by MARGGRAFE after his expedition to this territory.[127] The text describes not only in detail the route from place to place, and where along the way grass and drinking water for the horses could be found, but the document also states how much time it took travelling from one point to another. In an almost mathematical approach, the travel time required for each segment is given in units of a whole, half, quarter, and even eighth parts of an hour.

TEENSMA argues on the basis of Germanisms that the Dutch text was translated from a German original, of which the 'mathematician' MARGGRAFE could have been the author. However this may be, the congruence between MARGGRAFE's map and the *Wegwijser* is so striking that according to the cartographic historian MARTIJN STORMS it seems certain that MARGGRAFE has used the information of the *Wegwijser* during the construction of his map of the area:

The route in this manuscript can be easily followed on MARGGRAFE'S map. Virtually every river that has been crossed and every settlement or church that has been passed is marked on the map. It took a total of 96 hours, without breaks, to travel from Mauritsstad to Fort Mauritius on horseback. It is striking that the route through what is now the province of Alagoas almost completely followed the coastline. The inland area was probably less known here. The last part of the route to Fort Mauritius followed the so-called *Gouverneurspad.* A passage in the *Wegwijser* describes the route between Porto Calvo and the *Limoenenberg* [...], one of the few Dutch-language toponyms on the map [*HJZ: no. 3*], which mainly contains Portuguese place names (*fig. 46*).[128]

FIG. 46. Route described in the *Wegwijser*, drawn on the two maps of Southern and Northern Pernambuco (1647). Picture from Storms (2011).

Returned to Mauritsstad in November 1641, MARGGRAFE received a supplementary mission before the end of the month. This time he had to survey an area around the river Machioppe, a tributary stream of the river Capibaribe, northwest of Recife. Again we learn more details from the minutes of the High Council in Recife, where the order to assist him is clearly stated:

> The bailiff HONINCKHUIJSEN[129] of Iguaracú charges everyone to give the Mathematician MARCKGRAEF all the help that [is required to construct] a perfect map on paper of the lands of Machioppe and the Caruba belonging to the Company, in order that the partition and the limits [of the lands] with its size and location can be designated better, to be used when these lots are sold (which is intended), to solve every question, according to the usual specifications.[130]

So, on 10 December 1641, MARGGRAFE was on his way again, now heading north, still carefully mapping the areas he traversed. That he kept a keen eye on the natural environment is proven by his account of a dead iguana on which he stumbled. When he curiously dissected this animal, he was surprised to find in its stomach a stone the shape and size of a chicken egg.[131] He also described a curious stick insect in January 1642 and noted that a certain plant already flowered in February 1642.[132] But when he reached his outer limit north, the fortress Ceulen in Natal, at the mouth of the Rio Grande do Norte (*fig. 47*), he would and should still pay attention to his original astronomical mission. For in the evening of 14 April 1642, a lunar eclipse was predicted, so MARGGRAFE took his portable sextant to observe the event, which would have been entirely visible from that location. However, a little later the sky became clouded and it started to rain. But luckily, 'an excellent fine weather' returned, so he could make some measurements which also would come in handy for a better determination of the geographical location of the fortress.[133]

PROCESSING GEOGRAPHICAL INFORMATION

When MARGGRAFE returned to his base on 16 June 1642, he had gathered a load of information about all of the territories that he had crossed.[134] The challenge now was to organize all the information in a number of handy maps. But to combine all the small maps he made

FIG. 47. *Fort Ceulen* near Natal (the former Portuguese fortress *Dos Reis Magos,* renamed in 1633), where MARGGRAFE observed the lunar eclipse of 14 April 1642. (Nationaal Archief , Den Haag, Cat. Leupe).

into larger ones, he needed a pantograph, an instrument to copy, reduce or enlarge diagrams and maps from one sheet to another. It was a device invented by the German astronomer CHRISTOPH SCHEINER, who had published about it in his book *Pantographice seu ars delineandi res quaslibet per parallelogrammum lineare seu cavum* (Rome 1631). In Leiden, MARGGRAFE had used SCHEINER's method to project sunspots, so he was well are of SCHEINER's work. But such a precision instrument had to be ordered from Holland, most likely from the Leiden-based firm of the mathematical instrument makers JAN and JACOB DAVID. Apparently this tool arrived in February 1643, as the surviving account of that month records that the ensign-bearer JOOS COPPIJN received a hundred florins for the delivery of 'an instrument and some paint to construct some maps for the Noble Gentlemen of the High Council'.[135] So, now the final maps could be made. This result, drawn on four partly overlapping sheets, reveals the terrain MARGGRAFE explored.[136] The charted area comprises a relatively narrow coastal strip between the Rio Grande do Norte and the city of Natal in the north and the Rio Vaza Barris near Aracaju, in the southwest. It is a route about 800 kilometres long and 80 kilometres wide. A few tracks go further inland, especially around the river São Francisco.[137] The mere fact that the northern territory of Ceará is *not* included in his map is in itself a proof that MARGGRAFE's inclusion in the earlier expedition to Ceará was for reasons other than surveying.

MARGGRAFE's four maps were largely based on his own information. The route taken in the various areas is expressed on the maps by small but clearly visible dotted lines. In his discussion of MARGGRAFE's general map of the colony, ZANDVLIET has alleged that he used also information from other expeditions, namely:

(1) an exploration of salt pans in the Capitania Rio Grande in the summer of 1641 and (perhaps) the spring of 1642;
(2) information provided in May 1641 about the newly built fortress in Sergipe del Rey, south of the Rio São Francisco, the 'official' southern frontier of the colony;
(3) the expedition lead by ELIAS HERCKMANS to the inland of the Capitania Paraiba in September, October and November 1641.[138]

However, after a closer inspection of the sources to which ZANDVLIET referred, only the detailed information provided by HERCKMANS about Paraiba seems to have contributed to MARGGRAFE's map.

Concerning the salt fields 'in the Capitania Rio Grande', ZANDVLIET refers to the minutes of the Political Council of 15 March 1641, which mention an invitation of the commander of the Fortress in 'Siara' to explore an area near the river Upanema where he had found these salt pans.[139] The 'second exploration of the salt pans in Capitania Rio Grande' concerned this same field of dry salt in the Capitania Ceará.[140] Evidently, as stated above, this part of Dutch occupied Ceará is *not* included in MARGGRAFE's map.

With respect to information about the newly built Dutch fortress in Sergipe del Rey, it is indeed stated in the minutes of the Political Council that ELIAS HERCKMANS, who had inspected in May 1641, brought with him a – now lost – map of the surroundings of this newly built Dutch fortress, but on MARGGRAFE's map of this area no fortress can be observed.[141] So, this information evidently was not used in MARGGRAFE's cartographic project.

On the other hand, HERCKMANS indeed provided valuable information to MARGGRAFE from his expedition to Copaoba (named by the Dutch 'Copaiba'), deep inland in Paraiba. This trip took place between 3 September and 4 November 1641. For a better understanding of the way in which MARGGRAFE's map was built up, it is important to know that he, after his return in November 1641, promised the Political Council to draw a 'pertinent map' of the land he had crossed, so that he better could explain what he had seen and experienced.[142] This HERCKMANS map, which is now lost, must have be the basis for the central part on the sheet showing the *Praefectura de Paraiba, et Rio Grande*, recognizable by the toponym C O P A | I B A (*fig. 48*).

HERCKMANS knew Paraiba well. He had been governor of the provinces Paraiba, Itamaracá and Rio Grande between 1636 and 1640. In that period, he made an extensive report about the coastal part of Paraiba, which since its publication in 1879 has been hailed by many scholars as being extremely important for its many geographical and ethnographical details about seventeenth-century Brazil.[143] A report of HERCKMANS's 1641 journey to Copaiba also survived, because it was included in BARLAEUS' *Rerum per octennium in Brasilia*.[144] As no itinerary of MARGGRAFE's geographical expeditions has come to us, it is important to dwell on this report more in detail, because it also shows in a more general sense how such trips were conducted. BARLAEUS, for instance, provides us with the official commission of the expedition, namely 'to search diligently the lands and the deserts of Copaiba, and investigate what this land has to offer and what could be enjoyed from it'.[145]

FIG. 48. Map of the *Praefectura de Paraiba, et Rio Grande*, with the long central area surveyed by ELIAS HERCKMANS in his 1641 expedition to the inland of Paraiba. The inset in the upper right corner depicts his expedition (see *fig. 49*), engraved after a picture designed by FRANS POST. (Koninklijke Bibliotheek, Atlas Van der Hagen).

On his expedition HERCKMANS started with a company of 53 soldiers, 60 Indians, some 'Brazilian' women, and 3 or 4 volunteers, who were curious to see the land (*fig 49*). But, already in the first days of the expedition 13 soldiers and 24 Indians left because of illnesses. The travel route, as provided by BARLAEUS, can be followed very well on MARGGRAFE's map. Starting in the bay of the river Guaramama, where HERCKMANS went ashore, a dotted line goes to the small Dutch settlement of Frederica (where the Paraibian reverend CORNELIUS VAN DER POEL joined the train),[146] and then up to the farthest point, a high mountain, named by HERCKMANS *De Steenen Keerberg* ('stone mountain of turning around'), to mark his point of return.[147] Several toponyms on the map are slightly different from those in BARLAEUS, but overall the two sources can be linked. The village 'Tamoatamer' in BARLAEUS is, for instance, 'Tamoatâmirî' on the map.

FIG. 49. The expedition to the inland of Paraiba led by ELIAS HERCKMANS in 1641, as illustrated on the map of the *Praefectura de Paraiba, et Rio Grande* (*fig. 48*) HERCKMANS is depicted at the front of the train, as the only European man present (inset). Left in the background a local church.

It is interesting that on several occasions the group stayed overnight at a Portuguese mansion, for instance at the sugar plantation of DUARTE GOMES DE SYLVEIRA, whose name also can be found on MARGGRAFE's map as 'Duarte Gomes'. HERCKMANS's dependence on the original Portuguese settlers who formed a majority of the farmers who exploited the cattle farms and sugar plantations in the inland of Pernambuco and Paraiba demonstrates the complexity of the Dutch occupation of the country. The ninety years old DUARTE GOMES, for instance, was one of the original Portuguese explorers who in the late sixteenth century had conquered the land. It had been under his command that the native town of Arazembeam was destroyed (called 'Ararembe' on MARGGRAFE's map), and its inhabitants had been forced to move to Portuguese-controlled settlements near the coast.[148] After the Dutch conquest of the land in 1635, DUARTE GOMES DE

Sylveira officially had renounced his allegiance to the Spanish king, but covertly he supported the Portuguese guerrilla resistance against the Dutch, a fact for which he was arrested and almost deported in 1639.[149] But given the fact that the West Indian Company desperately needed the yield of the Portuguese-owned sugar plantations, Duarte Gomes – as the most prestigious man among the Portuguese settlers – was eventually allowed to return to his mansions.[150] A story like this provides context to the statement in Marggrafe's *Vitae* that he got along well with 'the Portuguese enemies', especially with the Jesuit friars whom he met along his journey.[151] This enjoyable social behaviour was not courtesy; it was a necessity for being able to travel in the inland of so-called Dutch (but in fact Portuguese inhabited) inner parts of Brazil.

Herckmans's itinerary further reveals an interesting ritual that seems not incidental, because it was performed twice, namely the erection of column with a shield bearing the coat of arms of the West Indian Company at some landmarks beyond the border of the 'official' Dutch territory. This 'monument' was evidently placed not only as a token to signify the ownership of the newly entered land, but as Barlaeus explained, it was also to mark the entrance of (European) civilisation into these lands of 'Barbaric nations'.[152] One can imagine that such rituals were also performed during Marggrafe's expeditions, as his overall map served similar goals.

Herckmans's expedition report of 1641, including his now lost map, was obviously not the only source Marggrafe used for his depiction of Paraiba. He also must have used the information from Herckmans's earlier report of 1639, as is demonstrated by the fact that most sugar plantations named in that document are carefully drawn on his map of 1643.[153] But even then, Herckmans's sources can account only for the part of Marggrafe's map around the river Paraiba. The other parts have to be drawn either from Marggrafe's own experiences or from those of unnamed others, as is evidenced by many dotted lines on the map, indicating further exploration routes.

Zandvliet has pointed to the fact that Marggrafe provides a wealth of information without cluttering his maps. He achieved that by designing a system of standardized symbols, explained in a key (*fig. 50*). This was quite an innovation, as Ormeling has observed: explanatory legends on maps are in the seventeenth century more the exception than the rule.[154] In order to convey all the details Marggrafe used 18 symbols: three for distinctive (European) settlements such as a city, a village, a fortress or a church; two to differentiate between the villages of domesticated 'Indians' (the Tupi's) and those of the more independent Tapuyas; four for various types of sugar mills (indicating also the presence or absence of a local chapel); and several others to indicate remote houses (differentiating between inhabited and deserted ones), cattle corrals, sources of water, salt pans, bushes or meadows and roads. In doing so MARGGRAFE gave a detailed impression of the economic and social value of the surveyed areas.

After all, the map foremost had to serve the commercial and military interest of the WIC. Or, to quote BARLAEUS:

> They seriously strived to gain money and property for the West Indian Company, with arms, with the advancement of commerce by searching for value in – and digging through – remote foreign lands. No forests, no wilderness, no rocks, no cliffs, no rivers, no lakes and seas, nothing stopped them [*HJZ: the explorers*] from achieving that goal.[155]

A MAP WITH ASTONISHING QUALITIES

MARGGRAFE's four maps were drawn in the first half year of 1643, but printed for the first time in BARLAEUS's *Rerum per octennium in Brasilia* of 1647. A large wall map, combining the four sheets into a large one, followed the same year.[156] Fortunately, one of the original manuscript

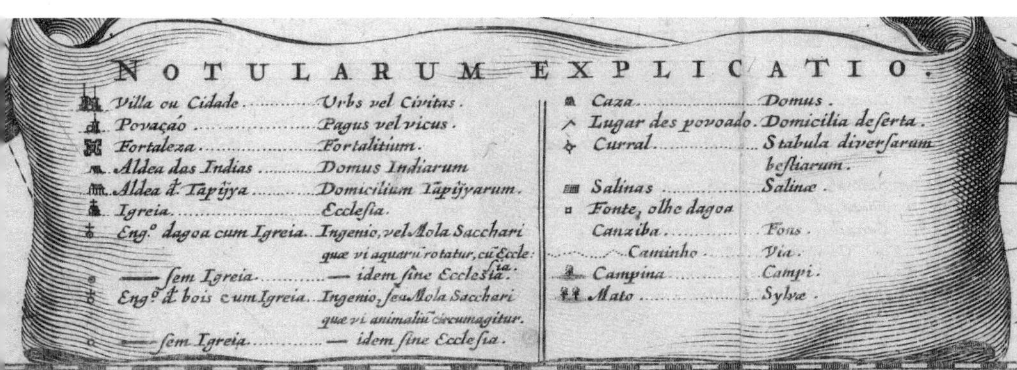

FIG. 50. MARGGRAFE'S *Notularum Explicatio* (explanatory notes) in the legend of his map.

maps has been preserved (*fig. 51 & fig. 52a*), allowing control of the crucial process of transferring the cartographic information to the copperplate.[157] ZANDVLIET already noted that 'the very manner in which MARGGRAFE drew up his map inspires confidence'.[158] The map shows only what was known to him. MARGGRAFE left the interior of the map blank, instead of filling it up with imaginary mountains, rivers, or fancy animals. But this approach was obviously too much for the commercially oriented JOHANNES BLAEU in Amsterdam, who had a business to run. So, he added ships at sea, and beautiful pictures by FRANS POST, depicting various scenes of daily life in Dutch Brazil (*figs. 44, 48, 49, 53*).[159] Nevertheless, as *fig 52b* shows, BLAEU respected the essence of MARGGRAFE's map. The additions do not interfere with MARGGRAFE's cartographic achievements and the engraving carefully follows the manuscript. The resulting map was highly appreciated by contemporaries, such as CASPAR SCHMALKALDEN, a surveyor who had served in Brazil as a soldier between 1640 and 1645. He found the four engraved plates with illustrations 'so splendidly done' that finding better depictions was sheer impossible.[160]

The idea to combine the cartographic information with nice pictorial stories from daily life in Dutch Brazil could also have come from JOHAN MAURITS personally. After all, after his return to the Netherlands, the count seems to have paid for most of the luxuriously produced publications, such as BARLAEUS's *Rerum per octennium in Brasilia* (1647), PLANTE's *Mauritiados Libri XII* (1647) and DE LAET's edition of the *Historia Naturalis Brasiliae* (1648). As BOOGAART has argued in his meticulously analysis of POST's images, BLAEU's edition of MARGGRAFE's map, which was accompanied by BARLAEUS's text, can only be seen as 'a paean of praise for JOHAN MAURITS's administration'.[161] BOOGAART sees the map as a civilizing project, in which the Brazilian colony was deliberately presented as 'a society of good standards of morality and respectability'. Or to quote BOOGAART,

> Colonization is portrayed here as a superior economic order, [...] The plantation colony showed that white people with their knowledge and skills brought from Europe and African slaves with their capacity for hard labour in the tropics could transform the wilderness of tropical America into a prosperous landscape.[162]

However it may be, during the preparation of the copper plates for BARLAEUS' book, BLAEU or JOHAN MAURITS must have seen the opportunity to combine the four sheets into one large wall map,

FIG. 51. Heading of the manuscript map (*fig. 52*) from 1643, with the correct Latin way of writing of his name 'GEORGIUS MARGGRAFIUS'. Although MARGGRAFE's signature has a different shape for the letter 'G' in his name, the mere fact that on a map letters are carefully drawn instead of fluently written makes it plausible that the map is indeed by his own hand.

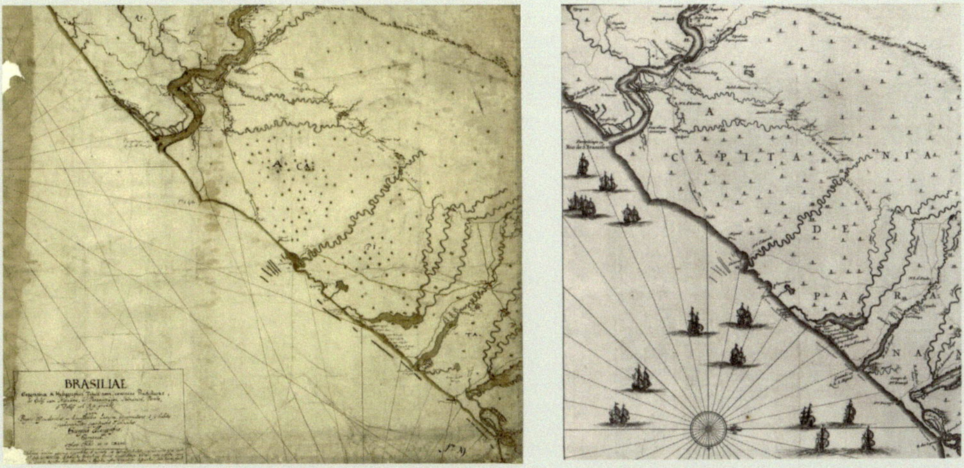

FIG. 52. (a) Detail of the only surviving manuscript map by GEORGE MARGGRAFE (1643 – Nationaal Archief, Den Haag, 4.VEL 695), combined with (b) the corresponding engraved part (1647) [= detail of the map 'Praefectura Paranambucae pars Meridionalis'], which has different compass lines, but otherwise exactly follows the original.

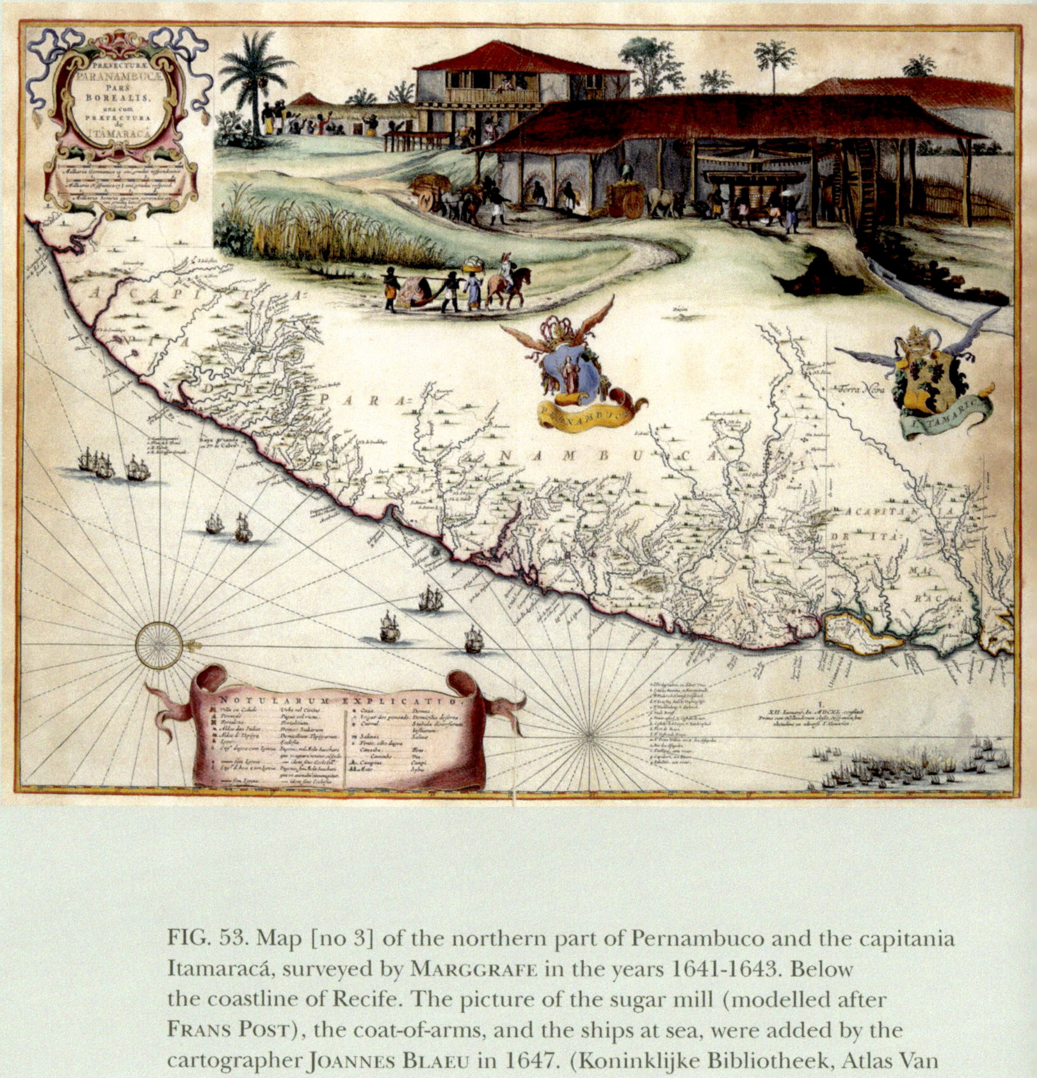

FIG. 53. Map [no 3] of the northern part of Pernambuco and the capitania Itamaracá, surveyed by MARGGRAFE in the years 1641-1643. Below the coastline of Recife. The picture of the sugar mill (modelled after FRANS POST), the coat-of-arms, and the ships at sea, were added by the cartographer JOANNES BLAEU in 1647. (Koninklijke Bibliotheek, Atlas Van der Hagen).

FIG. 54. Wall map constructed from the four sheets drawn by GEORG MARGGRAFE in 1643. The north side is at the right. (British Library, London, Klencke Atlas; The text below the map has been deleted in this picture).

filling the empty spaces with the earlier mentioned scenes by FRANS POST (*fig. 53 & fig. 54*). After all, POST and GEORG MARGGRAFE had known each other well. A list from April 1643 shows that they dined together at JOHAN MAURITS's court, accompanied by the second painter in the count's service, ALBERT ECKHOUT. At that moment MARGGRAFE was no longer denoted as an astrologer (as in 1641), but instead as a cartographer.[163] This renaming of his function is no surprise, because at that moment MARGGRAFE and his assistant – *een jongen* ('a boy') – were fully engaged in the process of transferring the gathered information into the overall map of the Dutch-occupied territories.

Wall maps as constructed by BLAEU were very much in vogue at the time, being used as decorations in fancy houses, as illustrated, for instance, in the painting by the Delft artisan CORNELIS DE MAN, showing the map of Dutch Brazil, made in 1648 by CORNELIS GOLIATH,

FIG. 55. A scholar in his studio, ca. 1670. Painting by CORNELIS DE MAN from Delft, showing on the wall the map 'Olinda, Maurits-Stadt ende 't Reciffo' by CORNELIS GOLIATH, 'Caertmaker van sijne Excellentie JOHAN MAURITS VAN NASSAU'. (Painting sold by Sotheby's, 2007, no. 31). The engraved map was published in 1648 by CLAES JANSZ. VISSCHER. (Scheepvaartmuseum, inv. nr A.3143-03).

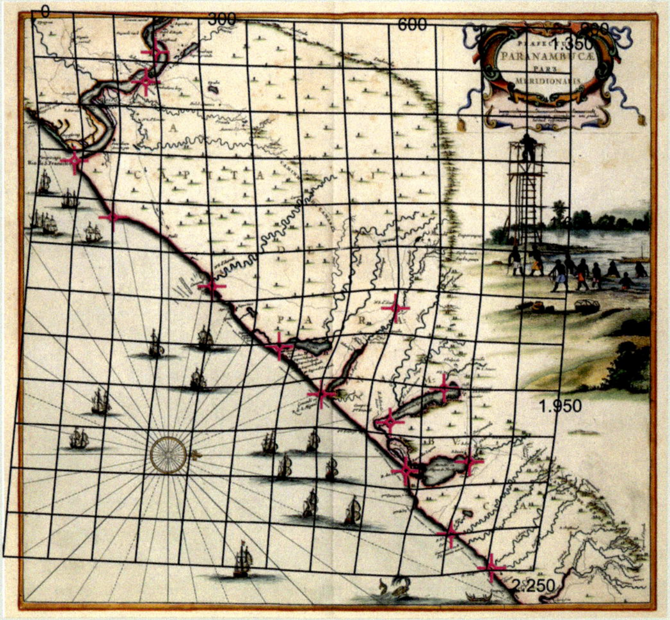

FIG. 56. Map of southern Pernambuco from the map of MARGGRAFE (1647) with deformation grid. Picture from Storms (2011).

'Caertmaker van sijne Excellentie JOHAN MAURITS VAN NASSAU' (*fig. 55a & fig. 55b*).[164]

Today, every commentator agrees that MARGGRAFE's map of Brazil belongs to the very best of seventeenth-century Dutch cartography. According to WHITEHEAD it is 'one of the most elegant Dutch maps of that period', and 'the best guide to north-eastern Brazil' for over a century.[165] ZANDVLIET called it a map with 'a monumental appearance, corresponding to Dutch painting and literature of that period', also 'because of the strip-like illustrations, engraved after designs by artists who had accompanied JOHAN MAURITS in Brazil'.[166] STORMS agrees by saying that GEORG MARGGRAFE 'made a monument for himself' with his map of Dutch Brazil: 'a map praised for its metric accuracy and topographic detail'.[167] Although the area mapped by MARGGRAFE is only a relatively narrow strip along the coast, the map is indeed strikingly detailed. STORMS even investigated this claim of metric accuracy by analyzing the maps with two digital tools. First, he linked a number of locations on the MARGGRAFE map with corresponding points on a modern map. His tool – MapAnalyst – then calculated a deformation grid. For a modern map, this grid would be a rectangle. Applied to the map of MARGGRAFE, it appears that only minor distortions can be observed, from which it can be concluded that the map is indeed rather accurate (*fig. 56*).

Secondly, STORMS used displacement vectors. These vectors show in which direction the inaccuracy manifests itself. The length of these vectors, which are also represented by circles, show the degree of inaccuracy: the larger the circle, the less accurate the position on the map. On MARGGRAFE's map, these vectors show a clear pattern. STORMS concluded from this test that the accuracy of MARGGRAFE's map is better along the coastline. Locations more inland are less accurate. It is striking that from the direction of the displacement vectors, it can be concluded that different sets of data were used. These differences indicate that MARGGRAFE compiled the map from data collected during various expeditions (*fig. 57*).[168]

This same method has led to a controversy between BENJAMIN TEENSMA and MARTIJN STORMS, because where TEENSMA was convinced to have identified MARGGRAFE as the maker of an anonymous coastal atlas known by the prosaic name 4.VEL-Y, STORMS has shown that although this atlas exactly follows the coastline of MARGGRAFE's map, its accuracy is far less, so the question of the identity of its maker, who evidently has used MARGGRAFE's map for filling in some details, remains unsolved.[169]

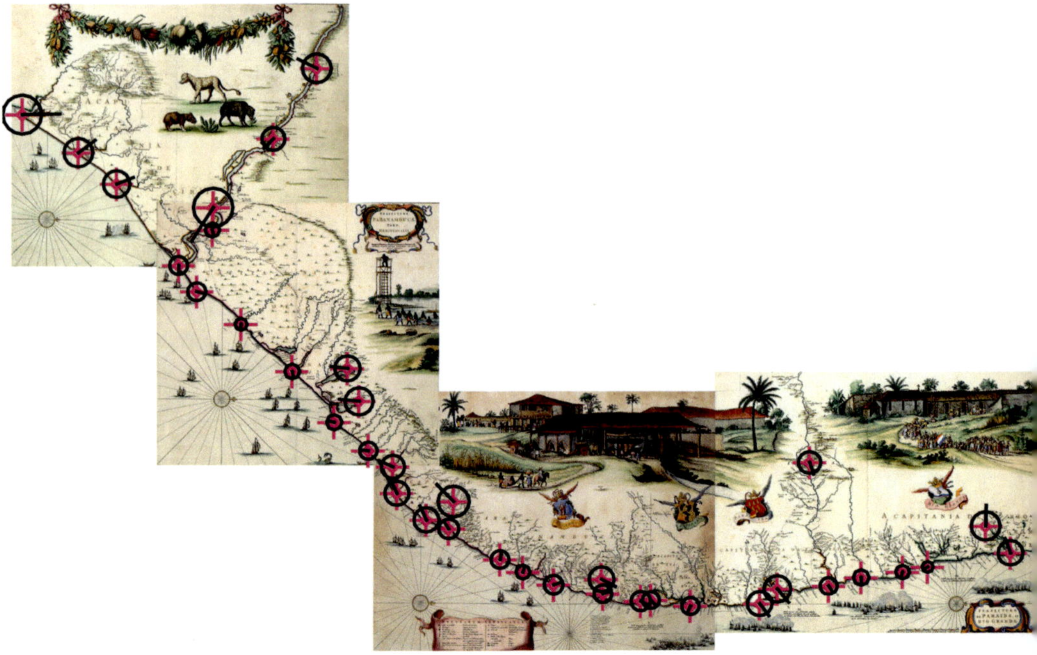

FIG. 57. MARGGRAFE's map with transposition vectors, demonstrating the use of different data sets. For instance, in the northern sheet of the *Praefectura de Paraiba, et Rio Grande*, the two rightmost transposition vectors have a different, upright orientation compared with the others on this sheet which have a downward orientation. Picture from Storms (2011).

An obvious reason for the accuracy of MARGGRAFE's maps is that he supplemented the available surveying techniques with astronomical measurements. An astronomical assessment of MARGGRAFE's maps, performed by the Brazilian scholars JORGE PIMENTEL and LEVY PEREIRA, corroborates this supposition.[170] According to their findings, MARGGRAFE used as his astronomical reference point the island of Ferro in the Canary Islands, at that time regarded as the prime meridian. They also concluded that the coordinates he used for the island of Antonio Vaz were based only on his observation of the moon eclipse in the night of 20/21 December 1638. They further hypothesize that MARGGRAFE designed a fine grid of meridians and parallels, based on astronomical observations, and fixed to these points of reference, which he then could fill in with specific features such as rivers, paths, land contours, cities, fortresses, etc.

PIMENTEL and PEREIRA's comparison of the astronomical coordinates on the map with those accepted today also led to the conclusion that the map has an unprecedented accuracy for that time. Moreover, the usage of about 1,800 geographic entities in mixed Portuguese, native and Dutch languages, shows a level of amazing intercultural cooperation.

ASTRONOMICAL AND METEOROLOGICAL ACHIEVEMENTS

Although astronomy had been his greatest ambition and *the* reason for coming to Brazil, MARGGRAFE was too busy with other assignments to fully fulfil his dream of becoming the southern TYCHO BRAHE. As the timeline for MARGGRAFE's Brazilian sojourn shows (*fig. 58*), most of the time he was unable to perform regular astronomical observations.

Still, what he accomplished in this field, given the time available, is astonishing and commands respect. We will discuss the details of his astronomical achievement elsewhere in this volume, but one

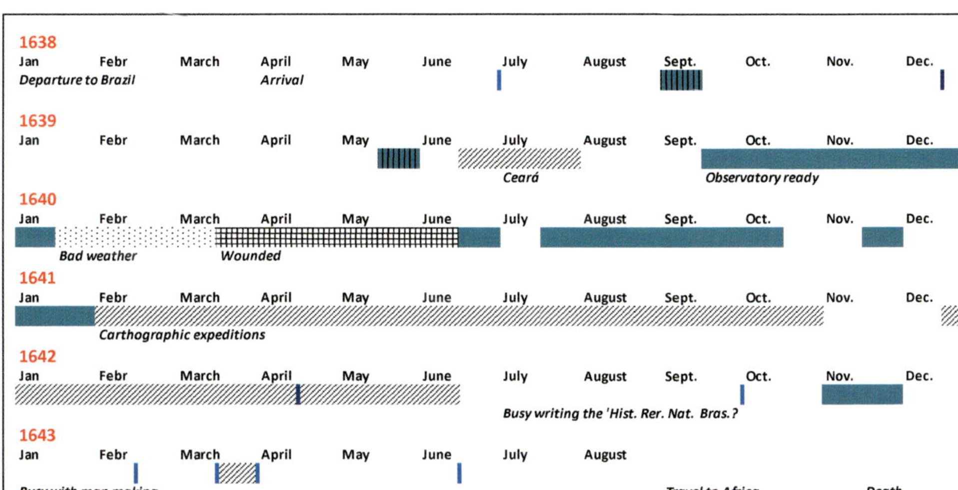

FIG. 58. Time line: *Blue* – period of astronomical observations; *shaded vertically* – time before the observatory; *shaded diagonally* – period of expeditions to inland Brazil; *chequered* – disabled due to shoulder luxation. In 1643 MARGGRAFE's time was for the largest part occupied for the assignment of drawing his maps of Dutch Brazil. He left for São Thomé and Angola at the end of August 1643 and died before the end of the year.

pioneering aspect related to his astronomical work deserves our attention here. It is an understatement that every astronomer needs a clear sky to see the heavenly bodies. But as MARGGRAFE was regularly confronted with bad or cloudy weather, his mathematical mind, used to tables and calculations, started to register the state of the weather. In the years 1640, 1641 and 1642 he kept daily meteorological records, noting wind direction, precipitation, and the incidence of thunderstorms (*fig. 59*). These are the oldest meteorological records made on the South American continent. From these records he even drew some conclusions about the seasons of north-eastern Brazil, for example that May was the wettest month of the year and November the driest.[171] It is interesting to note that these records carried on during MARGGRAFE's absences from Recife; he evidently continued these observations during most of his expeditions. Only at some moments do the meteorological records begin to tail off, for instance, during the expeditions of November and December 1640. But on his return to Recife, they instantly resumed.

TYPVS ANNI cIɔ Iɔc xL.

	Ian.	Feb.	Mart.	Apr.	Mai.	Iun.	Iul.	Aug.	Sept.	Octob.	Nov.	Dec.
1	p.	N.O	S.O	S.O	p.N.O	S.	p.S.O	p.S.O		S.O	p.	O.S. O.
2	S.O	N.O	S.O	p.N.O	p. S.	S.O	p.	p.S.O	S.O	O.S O.		p.N.O
3	S.O	N.O	S.O	N.O	p. S.	p.S.O	p.S.O	S.O	S.O	p.S.O	p.	p.N.O
4	S.O	N.O	p.	p. V.	p. O.	p.S.O	p.S.O	p. S.	p.	p.S.O	p.	p.S.O
5	S.O	p. V.	p.N.O	p. V.	p. S.	S.O	p.S.O	p. S.	p.S.O	S.O		S.O
6	O.	S.O	p.	N.O	p.S.O	S.O	p. S.S O.	p.S.O	S.O	S.O		S.O
7	p. O.	p.S.O	p.S.O	N.O	p.S.O	S. S. O.	p.S.O	p.S.O	p.S.O	O.		O.
8	O.S. O.	S.O	S.O	p. V.	p.S.O	p.	p.S.O	S.O	S.O	O.N O.		p.N.O
9	id.	p. O.	p.S.O	p.	p.S.O	p.	p.S.O	p. S.	p. S.	O.N O.	N.O	•
10	id.	p.	S.O.	p.	S.O	p.	p. S.S O.	p.S.O	S. S. O.	id.	p.N.O	p.N.O
11	S.O	p. O.	p.S.O	p.S.O	S.O	S.	p. S.	p.S.O	p. S.	O.S O.	N.O	N.O
12	p.S.O	p.S.O	p.S.O	p.S.O	S.O	p.	p.	p.S.O	p.S.O	p.S.O	N.O	N.O
13	p.S.O	p.	S.O	p. O.	S.O	p. O.	p.S.O	p.S.O	p.	p.S.O	N.O	
14	p. S.	p.N.O	S.O	p.S.O	p.	S. S. O.	p.S.O	p.	p.S.O	p. S.	N.O	
15	O.	p.N.O	p.	p.S.O	p.	S.O	p.S.O	S.O	S.O	S.O	N.O	
16	O.S. O.	p. O.	p.S.O	p. S.	p.S.O	S.	p.S.O	S.O	p.S.O	N.O	O.	p.
17	S.O	O.	p.S.O	p.	p.N.O		S. S. O.	S.O	S.O	N.O	p.S.O	
18	p.S.O	O.	•	p. V.	p.N.O	S.		p.S.O	S.	p.N.O	p.S.O	
19	p.S.O	p. O.	p.S.O	p. S.	S.	p. S.	S.O	p.S.O	N.O	N.O	N.O	
20	S.O	O.	N.O	p.S.O	S.	p. S.	p.S.O	p.S.O	N.O	N.O	N.O	O.
21	p.S.O	p. O.	S.S. O.	p.	p.	p. S.	p.S.O	p.	N.O	p.N.O	O.	O.
22	p.S.O	p.S.O	S.O		p. S.	p. V.	p.S.O	p.S.O	N.O		p.	O.
23	p.	O.	p.S.O	p. O.	p. O.	O.S. O.	S.O	p.S.O	O.N	p.N.O	N.O	O.S O.
24	S.O.	O.	p.N.O	p.	p. S.S. O.	p.S.O	p.S.O	p.S.O		N.O	p.N.O	
25	S.O.	N.O	p.N.O	p.	p.	p.S.O	p.S.O	p.	O.	p.	p.S.O	p.S.O
26	S.O.	N.O	p.N.O	O.	p. S.	p.S.O	S.O	S.O	S.O		p.S.O	S. S. O.
27	p.S.O.	p. O.	p.N.O	S.O	p.S.O	p.S.O	p.S.O	p.S.O	p. S.	p.N.O	S.O	p. id.
28	S.O.	S.O.	p.N.O	p. S.	p. S.	p.O.S	p.S.O		p.	p.N.O	O.	p.S.O
29	S.O.	S.O.	p.S.O	S.O	O.	p.S.O	p.S.O	p.S.O	S.O		O.	S.O
30	p. O.		p.N.O	p. O.	p. O.	p.S.O	p.	S.O	p.S.O	O.N O.	O.	p.N.O
31	O.		p.S.O		p.	p.	p. S.	S.O		N.O		N.O

FIG. 59. Meteorological table for the year 1640, as published in the *Historia Naturalis Brasiliae* (1648), 265 and the *Indiae utriusque re naturali et medica* (1658).

Dutch strongholds on the West African coast

In May 1643, the States General wrote Johan Maurits that his rule over Dutch Brazil had come to an end. He was requested to return to the Republic. For many in the colony, this message was experienced as a severe shock. But a petition from a large number of citizens and merchants to maintain the count as governor in Dutch Brazil was ineffective.[172] So, slowly but surely, preparations were made for a return to the Dutch Republic. Marggrafe, too, planned to return to Europe. By the summer of 1643, he 'had packed up all his goods', as he wrote to his former roommate Samuel Kechel, 'expecting that, by God's assistance, he might with his highness the Prince return into his own country'.[173] However, Johan Maurits would not part from his dear colony until May 1644, so although Marggrafe wanted to return to the Netherlands together with the count, there was still plenty of time for a new assignment. Such a new commission came when Marggrafe had just finished the manuscript of his four maps of Dutch Brazil. He was assigned to chart the conquests of the WIC that had not yet been mapped, namely the newly conquered African territories, which for their slave trade were so vital for the economy Dutch Brazil (*fig. 60*).

Already in the time of the Portuguese rule of Pernambuco, the functioning of the colony had been highly dependent of the presence of slave labour. When the Dutch took over the control of the territory in 1630, for a short time they had been rather reluctant towards slavery, but soon everyone accepted the notion that for a profitable exploitation of the Brazilian sugar plantations and cattle farms, slave labour was absolutely a requirement. Or to quote a document from 1638:

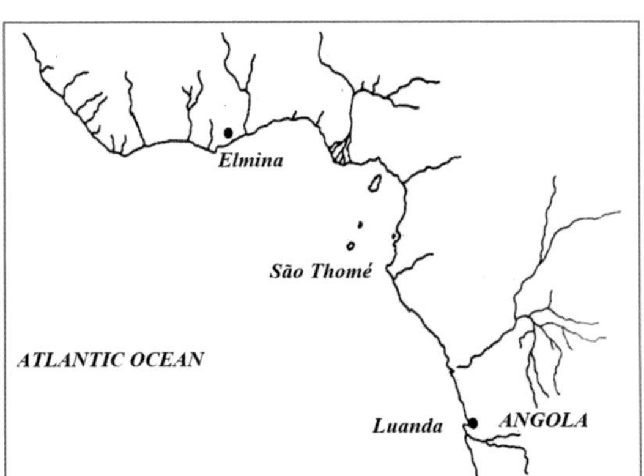

Fig, 60. The three Dutch conquests in Western Africa in the years 1637-1641. Adapted from Den Heijer, *De geschiedenis van de WIC* (2002), 70.

> Without such slaves it is not possible to accomplish anything in Brazil: without them no sugar mills can grind and no land can be cultivated. So, it is necessary that in Brazil there must be slaves. There is no other way, and if someone would feel burdened by this, that is just unnecessarily tormenting your conscience. Thus, since Brazil cannot be built without black slaves, and a large number of them is needed, and because there is a constant complaint about the lack of such slaves, it therefore is imperative that every possible means is used to trade in black slaves at the coast of Africa. This is of the utmost importance to the Company, for in addition to the fact that the Company sells them for a lot of money, the Company also annually receives from each black one third of his performance, so that he works for both the Company and its owner.[174]

In order to guarantee the undisturbed supply of new slaves from Africa, both the High Council in Brazil, as well as the *Heeren XIX* in the Netherlands, desired to have their own foothold on the West African coast. The first Dutch stronghold became the fortress São Jorge da Mina (in short 'Elmina') on the Gulf of Guinea. The conquest of the fortress in August 1637 was JOHAN MAURITS's first feat after his arrival in Recife. Fortress Elmina, the local centre for slave trade, had been the first Portuguese settlement in Africa, founded in 1482.

But this was not sufficient. The above-mentioned report of January 1638 refers to the experience that the most useful slaves came from Angola, 'partly because they want to work better, partly because, when newly arrived, they learn better from the old negroes because they understand each other's language'.[175] So, at the end of May 1641 the decision was made to conquer the city of São Paulo de Luanda in Angola, the centre of the Portuguese slave trade. As a second target, the island of São Thomé was chosen, as a strategic site near the Congo river, another fruitful place for the trade of slaves. A large fleet of 21 ships with some 3,000 sailors and soldiers was prepared in Recife, and in August 1641 Luanda was indeed added to the dominions of the Dutch West Indian Company, to be followed by the island of São Thomé in October 1641.[176] Both conquests, however, were very precarious, being faced with fierce resistance from the Portuguese side. By December 1642, for instance, on São Thomé only the fortress São Sebastião was still in Dutch hands.[177]

The inevitable end

In spite of the precarious situation in both the island of São Thomé and the city of Luanda, Marggrafe was nevertheless commissioned to chart these newly conquered lands. In August 1643 the *Hoge en Secrete Raad* of Brazil dispatched a letter to the WIC directors on the coast of Africa with the following message:

> We recommend that you give to the person George Markgraef all help and affection; he is coming there in order to make a pertinent description and map of this area.[178]

For this task 'to describe the country there in a map', the count's treasurer Johannes Grevingh paid him 200 guilders, probably on the day of his leave, on the 20th of August.[179] Marggrafe's last recorded observation in his 'Historiae Rerum Naturalium Brasiliac' is indeed from that same month.[180] It was the fast yacht *De Brack* that would bring him to his new destination, together with victuals and ammunition.[181]

Unfortunately for Marggrafe, the island of São Thomé was ill-famed for its 'unhealthy air'.[182] This was an understatement for the bare fact that of the invasion fleet hundreds of sailors and soldiers died from what then was called the *lantsieckte* ('land sickness'), today interpreted as *malaria quartana*. It was said at the time that, on shore, the mortality rate was between 80 and 90 %.[183] The symptoms of this disease were evident. The sick experienced a feeling of great fatigue, and despite the oppressive heat, the victims shivered from the cold. A severe headache, sometimes reminiscent of delirium, overtook them, followed by a severe attack of fever that lasted for several hours. If a patient survived this first attack of fever, a second attack followed after about four days, and after another four days there was a third attack that usually broke the resistance of even the strongest.[184] Even the commander of the fleet, admiral Cornelis Jol, the army commander captain Joan Valet, and Jol's replacement, skipper Matheus Jansen, did not survive this disease. The latter requested on his death bed to be discharged at sea, instead of being buried on São Thomé, this 'cursed human-destroying island'.[185]

The inevitable indeed happened. According to a testimony of Willem Piso, made long after his return to Holland, Marggrafe survived in Africa only 'four to six weeks'.[186] Given the fact that a crossing from Recife to the African island of St Thomé took about two weeks, MARGGRAFE arrived there well before mid-September 1643.[187] He probably stayed on the island only for a week or so. His yacht *De Brack* had to continue its trip to Angola relatively quickly, due to the required fast delivery of the letter of the High Council of Brazil to the local administrators in Luanda. This timeline implies that GEORG MARGGRAFE died long before the new year of 1644.

Despite this short time span of 'four to six weeks', it seems that MARGGRAFE still managed to finish his surveying task. In BARLAEUS's *Rerum per octennium in Brasilia*, a map of (only!) the fortress on São Thomé, as well as a map of the peninsula at St Paulo da Luanda is present (see *fig. 61* and *fig. 62*).[188] These engravings are identical to two neat manuscript maps constructed by JOHANNES VINGBOONS (ca. 1616-1670), a cartographer who never left the Netherlands.[189] So, it seems that cartographic drafts made by MARGGRAFE were processed by VINGBOONS into the maps published in BARLAEUS's *Rerum.* As both São Thomé and Angola were lost for the West Indian Company as early as 1648, and because JOHAN MAURITS – the driving force for the cartographic exploration – returned to the Netherlands in 1644, who else could have charted these extremely unhealthy locations?[190] Anyway, a deadly disease (unfortunately avoidable if he had stayed in Brazil) brought GEORG MARGGRAFE's wanderings to an end. His costly legacy, however, survived the ages, as we will see in the next chapter.

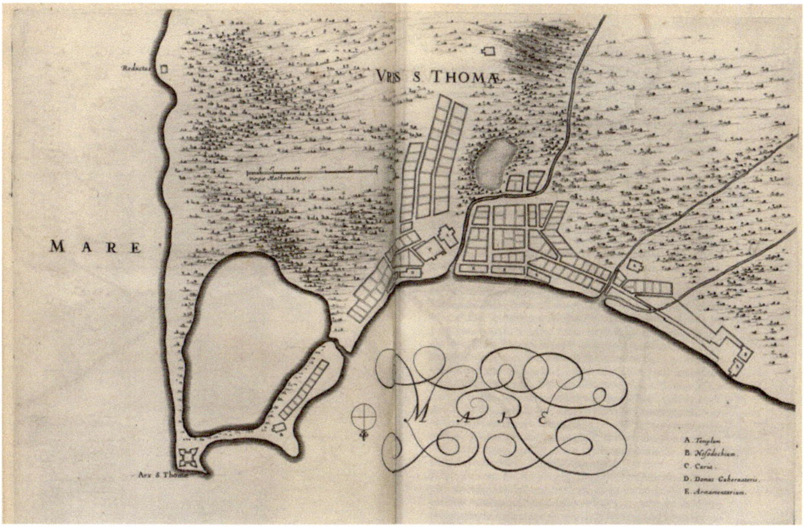

FIG. 61. Map of the fortress São Sebastião at the island of São Thomé, as issued In Barlaeus' *Rerum per octennium in Brasilia* (1647). (Rijksmuseum, Amsterdam).

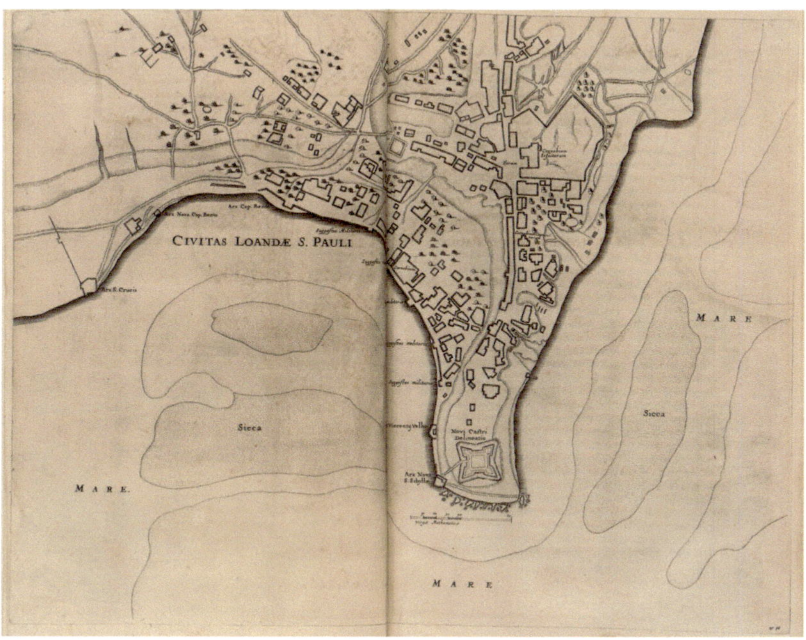

FIG. 62. Map of the city St Paulo de Luanda, as issued in Barlaeus's *Rerum per octennium in Brasilia* (1647). (Rijksmuseum, Amsterdam).

FIG. 63. The remnants of JOHAN MAURITS's ivory furniture from his mansion *Vrijburg*, damaged in WW II, together with two canvasses by an unknown painter, with views of Recife and Mauritsstad. Originally the set contained, next to the bench, two big chairs with elbow rests, two smaller ones, a table with corresponding chairs, a cabinet with several drawers, decorated with pilasters and festooned with the 12 apostles carved in it, two gueridons with candlesticks, a mirror and some smaller items. (Schloss Oranienburg, Berlin; Photo Hans Christian Kraas).

CHAPTER 6

LEGACY

The spread of Marggrafe's paper legacy

When Johan Maurits finally left Dutch Brazil in May 1644, many of his loyal followers went with him in this large fleet home. As the prime commander, Johan Maurits travelled on the flagship, the *Amsterdam,* which carried in its hold the most valuable cargo, such as various paintings, Brazilian curiosities and the beautifully carved ivory furniture from his personal cabinet in *Vrijburg* (the remnants of which are currently preserved in *Schloss Oranienburg* in Berlin, *fig. 63*). Less eye-catching in the hold of the *Amsterdam* was 'a suitcase with astronomical instruments', along with three trunks of an unidentifiable person (his name was accidentally ripped off the cargo list).[1] So, evidently the astronomical observatory on the island of Antonio Vaz was demolished after the news of Marggrafe's death and his belongings went to the Netherlands together with the count. After arrival, Marggrafe's trunks with his costly papers were delivered in Leiden to the university. From there their contents were distributed to the house of WIC-*bewindhebber* Johannes de Laet, and to the houses of Marggrafe's former professors Jacob Golius and Adolphus Vorstius.

The sequel is known. Count Johan Maurits gave both De Laet and Golius the explicit instruction to publish the relevant content of the papers at his expense.[2] De Laet indeed edited the material he received into the impressive illustrated folio book *Historia Naturalis Brasiliae,* published in 1648 in a joint effort of the Leiden publisher Franciscus Hackius and the Amsterdam publisher Lodewijk III Elzevier. De Laet also took care that the cartographical information was handed over to the Amsterdam publisher Johannes Blaeu, who in 1647 produced the earlier-mentioned set of four engraved maps for Barlaeus's *Rerum per octennium in Brasilia* and combined these later into the large wall map. But, as reported in the introduction,

Golius's astronomical publication was never realized. At the time, only Marggrafe's observation of the 1640 solar eclipse was published in Barlaeus's *Rerum*.

The genesis of the *Historia Naturalis Brasiliae*

What happened during the process of transforming Marggrafe's field notes into the codified text of the 'Historiae Rerum Naturalium Brasiliae', the second and largest part of the *Historia Naturalis Brasiliae*? Which steps were followed to compose this large folio volume? And why did this book achieve its long-lasting authority in the field of natural history? Thanks to preserved correspondence as well as some additional information, we are able to follow the steps taken in detail.

Processing the raw material

At first, it must be noted that Johannes de Laet derived the title of the book from a design he had found among Marggrafe's papers, which small sheet of paper – now torn – survived among the documents in the Leiden archive (*fig. 64*).[3] This sheet also shows that in the printed edition De Laet did not follow Marggrafe's scheme, but rearranged the text and also introduced some new material.[4] But before De Laet could start with the editorial process, he had to cope with the problem that not only Marggrafe's ordinary handwriting was rather difficult to read, but that some of the notes were written in cipher. De Laet explains this in his 'ad lectores' in the *Historia Naturalis Brasiliae*:

> When his [=Marggrafe's] comments, though imperfect and disordered, were handed over to me by the illustrious Count Johan Maurits (with whose support, favour and financial expenses they were made), there soon appeared what was not a little obstacle. The author, fearing that others would vindicate his work as their own, if perhaps something would happen to him before the publication of his book, wrote a large part of his work and especially the most important passages, with signs he invented and that had to be interpreted and transcribed first, following a secret alphabet, much to the disturbance of those who would want to work with an unfamiliar text.[5]

Why did Marggrafe write his natural history notes in cipher? The best assumption is that after 1640, his relationship with Piso was

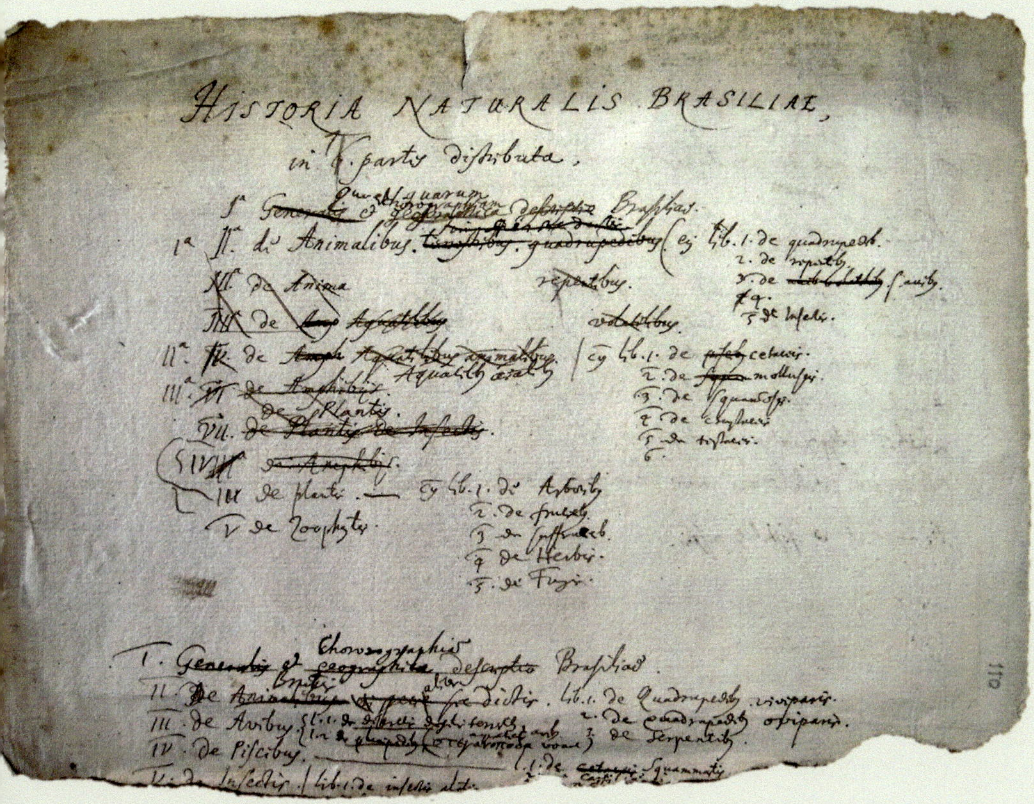

fig. 64. MARGGRAFE's draft of the contents of the *Historia Naturalis Brasiliae* (ELO).

seriously disrupted, and that in this way he wanted to prevent that PISO would run off with his work. In the *Vitae*, his brother CHRISTIAN writes at least that he had heard 'from many' that in Brazil 'there was no entire friendship between MARGGRAFE and PISO'.[6] Anyway, DE LAET, who evidently possessed the key to the secret code, not only had to edit MARGGRAFE's text, but first he had to decipher it, which took a lot of time. Then he had to order the raw material in a systematic way.

SELECTING THE PROPER IMAGES

A further step was to combine MARGGRAFE's descriptions with the available pictorial information. As BRITT DAMS has put it, 'It is important to emphasize the fact that the *Historia Naturalis Brasiliae* is more than a collection of mere textual descriptions. [...] The images do not merely support the textual descriptions, they also function as an extra tool to comprehend the described phenomena ...'.[7] That feature, combined with MARGGRAFE's businesslike approach of describing only what he himself had observed, made the book a novelty of its kind.

What sources did DE LAET have at his disposal for the illustrations? As WHITEHEAD, BRIENEN and SCHARF have pointed out, DE LAET had quite diverse material that could be used for the 533 illustrations in the *Historia Naturalis Brasiliae*, of which 422 are placed in the botanical and zoological parts written by GEORG MARGGRAFE.[8] In Brazil, MARGGRAFE had collected not only field studies (basic drawings in a now-lost sketchbook), but also neat watercolours, animal skins and herbaria containing dried plants and insects. 'Figures of some plants that were missing' were indeed 'drawn according to the dried herbs preserved', DE LAET told his readers.[9] This use of MARGGRAFE's – still-preserved – herbarium for the manufacture of 15 woodcuts in the *Historia* is indeed confirmed by the study of ANDRADE-LIMA (see also hereafter, *fig. 76*).[10]

But most of the woodcuts were made after the available pictures. Examples of MARGGRAFE's botanical field drawings (probably from MARGGRAFE's lost Brazilian sketchbook), can still be found in a manuscript from the HANS SLOANE collection, now in the British Library.[11] This manuscript is composed by DE LAET and is an earlier draft version of the text in MARGGRAFE's botanical part of the *Historia*. It contains basic, uncoloured drawings identical (also in size) to the woodcuts in the *Historia*.

For other woodcuts, especially the zoological ones, De Laet relied on other sources. The probably most important of these consists of two volumes with watercolours of – foremost – animals, which books were used by Count Johan Maurits personally to write down his own observations about Brazilian nature. These two so-called *Libri Principis* – also known as the *Handbooks* – were already made in Brazil, most likely – as shown in a meticulously executed study by Claudia Scharf – after illustrations recorded in previous notebooks and sketchbooks.[12] The watercolours in the two surviving *Handbooks* are not signed. Nevertheless there is consensus between most recent authors that, although the hands of six different artists can be distinguished, the majority of the pictures, especially those in *Handbook I,* can be attributed to Georg Marggrafe.[13] The fact that *Handbook I* follows the same order Marggrafe originally intended for his part of the *Historia Naturalis Brasilae* underpins this attribution.[14] More about this in the next paragraphs.

Ad vivum?

With respect to the way of representation of the animals in the *Handbooks* (and thus also in the *Historia*), it is interesting to note that Marggrafe, in his dedication to Count Johan Maurits, states that he made all the illustrations *ad vivum.* This is underscored in an inventory of the books, made in 1652, in which it is also noted that all the images are 'artfully represented after life'. However, the static way of representation of the animals makes it nonetheless clear that these pictures do not record encounters with wild animals in their natural environment.[15] One drawing of a monkey even shows that he was chained to a wall.[16] So, most animals are probably portrayed in the count's menagerie at the *Vrijburg* palace, or from stuffed specimens in his cabinet of curiosities at the residence *Boa Vista.*[17] Thus, to cite Neil Safier, 'what appears in the *Historia Naturalis Brasiliae* as an encyclopaedic menagerie of Brazilian animals and botanical species was in fact (at least in part) a constructed vision of the region, collected by Johan Maurits and his minions, but presented in such a way as to emphasize the on-the-spot character of the observations Marggrafe had obtained'.[18]

The transmission of visual evidence into woodcuts

Having selected the required pictures, De Laet had to organize the reproduction of these images. As so many illustrations were required for this transmission process, woodcuts were chosen for the printed edition. Woodcuts were much cheaper to produce than copperplate engravings, but unfortunately, in woodcuts a lot of detail is lost. Moreover, according to JOPPIEN and WHITEHEAD, many woodcuts in the *Historia* are rather crudely executed, which sometimes has led to a loss of crucial information, often causing misinterpretations.[19] A case in point is the picture of the *Tamandua* or anteater. Here the information about the dark skin of the animal is completely lost in the woodcut in the *Historia Naturalis Brasiliae* (*fig. 65* and *fig. 66*).

FIG. 65. MARGGRAFE's watercolour of the large *Tamandua*, or anteater, from *Handbook I*, fol 84. (Jagiellonian University Library, Kraków)

FIG. 66. Woodcut of the large anteater from the *Historia Naturalis Brasiliae*, part 2, page 225.

However, this loss of information about the skin of the anteater is recovered in the hand coloured copies of the *Historia Naturalis Brasiliae*, of which at present six copies are known (*fig. 67*). One of these hand-coloured editions of the *Historia* was once owned by JOHAN MAURITS, who carefully copied his own remarks from the *Handbooks* into this folio before he donated the two *Libri Principis* to the Elector of Brandenburg.[20]

Unexpectedly, this same depiction of the large anteater confronts us with a new, but interesting problem. It appears that ZACHARIAS WAGENER (who left the Dutch colony in April 1641) copied this very image in his *Thierbuch*, now preserved in Dresden (*fig. 68*).[21]

FIG. 67. Woodcut of the large anteater from a coloured copy of the *Historia Naturalis Brasiliae*, part 2, page 225 (Leiden University Library).

FIG. 68. Watercolour of the large anteater by ZACHARIAS WAGENER in his *Thierbuch* (fol. 82), made before his departure from Brazil in April 1641. (Kupferstich-Kabinett, Staatliche Kunstsammlungen Dresden, Photo: Herbert Boswank).

A comparison of WAGENER's image with the ones mentioned above first confirms that WAGENER and MARGGRAFE interacted in Brazil, but this comparison also leads to the possibility that DE LAET did not use *Handbooks* after all. Instead he could have employed a – now-lost – sketchbook as his prime source for the woodcuts. After all, WAGENER's representation is not only in reverse (as is often the case compared to a woodcut or engraving), but both the tree, as well as the cut-off tail of the animal, resemble the woodcut better than the picture from *Handbook*. In that case the *Libri Principis* contain neat copies from earlier sketches. This is in line with the recent study by SCHARF, who has concluded that the watercolours in the *Handbooks* are derived from now lost notebooks.[22] This example demonstrates the complexity of reconstructing a process of transformation of knowledge acquired in the field to knowledge presented in a printed edition.[23]

FRAMING A FRONTISPIECE

One of the last required steps before the publishing of the *Historia Naturalis Brasiliae* was the design of its title page. In the past, large natural history books often referred in their frontispieces to previous scholars, but this time the publishers, or – more likely – JOHAN MAURITS, decided on another approach. After all, only the count's name figures in this title page. Remarkably, this engraving does not correspond at all with the businesslike no-nonsense approach of the content of the book. On the contrary, in the frontispiece Brazil is framed as a new garden of Eden (*fig. 69*). Or, to quote the study of DA SILVA and ALCIDES:

> Not only its written contents, but also its iconography confirm the obsession [JOHAN] MAURITS had with personally building an Earthly Paradise. Its frontispiece shows a Brazilian version of ADAM and EVE, with serene and almost smiling faces. They are in the foreground of a perspective composed of trees, some palm trees. Tamandua, fish, turtles and crabs appear in the lower part of the drawing, while in the upper part snakes coil around the trees in which both bearded and unbearded monkeys and a sloth appear. EVE does not carry an apple but a bunch of caju fruit (*Anacardium occidentale L.*) and a stalk of sugarcane. ADAM holds a set of spears. Between the couple are the star-shaped leaves of the manioc and a magnificent pineapple. The scene is completed by representations of several different kind of birds and fruit, and in the far background by a group of natives dancing. This work could well have as

FIG. 69. Title page of the *Historia Naturalis Brasiliae*, probably designed and engraved by Theodor Matham, who perhaps also was responsible for the hand colouring of the plate. (Leiden University Library).

its title the well-known expression by Maurits: '*Ce beau Pay de Brasil, lequel n'a pas son pareil sous le ciel*'.[24]

The fascinating frontispiece was probably designed and engraved by the skilled artisan THEODOR MATHAM from Amsterdam, whose name is also suggested as the artisan – in Dutch, the *afsetter* – responsible for the coloured copies of the *Historia Naturalis Brasiliae*.[25] Whoever that may be, SCHARF noted that the artist who polychromed these copies was very skilled. Even in very small areas, the applied paint seldom crosses the lines of the engraving.[26]

DE LAET'S ADDITIONS

DE LAET was educated as a classical scholar. This is noticeable, for example, in his comments to MARGGRAFE's observations and illustrations (almost a hundred), which he carefully kept apart from MARGGRAFE's own text by putting them every time under a separate heading, called *Annotatio*. In these remarks DE LAET frequently refers to classical literature, to which he was clearly attached. One time, he can even be caught making a deliberate mistake. When in Chapter 6 DE LAET started editing the description of the Brazilian quadrupeds, the first animal he encountered in MARGGRAFE's notes was the sloth, or in Dutch the *luiaard*. MARGGRAFE had made a nice watercolour of this animal in *Handbook I*. However, the sloth was an animal that already had an iconic status in Europe, because early examples had been brought to this continent in the late sixteenth century. Around 1600 already several drawings of the sloth circulated and one of these was published in the *Exoticorum libri decem* (1605) written by the Flemish scholar CHARLES DE L'ÉCLUSE, better known as CAROLUS CLUSIUS. It was a book DE LAET owned in his library, and he evidently preferred the iconic woodcut from that book, instead of MARGGRAFE's depiction, which watercolour nevertheless got a place in the frontispiece of the *Historia Naturalis Brasiliae* (*fig. 70* and *fig. 71*).[27]

Another intervention in MARGGRAFE's part of the *Historia Naturalis Brasiliae* was DE LAET's drastic addition to the last, eighth chapter. It was an understandable adjustment, because MARGGRAFE had actually left little more than an outline of this chapter, in which he wanted to discuss the region, its rivers and climate, as well as the indigenous people who originally lived there, their language and their habits. Having found this outline, DE LAET felt the necessity to compose this chapter himself, adding to MARGGRAFE's fragments information from various sources. In this way he included

FIG. 70. De Laet's choice for the depiction of the Brazilian *Luiaard* (Sloth) in the *Historia Naturalis Brasiliae* (1648), part 2, page 221, copied after Clusius' *Exoticorium Decem libri* (Antwerp: Plantin, 1605), page 373. (Leiden University Library).

FIG.71. Marggrafe's watercolour of the Brazilian *luiaard* (sloth) in *Handbook* I (*Libri Picturati*, A36), fol. 112 and its application in the frontispiece of the *Historia Naturalis Brasiliae* (1648) (Jagiellonian University Library, Kraków / Leiden University Library).

contributions on Tupi and Tapuya vocabulary and grammar. 'Because of the similarity of subject' De Laet concluded with an appendix about Chile and its inhabitants, based on information from the WIC expedition led by Elias Herckmans and Hendrik Brouwer in 1644. This intervention gives this last chapter a quite different character than the rest of the *Historia*. Where the rest of the book provides new empirical information, this section is only a compilation of already circulating information.[28]

The last phase of editing and printing

With all the preparatory work in deciphering, arranging, selecting pictures and combining these with the corresponding text, plus making additional notes and performing other editorial tasks, it is not surprising that De Laet, in his correspondence with the Danish collector of natural curiosities Øle Worm, did not mention the *Historia Naturalis Brasiliae* before July 1646. But after more than a year of hard work, he finally could write the following to Worm:

> From Brazil, I received a description of all plants and animals, both on land and in the sea; it is made by the highly educated young man Johannes [Georg] Marcgravius from Meissen and is provided with finely drawn images of most of them. I am eager to publish them, but hampered by the lack of artisans who can carve the pictures into wood, for there are almost five hundred images; however, in time I hope to be able to overcome these difficulties, so that I can communicate at least something, if not the whole, then the most important of it.[29]

When three and a half months later Worm received this letter, he was extremely enthusiastic.

> There is no doubt that you will perform a very hospitable act for all lovers of nature by speeding up the publication of your report on Brazilian plants and animals. This work makes my mouth water as I think its author will reveal a lot of new rarities. Good luck! [30]

During the following months, De Laet worked steadily further on the future folio book. In November 1646 the text was ready, so the printing could start.[31] We can safely assume that this was done on the two printing presses of the Leiden bookseller, publisher and printer François Hackius (c. 1605-1669) in the Pieterskerkkoorsteeg. His name is on the title page as the first publisher of the *Historia Naturalis Brasiliae*, to be followed by the Amsterdam bookseller LODEWIJK III ELZEVIER (1604-1670). The latter was a full cousin of the Leiden Elzeviers, in which publishing house he had worked until his move to Amsterdam in 1637. Why DE LAET chose HACKIUS is not known. He had never dealt with HACKIUS before. Until 1642 DE LAET had issued books only via the Leiden academy printers BONAVENTURA and ABRAHAM ELZEVIER. But from 1643 onwards, DE LAET started using other publishers. He then issued two small books via JOHANNES LE MAIRE and FRANCISCUS HEGERUS (both from Leiden) and another one through LODEWIJK III ELZEVIER in Amsterdam. But for his *Iaerlijck verhael van de verrichtinghen der geoctroyeerde West-Indische Compagnie* (1644), DE LAET returned again to the Leiden Elzeviers. Perhaps LODEWIJK ELZEVIER was the link. HOFTIJZER supposes that HACKIUS was also an apprentice from the Leiden Elzeviers. If so, he was a former colleague of LODEWIJK.[32] It is known that LODEWIJK ELZEVIER, for his own publications, often used the Leiden printing presses of HACKIUS.[33]

For DE LAET, it was indeed convenient to have a printing press close by. This made it easier to correct the proofs of the book. As so many woodblocks had to be combined with the text, it may be assumed that proofreading was required several times. In May 1647, when DE LAET's English correspondent JOHN MORRIS received DE LAET's newly printed book about gems and gemstones, the book was wrapped in proof-sheets of MARGGRAFE's text.[34] Anyway, in July 1647, DE LAET could finally report the following to his correspondent ØLE WORM:

> The book about Brazil's natural objects is progressing rapidly, but I cannot promise it will be ready before winter. Three books about the plants have now been printed, as well as one about the fish another one about the birds; the chapter that deals with the four-footed animals is now on the press; after it follows one about insects and finally there will be a chapter about the people of the country. I think there will be over 500 pictures, all of them beautiful wood-cuts. Due to the small available number of this kind of artisans, this causes a great delay. But at the moment almost four hundred have been completed. I have no doubt that you will enjoy reading the book.[35]

This complaint about the scarcity of woodcarvers and engravers was widely heard in the seventeenth century. That's why the *Historia Naturalis Brasiliae* was finished only in February 1648.[36] WORM desired to get the book 'at all costs' and commissioned his cousin ERASMUS BARTHOLIN, who at the time studied in Leiden, the task of providing it. A leather-bound copy stamped on both covers with a pattern characteristic for Dutch books was sent to WORM in April 1648.[37] Having finally received the folio, WORM's response to DE LAET was heart-warming, as he wrote in May 1648:

> The very fine work, the *Historia Naturalis Brasiliae*, which you have generously given me, has recently been sent to me by the learned young man ERASMUS BARTHOLIN. It is indeed a golden book, highly pleasing and useful to all who feel any interest in strange foreign objects. I only can say, thank you very much for this gift. I wish there was something in my possession that would allow me to pay you back in a dignified manner. All scholars will acknowledge with gratitude that the work you have put in reviewing, arranging, and completing MARCGRAVIUS's work has been excellently applied.[38]

INNOVATIVE ASPECTS OF THE *HISTORIA NATURALIS BRASILIAE*

It is striking that DE LAET and WORM in their letters never mention WILLEM PISO. Still, the *Historia Naturalis Brasiliae* starts with PISO's part, entitled *De medicina Brasiliensi* (four chapters with 132 pages), followed by MARGGRAFE's part, entitled *Historiae rerum naturalium Brasiliae* (eight chapters with 303 pages).[39] It was not that the subjects addressed by PISO (tropical botany and the medicinal use of Brazilian plants) did not matter to them; no, WORM was right, MARGGRAFE had made a kind of text never seen before. MARGGRAFE was *the* real innovator. So, what was so innovative in his approach? What sparked WORM's enthusiasm for his text?

Firstly, because MARGGRAFE only reported what he himself had witnessed. In his descriptions he avoided all rhetoric and adornments. It was the same empirical approach that he followed with his maps, 'don't fill in what you haven't seen'. While earlier famous naturalists, such as the Swiss CONRAD GESSNER (1516-1565), the Italian ULISSE ALDROVANDI (1522-1605) or the Flemish CAROLUS CLUSIUS (1526-1609), included numerous textual references as an integral part of their descriptions, such remarks are almost absent in MARGGRAFE's part of the *Historia*. When earlier literature was cited, this was done by DE LAET as editorial comments.

Secondly, MARGGRAFE's approach was innovative because – as DAMS has put it – with MARGGRAFE 'mapping [of nature] has finally given way to classifying'.[40] MARGGRAFE described the Brazilian plants and animals with scrutiny, and he meticulously categorized the acquired knowledge in taxonomic groups.[41]

Innovative, too, was that MARGGRAFE, for most of the local plants and animals described, used names derived from local Portuguese settlers or native Indians. Even today names of plant species from the *Historia Naturalis Brasiliae* can be found in contemporary Brazilian markets.[42] That ethnographic aspect, revealed in the nomenclature applied, was not only new, it was possible only thanks to the role of go-betweens who could transfer and translate the relevant information from one language to the other. Such interaction with the local population also led to the inclusion in the *Historia* of indigenous knowledge about the therapeutic value of exotic herbs, and the habits of animals.[43]

Finally, but not least, MARGGRAFE was a pioneer in the use of a magnifying glass or simple microscope in his research of natural specimens. In the world of scholars from the Dutch realm, he is the very first known to have done this.[44] The canvases of Dutch still life painters from the early seventeenth century, in which insects are often accurately depicted, suggest that among them the use of a magnifying glass was already rather common.[45] But, until now, the first Dutch naturalist known to have used such a magnifying instrument is the painter-entomologist JOHANNES GOEDAERT from Middelburg, who published about his research in the *Metamorphosis Naturalis* of 1660.[46] MARGGRAFE used a *megascopium*, as he called it, almost twenty years earlier. He used his magnifying glass for several purposes. In the *Historia* he firstly reports about his investigation of the pistil and stamens of a plant called the *Caapomonga Brasiliensibus*. Unfortunately, his drawing made with this magnifying glass is lost. DE LAET made a mistake when he placed the woodcut of the wrong plant next to this text.[47] But fortunately, MARGGRAFE also applied his *megascopium* in his study of Brazilian insects.[48] This time, the correct woodcuts accompany the text. The first insect he drew was the *Nhatru Brasiliensibus*, a mosquito, called in Dutch the *langpootmug*. The original drawing has not survived, but even the woodcut shows a rather detailed picture (*fig. 72*).

Far more interesting is MARGGRAFE's report of a black bug which was 'barely visible' with the naked eye. But with the help of the *megascopium* he could establish it was perfectly round, and that it possessed

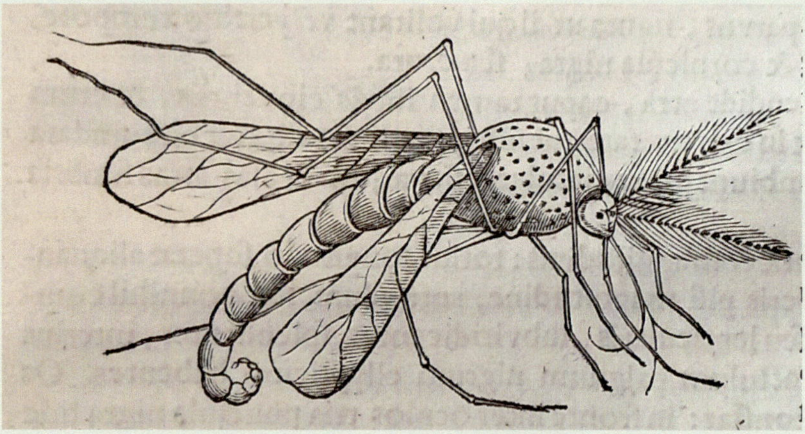

FIG. 72. Woodcut after MARGGRAFE's drawing of a mosquito, made with his magnifying glass or simple microscope, a device he called *megascopium*. Note the small human-like face that MARGGRAFE thought to have seen. *Historia Naturalis Brasiliae* (1648), part 2, 257. (Leiden University Library).

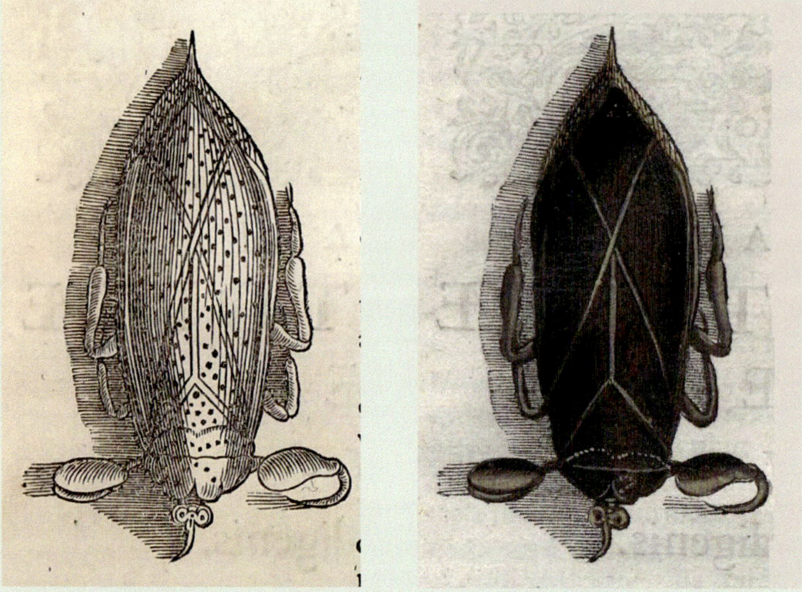

FIG. 73. Woodcut after MARGGRAFE's drawing of an unknown bug, made with his *megascopium*. This insect, of which MARGGRAFE did not give a name, was dark yellow in colour, so the black colour of the illuminator is a mistake. Picture from an uncoloured and coloured copy of the *Historia Naturalis Brasiliae* (1648), part 2, page 259. (Leiden University Library).

a black, shiny shield and six protruding legs. With this magnifying instrument, MARGGRAFE also observed a bug with multiple legs. 'It turned out to have two horns and six legs', he wrote. The body and its wings were dark yellow in colour, with black transverse streaks (*fig. 73*).[49] The corresponding woodcut in the *Historia* demonstrates that the lens MARGGRAFE used had a rather high magnification. But the coloured copy of the *Historia* also shows that the original drawing must have been in black and white, because in this case the Dutch *afsetter* (illuminator) used the wrong colours. He did not notice that in this text MARGGRAFE reported about two different bugs. So, that's why the second yellow bug received the black colour of the first.

THE RECEPTION OF THE *HISTORIA NATURALIS BRASILIAE*

The *Historia Naturalis Brasiliae* has been described as one of the works that marked the turning point in the way natural history was practiced in the early modern period.[50] No wonder that the book was soon studied by other naturalists. ØLE WORM, for instance, used the *Historia* as a model for his own museum catalogue, issued in 1655. WORM even wanted his book to be issued by the same publisher, using a similar typeset and, if possible, with illustrations made by the same woodcutters. WORM's book contains even some of the original woodcuts of the *Historia,* in places where these could illustrate specimens in his museum.[51] Others soon followed WORM's example of using the *Historia* as a work of reference. In the seventeenth century the book was cited and used by naturalists throughout Europe, for instance by JAN JONSTON from Poland, JOHN RAY from England, CHARLES DE ROCHEFORT from France, ERASMUS FRANZ from Germany and ADAM OLEARIUS from Sleswig.[52] And, as BRIENEN has pointed out, MARGGRAFE's inclusion of additional information regarding food and habitat in his representations of quadrupeds would become standard in the eighteenth century.[53]

Why did the *Historia* obtain such great authority? According to NEIL SAFIER, who studied this question, this can be attributed to a number of reasons. At first, the approach chosen by MARGGRAFE to present himself as firsthand observer 'earned him the right to speak authoritatively about what he had seen, differentiating him and his work from earlier armchair natural historians who had merely compiled travellers' accounts they had received from abroad'.[54] A second strong point of the *Historia Naturalis Brasiliae* was that DE LAET, in his preface, had presented the book not only as a description of the nature of northern Brazil, but also as useful for other regions throughout the Americas. Further, MARGGRAFE repeatedly used a strategy of comparing the things he described with animals and plants that his future readers were already familiar with. Finally, MARGGRAFE emphasized the information he gained from the indigenous people, which rhetorical gesture enhanced the authority of the knowledge presented. This all contributed towards the reputation of the *Historia Naturalis Brasiliae* as a reliable source, quoted until this very day. This in spite of the fact that, as TEIXEIRA has demonstrated, after close reading, it appears that the *Historia* was written with a utilitarian view that omitted less useful species and also included some elements foreign to Brazilian nature.[55]

PISO'S 1658 REWRITING OF THE *HISTORIA NATURALIS BRASILIAE*

The publication of the *Historia* in 1648 had crossed WILLEM PISO's original plans. As MARGGRAFE's letter to DE LAET, cited in Chapter 2, has shown, PISO was already busy composing a text on Brazilian medicine as early as February 1640. After his return in Holland, he settled in his parental house in Leiden and in March 1645 he registered himself at Leiden University with the intention to edit his own work.[56] But as JOHAN MAURITS urgently requested PISO to follow him to his new residence in Cleve, he reluctantly obeyed and left his material at the disposal of JOHANNES DE LAET, who included PISO's findings in the *Historia*.

But that course of affairs left PISO unsatisfied. In a letter written to CASPAR BARLAEUS from Cleve in September 1645, he informed the Amsterdam professor about his intention to settle in Amsterdam, now that his father has died.[57] He hoped to establish himself there as a physician, being able to start a family.[58] Ten years later, when PISO indeed was married and lived in Amsterdam as a respected medical doctor, he saw his chance to cash in on his Brazilian fame after all.[59] In 1658 he issued a revised edition of the *Historia*, with the title *Indiae utriusque re naturali et medica* ('On the natural history and medicine of both Indies'). In this book he combined his – and MARGGRAFE's – observations in Brazil with a smaller book by the late medical doctor JACOBUS BONTIUS (1592-1631), who had worked in the East Indies. By combining the medical experiences from the two indies, PISO wanted to compose the ultimate reference book for tropical medicine.

BONTIUS's posthumous book *De medecina indorum* (in which he described several tropical diseases for the first time) was published by FRANCISCUS HACKIUS in 1642. This Leiden printer-publisher still possessed most of the woodblocks of the *Historia* as well as its only copperplate. As PISO's new book was issued by HACKIUS's colleague and frequent business partner LODEWIJK ELZEVIER, this eased the production of the new book in which much material of the *Historia* would be re-used. It is even plausible that the new edition was printed in Leiden on the printing presses of HACKIUS. For this 'revised' edition, the copperplate of the frontispiece was changed. In the adapted engraving, several areas were erased, on which flattened parts new subjects were engraved to highlight that the new book also contained information about the East Indies. The 'Brazilian Eve' (the Tapuya woman) was replaced by a native with a turban (a similar figure was depicted on BONTIUS's book from 1642), and the dancing Tapuyas in the background were changed into a dodo and a rhinoceros. Both were – non-Indonesian (!) – animals that PISO had added to BONTIUS's text.[60] The picture of the rhinoceros was a poor, reversed copy of the famous woodcut by ALBRECHT DÜRER of 1515.[61] Further were the monkeys around the book's title replaced by two swans, and the mythical figure in the foreground was changed into a Moluccan *babirusa* (a pig-like animal from Indonesia) together with an Indonesian leopard (*fig. 74*).

In a letter to the diplomat and poet NICOLAAS HEINSIUS, which accompanied a copy of the new book, PISO explained that its publication had been necessary because the first edition had 'an infinite number of errors', caused by the haste pursued by Mr. DE LAET and the printer, 'in my absence'.[62] PISO had inserted a similar note in the preface of his book. There, he explained to his readers that his original text had been written in haste and, due to his absence (in Cleve), he had not been able to give his text the shape he had wished. The *Historia* therefore contained errors he wanted to correct. He now claimed to have made a better text. Although PISO acknowledged that he obtained much information from his servants,[63] among whom was especially the 'very accurate' GEORG MARGGRAFE, 'candidate in mathematics and medicine', he excluded MARGGRAFE's name as a co-author, a choice he justified in the middle of his folio by writing:

> I received, on loan, from my great and very diligent assistant, MARKGRAVIUS, some pictures and notes observed in our travels. And I want to warn beforehand, so that no malevolent growl will occur, that I have decorated my writings with stolen colours. Because when the public service denied me the necessary leisure, I requested him to supply me with his work, so that I could investigate and experience more closely the nature and property of the species, and that

FIG. 74. Title page of Willem Piso's *Indiae utriusque re naturali et medica, libri quatuordecim* (Amsterdam: L. & D. Elzevier, 1658), adapted from the frontispiece of the *Historia Naturalis Brasiliae* (Leiden: F. Hackius / Amsterdam: L. Elzevier, 1648). (Leiden University Library).

> he should carefully draw the figures outside. I did so, because it does not matter that our discoveries are attributed to him or me, or that they come to light under one name or the other, as long as the findings dissolve some doubts and reveal errors and are of use to patients and doctors, in both the New and Old Worlds.[64]

Having written this disclaimer, PISO now boldly published all material from MARGGRAFE under his own name, except two small chapters, one on topography, meteorology and the sun eclipse of 1640, and the other on the indigenous people of Brazil and Chile (which part – as we saw earlier – was in fact almost completely written by DE LAET). Moreover, PISO did not mention where he had added, deleted or adjusted the original texts. The German botanist and zoologist MARTIN LICHTENSTEIN, who made a close comparison of the content of both the 1648 and 1658 editions, made the following scathing judgment about PISO's considerable alterations:

> PISO's descriptions are consistently shorter than MARCGRAVE's, and arbitrary omissions make them completely incomplete. Often, in order to avoid the appearance of a literal copy, the position of MARCGRAVE's words is changed and so the meaning of the sentence is frequently in contradiction with itself and the illustration. PISO's work therefore deserves only a casual mention, and is mainly only considered where it has given rise to errors that need to be refuted.[65]

One error that PISO corrected was the illustration of the sloth, or *luiaard*. At the proper place he deleted DE LAET's image, copied from CLUSIUS, and changed it for another picture, modelled after a drawing by ALBERT ECKHOUT in the *Theatrum rerum naturalium Brasiliae*.[66] To soften all the criticism about PISO, WHITEHEAD and BOESEMAN argued that 'in fairness to PISO', it must be said that 'he often introduced much more biological and ecological data than are in the *Historia*' and that 'all authors' agree that PISO is best remembered as an early pioneer in tropical medicine.[67] DAMS, too, remarked that PISO added numerous pieces of advice concerning hygiene and diet. His rules of conduct contributed in this way to better health conditions in the Dutch tropical colonies.[68]

However, PISO's manner of acting failed to convince posterity. The reaction of MARGGRAFE's brother CHRISTIAN, for instance, was furious. In the *Vitae* he wrote that he was convinced that everything

Piso had published came from his brother's notebooks. He also mocked Piso's ignorance, who for the determination of some Brazilian fishes and birds needed the assistance of surgeons and apothecaries from Amsterdam, who afterwards 'merrily jested at Piso's improving ye second edition of ye Brasile History. Had he been a party to this knowledge, he would not have been obliged to seek the terms elsewhere', Christian concluded.[69]

Many later authors have also accused Piso of plagiarism, criticizing him for his failure to give the deserved scholarly credits to the original researcher. Indeed, Piso went rather far with this conduct, as he even presented Marggrafe's observations with the *megascopium* as being his own.[70]

So, in the end Piso's work was marked by suspicion of plagiarism. A century later Carolus Linnaeus would name a new genus *Pisonia,* giving as his comment that the spines of this tree were as nasty as Piso's reputation. He loathed the memory of the man who after Marggrafe's death had seized all his results.[71] However, as Rodrigues has argued, at that time (and even until the twentieth century), it was quite accepted to use information collected by servants as being your own. After all, Marggrafe had indeed been in Piso's service in the years 1638-1640, and what Marggrafe had written after 1640 and outside Piso's expertise (such as the meteorological observations), Piso correctly had published under Marggrafe's own name.[72] So, in retrospect, not Piso's plagiarism, but foremost his many editorial mistakes are to blame for the bad reputation of his *Indiae utriusque re naturali et medica.*

Christian Marggrafe claiming the inheritance of his late brother Georg

In June 1652, four years after the publication of the *Historia Naturalis Brasiliae,* and two years after the death of its editor, Johannes de Laet, a new student from Liebstadt arrived in Leiden. It was Georg's brother Christian Marggrafe, 16 years younger. He probably came from the neighbourhood of Altenburg in Thüringen, where in 1651 he had contributed to a collection of funerary poems devoted to a young Lutheran *Pfarrer* from Grosstechau, who was murdered during a robbery on the streets of Copenhagen.[73] In Leiden, Christian registered as a student of medicine on 3 June 1652, giving as his address a student home in the Klocksteeg, called *De courvorstinne van Brandenborgch* ('The Electress of Brandenburg' – location no. 8 on the map in *fig. 19*).[74] This boarding house was not far from where his brother once lived. In July Christian already had visited Golius's house on the Rapenburg (no. 6 on the map in *fig. 19*), where he saw his brother's astronomical observations, about which he wrote that very month to the well-known astronomer Johannes Hevelius in Dantzig (as reported in the introduction).[75] At that first meeting, Golius probably gave him the *Itinerary*, the diary Georg had kept from the time of his departure from the parental home until the year 1640. Years later, Christian would use that document for the *Vitae*, the biography of his brother, published in 1685. Golius also told him that Count Johan Maurits had donated some goods from Georg to the university, mostly animal skins and stuffed animals. These objects were displayed to the public in the gallery of the *Theatrum Anatomicum*. On another occasion, Golius also gave him an Arabic manuscript that he had found in the chest delivered to him in 1644. So far, so good. In July 1654, Christian obtained the degree of medical candidate with a disputation that he gratefully dedicated to both Jacob Golius and his botanical colleague Adolph Vorstius (*fig. 75*).[76]

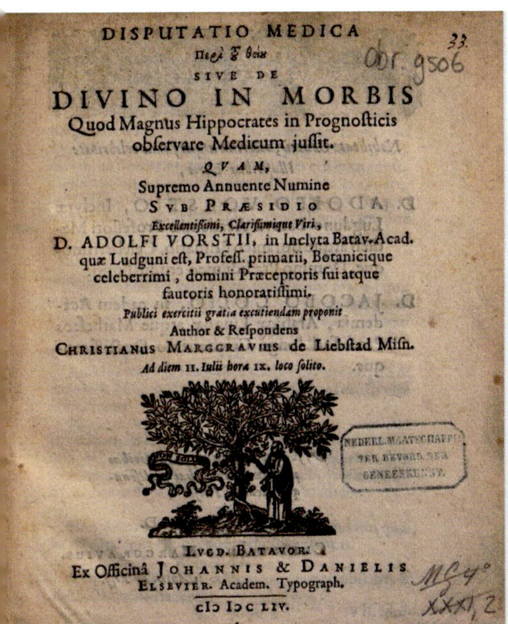
DISPUTATIO MEDICA
Περὶ τοῦ θείου
SIVE DE
DIVINO IN MORBIS
Quod Magnus Hippocrates in Prognosticis observare Medicum jussit.
QVAM,
Supremo Annuente Numine
SVB PRÆSIDIO
Excellentissimi, Clarissimique Viri,
D. ADOLFI VORSTII, in Inclyta Batav. Acad. quæ Ludguni est, Profess. primarii, Botanicique celeberrimi, domini Præceptoris sui atque fautoris honoratissimi.
Publici exercitii gratia excutiendam proponit
Author & Respondens
CHRISTIANUS MARGGRAVIUS de Liebstad Misn.
Ad diem II. Iulii hora IX. loco solito.
LVGD. BATAVOR.
Ex Officinâ JOHANNIS & DANIELIS ELSEVIER. Academ. Typograph.
cIↃ IↃC LIV.

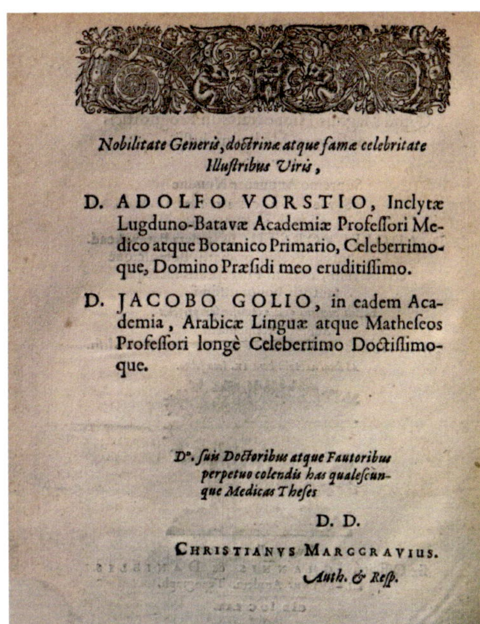
Nobilitate Generis, doctrinæ atque famæ celebritate Illustribus Viris,
D. ADOLFO VORSTIO, Inclytæ Lugduno-Batavæ Academiæ Professori Medico atque Botanico Primario, Celeberrimoque, Domino Præsidi meo eruditissimo.
D. JACOBO GOLIO, in eadem Academia, Arabicæ Linguæ atque Matheseos Professori longè Celeberrimo Doctissimoque.
Do. suis Doctoribus atque Fautoribus perpetuo colendis has qualescunque Medicas Theses
D. D.
CHRISTIANVS MARGGRAVIUS.
Auth. & Resp.

FIG. 75. Medical dissertation of Christan Marggrafe, defended 2 July 1654, dedicated to both Adolph Vorstius and Jacob Golius. (Amsterdam University Library).

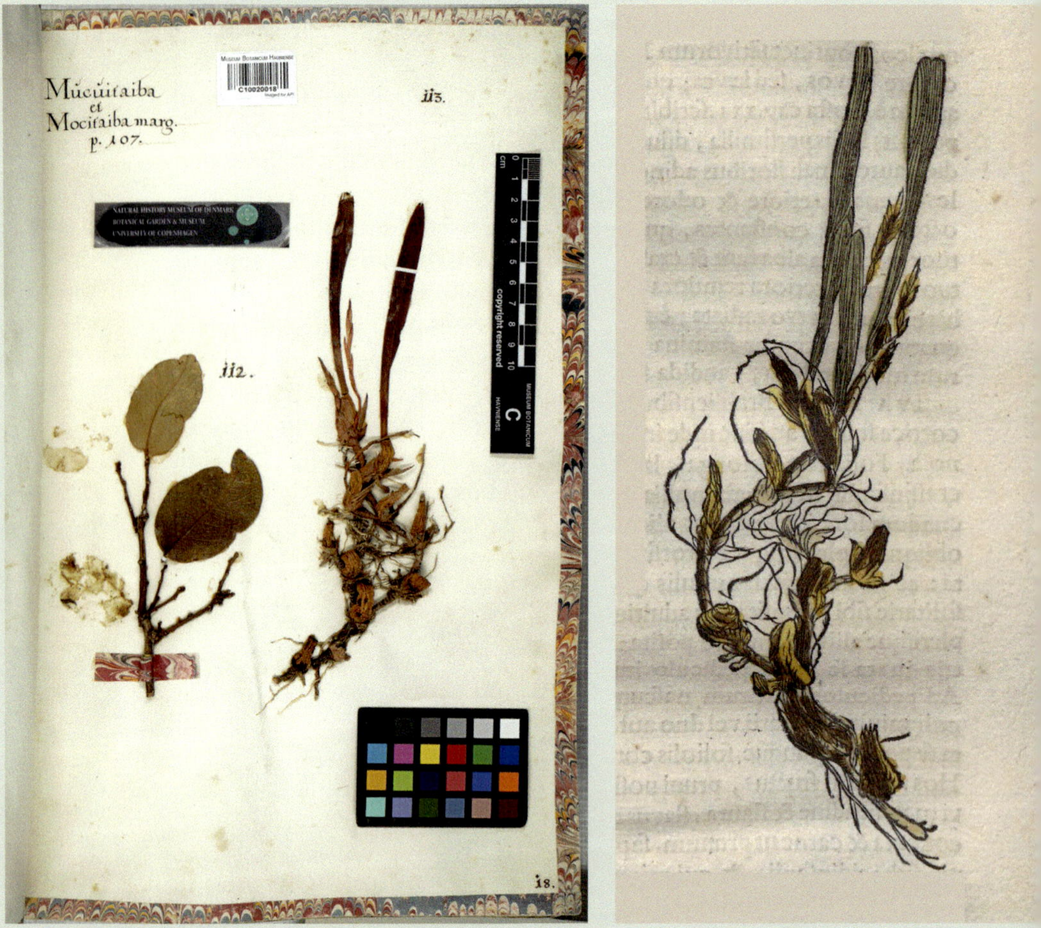

FIG. 76. A random page from GEORG MARGGRAFE'S *Herbarium Vivum*, bought in 1653 by VILLUM WORM (Reproduced with permission of the Natural History Museum of Denmark, Copenhagen), compared with the woodcut (in reverse!) made after this specimen in the *Historia Naturalis Brasiliae* (1648), part 2, page 107. (Leiden University Library). As Pickel (2008), 109 noted, this is *not* the *Mucutaiba* or *Mocitaiba Brasiliensibus*, described by MARGGRAFE (the tree that provided the very tough and heavy wood, called *Pao Santo*, of which his astronomical quadrant was made). Instead this sample concerns an orchid, sometimes found in its branches.

But shortly afterwards CHRISTIAN heard some disturbing rumours, prompting him to investigate more deeply what had happened to his brother's belongings. The news that the Danish student VILLUM WORM had been able to buy a *herbarium vivum*, full of dried plants collected by his brother, will have shocked him. But this collection was dispatched to Copenhagen before CHRISTIAN could take any action (*fig. 76*).[77]

To top it all off, GEORG MARGGRAFE's former Leiden roommate SAMUEL KECHEL claimed to have seen in Haarlem the sale of a book with dried Brazilian insects, with names 'all written in MARCGRAVE's hand'. According the *Vitae*, KECHEL told CHRISTIAN that the book yielded the enormous sum of 4,000 guilders (which in retrospect seems to be quite an exaggeration).[78]

By that time, the *Historia Naturalis Brasiliae* had been distributed throughout Europe, so the importance of GEORG MARGGRAFE's scholarly work had become known far and wide. CHRISTIAN now became aware of the monetary value of Brazilian curiosa, evidently also those of his late brother. In the spring of 1655, he heard that JOHAN MAURITS used parts of his Brazilian collection as diplomatic gifts to monarchs and other high-ranking individuals to raise his own status and receive favours back.[79] This was a strategy JOHAN MAURITS applied even before he returned to the Netherlands. Already in 1641, he had asked ZACHARIAS WAGENER to deliver *Schreiben, Mahlereyen und Papegoyen* (documents, paintings and parrots) to various recipients in The Hague, Haarlem, Delft, Rotterdam and Leiden.[80]

In 1652 JOHAN MAURITS made a large donation of Brazilian artefacts to his distant relative, the Elector of Brandenburg, FRIEDRICH WILHELM I, to be followed in 1654 by a gift of several Brazilian paintings by ALBERT ECKHOUT (with corresponding curiosities) to another distant relative, the Danish King FREDERIK III.[81] The go-between for this last gift was the Danish nobleman CHRISTOFFEL LINDENOV, then an Admiral in the Danish fleet, but before 1644 JOHAN MAURITS's *stalmeester* (equerry) in Recife.[82] LINDENOV was a man CHRISTIAN MARGGRAFE knew personally. He names him in the *Vitae* as someone he had spoken to about the time his brother lived in Brazil. As an educated mathematician and as independent collector of naturalia, LINDENOV must have had quite some common grounds with GEORG MARGGRAFE.[83] In Brazil they both shared the privilege of receiving a free meal at JOHAN MAURITS's court.[84]

Given these facts and rumours, it was time for action. In February 1655, CHRISTIAN went to The Hague and requested an audience with JOHAN MAURITS in his beautiful mansion, built there during his absence in Brazil. CHRISTIAN asked the count to write a letter of recommendation to the trustees of Leiden University to obtain clarification about what had happened to his brother's belongings. Was he – (who pretended to be) GEORG's only brother and sole heir – actually not entitled to receive these goods?[85] When these properties were missing, should he therefore not be compensated? JOHAN MAURITS indeed granted this request and wrote the following letter to the university's trustees:

> My Lords, when I returned here in the Fatherland from Brazil in January 1644, the belongings and furniture of my mathematicus, MARGGRAVIUS, who died in Angola, were also transferred [to the Netherlands]. Because at the time no one came forward as his heir, I gave everything to the laudable Academy in Leiden. But now the medical student CHRISTIAN MARGGRAVIUS, the real brother of the aforementioned MARGGRAVIUS, and the bearer of this letter, has come to this country. He has inquired about his brother's estate, and has requested that I give him a letter of recommendation for you, the honourable gentlemen Trustees of the illustrious Academy of Leiden. This in order that all furniture and goods which were put into the Academy's custody (listed in an inventory made at the time), could be given to him, or that he receives a reasonable and proper satisfaction.[86]

CHRISTIAN was really upset by this letter. Already in 1648, evidently shortly after the publication of the *Historia Naturalis Brasiliae,* he had written to professor VORSTIUS, asking him for his brother's belongings.[87] But CHRISTIAN never had received an answer. In 1652, after he had arrived in Leiden, professor GOLIUS indeed gave him some documents, but GOLIUS never told him about an inventory of his brother's 'furniture and goods', as stated in the count's letter. For that reason, CHRISTIAN now wrote again to VORSTIUS asking to provide him with this 'legitimately approved inventory' of his brother's chest, made at the moment it was delivered to the university.[88]

VORSTIUS again gave no response, but GOLIUS confirmed that the goods had not been donated to the university. But according to the professor, the value of the goods was no more than ten guilders; he remembered a herbarium, an Arabic dictionary, and some other writings, including two in Portuguese. He further claimed that at the time a large chest had been brought to VORSTIUS, and that he himself had kept only a small one, in which there was nothing but some seeds, various dried roots or fruits, some insects and the like. The Brazilian herbarium from his chest was also given to VORSTIUS.[89]

Shocked by this information, CHRISTIAN now submitted to the University's trustees a formal request, asking for clarification, return of all the goods, or – if missing – for compensation. He inquired in particular about the herbarium that should be in the possession of VORSTIUS, which to his knowledge was marked with his brother's personal Greek motto.[90] CHRISTIAN was especially angry that ten days before, in the medical auditorium of the university, GOLIUS publicly had protected VORSTIUS by denying that CHRISTIAN was entitled to the goods of his brother, because the count had donated them to the university.

CHRISTIAN's request stirred a lot of excitement among the university's trustees. They had indeed regarded GEORG's naturalia as a gift and these objects were now exhibited in the *Theatrum Anatomicum* and in the museum next to the *Hortus Botanicus.* Their reaction was fast. They immediately asked JOHAN MAURITS to write a second letter, downplaying the importance of GEORG MARGGRAFE's goods. In this second letter, the count argued there was indeed some misunderstanding. After its arrival, GEORG's chest was opened in the presence of several persons. They had found nothing in it other than some writings and observations about Brazil, 'which we had ordered him to record'. Further, there were some books and curiosities from Brazil, namely a few skins, horns and bones of some animals, and several dried plants, with the fruits thereof, 'all collected at our expense and to our benefit'. To be specific, the chest contained absolutely no valuables, no money, gold, silver or precious minerals. In short, there had been no objects that made it necessary to draw up an inventory.[91] Moreover, how could it be otherwise? Everyone from Brazil knew that GEORG MARGGRAFE always was short of money. Further, the count explained that it had been on his request that the goods were given to the university, 'to be displayed and preserved there'. He had ordered that the writings and drawings concerning the natural history of Brazil, as well as the descriptions and surveys of that country, be sent to DE LAET, and the astronomical observations to GOLIUS, 'to investigate and collect them at our expense, and to publish them, as has already been partially done'.[92] Having said this, the count still found that the younger brother should be given some of the preserved goods. But, as the value of MARGGRAFE's goods never exceeded the price of a pin, any monetary compensation for the younger brother should be out of the question. CHRISTIAN MARGGRAFE, however, did not agree with this proposal and requested the academic *Vierschaar*, the university's judicial court, to order that every item from GEORG's chests be handed over to him.

To prepare his defence in this legal procedure, GOLIUS now requested WILLEM PISO's assistance, reason why from PISO's hand two letters are preserved, providing more details about GEORG MARGGRAFE and his legacy. PISO declared that MARGGRAFE had left his service after two years, but by coincidence he had seen his chest in Brazil shortly before his departure to Africa. It then contained only shabby cloths, some student's notebooks, a striped seahorse skin, some old seed from Brazilian crops and various dried herbs or vermin.[93] He also recalled that MARGGRAFE always said that he was interested only in his observations with regard to astronomy and natural history, and that he did not care about money. GEORG was proud to cite the expression *Cantabit vacuus coram latrone viator*, in English 'the empty-handed traveller can sing, even in the neighbourhood of a robber', meaning a penniless man has nothing to lose.[94] GEORG MARGGRAFE was always short of money, PISO declared. If he had a penny left, he immediately spent it by drinking. MARGGRAFE's finances were so bad that when PISO returned to Leiden in 1644, he learned that MARGGRAFE had not even paid his debts to his shoemaker ARENT SYMOENSZ and his landlady GRIETTIE BAILLY.

Whether or not PISO made these statements deliberately to frame a penurious picture of MARGGRAFE is difficult to tell. But it certainly looks like it. As already has been remarked by WHITEHEAD and BRIENEN, MARGGRAFE's accurate scholarly work leaves little room for the frequent intoxication described by PISO. Or, to cite WHITEHEAD, 'night after night' MARGGRAFE 'was peering through his telescope, reading off divisions of less than a millimetre on his ruled scale, and meticulously noting his result [...] He cannot have been anything but sober the night before'.[95]

At GOLIUS's request another statement was made by ABRAHAM DE VRIES, who in Brazil had served as one of JOHAN MAURITS's treasurers. In his document the notary inserted in the margin the remark that in Brazil MARGGRAFE had been *behouftig* (destitute), and noted further that in Recife he had borrowed money from one of the count's lieutenants, a certain SETSKI, who according to DE VRIES currently served the count as his kitchen steward (*hofmeester*). This SETSKI can be identified as CHRISTOPHER SCHETSKI, one of the officers in JOHAN MAURITS's retinue receiving *costgelt* in Brazil. SCHETSKI, too, had never seen the return of his money, DE VRIES declared. However, upon closer examination of this statement, it appears that MARGGRAFE had not received SCHETSKI's money until after he had given him in return the claim against his arrears of salary, which would be paid to him after his return to the Netherlands. In Brazil, this procedure was quite normal. But the Amsterdam Chamber of the WIC refused to pay MARGGRAFE's overdue salary as court astronomer and surveyor. In 1637 the company had hired MARGGRAFE as assistant to the court physician WILLEM PISO. His later appointment as mathematician (with subsequent higher wages) was never ratified by the *Heeren XIX*. Therefore, the company's bookkeepers regarded this part of MARGGRAFE's work as a personal affair of Count JOHAN MAURITS, who as a consequence had to reimburse SCHETSKI for the overdue salary, instead of the WIC. Apparently JOHAN MAURITS had not accepted this view, which left SCHETSKI still waiting for his money.[96]

Apart from all the gossip, PISO's letters are in particular interesting for the fact that he gives more details about what happened with MARGGRAFE's astronomical papers. From his contacts with the Leiden publisher ELZEVIER, PISO had heard that MARGGRAFE's former Leiden roommate SAMUEL KECHEL AB HOLLENSTEIJN had composed a *Planisphaerium* of the southern stars, based on MARGGRAFE's observations. This work undoubtedly was done at GOLIUS's request, who always allowed KECHEL free access to the Leiden Observatory for making astronomical observations.[97] Evidently this map of the Southern Hemisphere was finished, but its publication was probably halted until GOLIUS had completed his work on Persian astronomy, as described in CHRISTIAN MARGGRAFE's letter to the astronomer HEVELIUS, mentioned in our introduction.

The legal procedures started in June 1655 and were finished by a first verdict in February 1656. CHRISTIAN could indeed obtain the astronomical observations and 'some Brazilian crops and other minor natural things, as these were present in a drawer of the defendant [GOLIUS]', to which also belonged the spindle of a globe with a brass meridian and a small instrument, as described in a newly made (but now lacking) inventory. Excluded from these goods, however, were the two volumes of GEORG's 'Mathematical Notebooks' (which probably contained his original astronomical observations) and the *Planisphaerium* of the southern stars that KECHEL had constructed from these observations. Count JOHAN MAURITS had ordered these goods to be returned to him because he had financed them.[98]

Christian however, refused to accept this verdict of the university's *Vierschaar.* He appealed to have the decision reconsidered, leading to another verdict in May 1656. The outcome was the same, but as an additional stipulation Christian was ordered to maintain 'an eternal silence' on these matters and he was convicted to pay all legal costs of the proceedings. But, because Christian still refused to accept this final verdict of the academic tribunal, all remaining objects and documents stayed in the custody of the university, some of them still being displayed at the *Theatrum Anatomicum* (*fig. 77*).

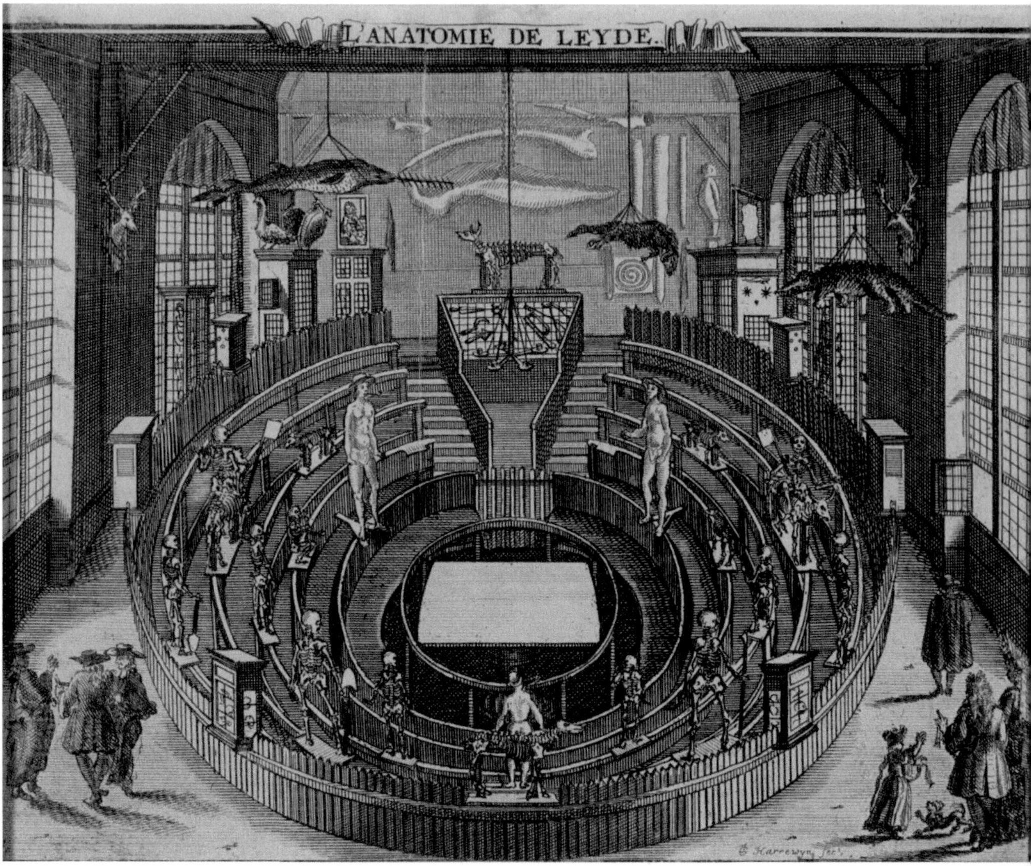

FIG. 77. The *Theatrum Anatomicum* of Leiden University on the Falide Begijhof (no. 4 on the map in *fig. 19.* In the showcases at the back of the room, all sorts of natural curiosities were exposed, including some of the objects collected in Dutch Brazil. Engraving published in 1712. (ELO).

'Maintaining eternal silence' was not something that CHRISTIAN MARGGRAFE adhered to. He continued to feel cheated. When in 1661 the Danish scholar OLE BORCH passed through, he told him that his late brother GEORG, 'the governor's mathematician in Brazil', always observed and speculated about the planet Mercury. But his inquiring mind had also led him to collect all kinds of rarities worth as much as 60,000 guilders in total, CHRISTIAN boasted. But these goods had fallen into bad hands after his brother's death. In the book recently published by PISO, the mathematical speculations of his brother were also neglected. Further, a gentleman by the name of WORM had bought his brother's herbarium 'for only three guilders', but CHRISTIAN thought that this was a book from the WIC *bewindhebber* DE LAET. That man had collected similar items from the soldiers who returned from Brazil. CHRISTIAN was convinced that some of the valuable goods of his brother were still in Leiden, BORCH noted in his journal.[99]

Some of MARGGRAFE's naturalia that remained with the university must indeed have been described among the objects in the *Catalogue of all the cheifest rarities in the publick theater and anatomy hall of the University Leiden* of 1678.[100] In this inventory, some naturalia are mentioned as donated by WILLEM PISO,[101] and others are attributed to JOHAN MAURITS. Other objects are listed without provenance, for instance, 'the skin of a Brazilian Beast, called *Tamandua Pepa*', which, in fact, may well have been MARGGRAFE's (*fig. 78*). He had dissected such an animal, noting that its skin was very thick.[102]

It is obvious that CHRISTIAN MARGGRAFE became a *persona non grata* at Leiden University after this affair with professor GOLIUS and the Leiden trustees. To finish his studies and to obtain his degree as

FIG. 78. *De Kleine Miereneter* (lesser ant eater), or *Tamaudua*. Watercolour on paper by GEORG MARGGRAFE in *Handbook I* of the *Libri picturati* A36, fol. 62. (Jagiellon University Library, Kraków). A skin of this animal was in the collection of the Leiden *Theatrum Anatomicum*. (Jagiellonian University Library, Kraków).

a medical doctor, CHRISTIAN had to travel to Franeker University, in the north of the country.[103] Nevertheless, he returned to Leiden and stayed there for the rest of his life. In Leiden, he built an impressive career as an independent physician and chemist, having a laboratory of his own. At home, he also gave academic lectures that were well-attended and appreciated by many students.[104] But CHRISTIAN's request to give these successful lectures at the university was politely declined.[105] His independent thinking in political and theological matters put him frequently in trouble with the local government or with the university professors. On several occasions he expressed as his opinion that the 'Lords of Leiden' had never done him justice. In August 1674, when he was 48 years old, the Leiden *schout* (sheriff) THEODORUS GOOL (a son of professor GOLIUS) jailed him for a short time on an accusation of blasphemy and contempt of the government of Holland.[106] Apparently shocked by his near-condemnation and associated almost-confiscation of all his properties, CHRISTIAN decided to end his life as a bachelor and to pursue a more regular family life. On 17 November 1674 a notice of marriage was registered between the 48-year medical doctor and the 19-year-old ADRIANA WILLEMSDR VAN GAMMEREN from rural Babyloniënbroek in Brabant.[107] A year later, their first child was born, to be followed by son GEORG MARGGRAFE III in 1677. In the next years five more children followed. Unfortunately, the young GEORG died in infancy.[108] CHRISTIAN himself passed away in January 1687, shortly before the publication of a lampoon aimed to ridicule his competitor in medical chemistry, the Leiden apothecary JACOB LE MORT (1650-1718).[109] This posthumous fierce attack on his competitor's credibility was in vain, because three years later LE MORT was appointed prefect of the newly erected chemical laboratory of Leiden University, becoming ordinary professor in 1695.

CHRISTIAN's widow, then aged 32, was left behind with five very young children, one of whom died within a few months after his father. Understandably, she remarried the following year.[110] CHRISTIAN's large library, which according to a newspaper advertisement reflected his scholarly reputation, was auctioned in May 1687.[111] As no catalogue has been preserved we cannot tell whether any manuscripts from his brother GEORG were among the items sold.

The fate of Georg Marggrafe's astronomical manuscripts

The lost mathematical notebooks and the accompanying planisphaerium

The two 'Mathematical Notebooks' and the *Planisphaerium* of the southern stars requested back by Johan Maurits in 1655 have never surfaced again. Nor are they mentioned in any inventory of the count's possessions after that date. It is highly likely that these papers are destroyed. Our guess is that they burned in 1665, with the house of Johan Maurits in Siegen.[112]

Surviving manuscripts in the Leiden archive

The documents preserved in the Leiden archive today are undoubtedly the remains of what was once disputed by Christian Marggrafe, but which was never handed over to him because of his refusal in 1656 to concede. They probably first stayed in the custody of the university and were transported at some moment to the Leiden city archive. Currently preserved is a set of 119 paper sheets, heterogeneous in size and content, referring to a variety of subjects.[113] All of them, except two letters by Dutch skippers, are written in Latin. Since Marggrafe spent almost ten years travelling before enrolling at Leiden University, some documents might even predate his arrival at Leiden. There are texts which appear to be class notes, covering astrological theory, astrological horoscopes, logarithms, trigonometric functions, etc. Other notes concern astronomical tables or astronomical observations. Many paper scraps contain sketchy notes, calculations (including one of the longitude of Recife based on an observed lunar eclipse), drawings of (parts of) instruments and plans of the observatory's ground floor in Recife, and finally a draft of the contents of his part of the *Historia Naturalis Brasiliae* (*fig. 64*). The Leiden manuscripts cover the astronomical observations made in Leiden rather well, but only a small part of those made in Brazil.

How and when these papers were transferred to the Leiden city archive is unknown. In modern times, the existence of draft letters written by Georg Marggrafe is first mentioned in 1855, in the historical journal *De Navorscher*, but without telling where these letters were kept.[114] Only insiders could guess that the 'R.E.' who had submitted the note was Willem Rammelman Elzevier, the keeper of the Leiden archives. This became evident only in 1864, when

RAMMELMAN ELZEVIER published an inventory of the city archive.[115] The first researcher to use and quote these preserved drafts was the medical doctor and historian BERNARD STOKVIS in 1883 in his essay *La médicine coloniale et les médecins hollandais du 17e siècle.*[116]

The fact that from MARGGRAFE not only letters, but also a bundle of astronomical calculations, observations and horoscopes, have survived was made public only in 1939, after the rediscovery of these papers by the Leiden archivist Mrs. L.J. RUYS. At her request, a short description of these findings was published in the history of science journal *ISIS* by CLAUDE AUGUST CROMMELIN, the first director of the *Rijksmuseum voor de geschiedenis der Natuurwetenschappen* (State Museum for the History of Science).[117] Unfortunately, after that time the drafts of MARGGRAFE's letters were incorrectly stored. When PETER WHITEHEAD, then head of marine fishes of the Leiden Museum of Natural History (currently Naturalis), wanted to consult these letters in the 1970s, they could not be found.[118] The letters remained untraceable until 2017, when INGRID POT-NOORDMAN, librarian of *Erfgoed Leiden en Omstreken* (the former Leiden City Archive), phoned one of us (HJZ) with the announcement that fortunately the drafts were found again in a folder with other completely unrelated material, such as medieval parchment strips found in the spines of book bindings.[119] It appeared then that drafts of the following nine letters are preserved:

1. to JOHAN MAURITS, in Portuguese, near San. Salvador, dated 15 May 1638.
2. to JACOB GOLIUS, in Latin, from 'San Antonio' (= Antonio Vaz), dated 26 June 1638.
3. to GEORG FROM, in Latin, no place, dated 26 June 1638.
4. to JACOB GOLIUS, in Latin, from Itamaracá, dated 4 October 1638.
5. to SAMUEL KECHEL, in German, from Mauritia (= Antonio Vaz), dated 18 May 1639.
6. to GEORG FROM, in Latin, from Mauritia, dated 18 May 1639.
7. to JEAN and ANTOINE VAN HUYCK, in French, no place and undated, but 18 May 1639.
8. to JOHANNES DE LAET, in Latin, from from Mauritia, dated 6 February 1640.
9. to JACOB GOLIUS, in Latin, from Mauritia, 16 February 1640.

Unfortunately, these texts are written in terrible handwriting and are messed up with with many strikethroughs, insertions and abbreviations that are often undecipherable. Nevertheless, some fragments could be read and used in this book.[120] From the other side of the correspondence, it appeared that in Copenhagen (Denmark), one draft letter to MARGGRAFE has been preserved:

10. from GEORG FROM, in Latin, from Leiden, dated 2 April 1638. This letter was transported to Brazil by BODECHER BENNING, and was received on 13 June 1638.[121]

In the above-named correspondence the following missing letters are mentioned:

11. from GEORG FROM, from Leiden, dated 13 March 1638, transported to Brazil with the ship *Amersfoort.*[122]
12. from JACOB GOLIUS, from Leiden, dated 22 June 1638, received on 3 October 1638.[123]
13. from SAMUEL KECHEL, from Leiden, received before 26 June 1638.[124]
14. to SAMUEL KECHEL, in Leiden, 1643. This lost correspondence is mentioned in the *Vitae*, published in 1685 by CHRISTIAN MARGGRAFE.

Finally, when in 2006 OSCAR MATSUURA requested digital scans of all documents that at the time were present in the Leiden archive, a curious surprise came to light. In a booklet with handwritten astronomical tables that MARGGRAFE had composed for his own usage, a coloured feather was hidden, probably of a tropical bird. It is the only remaining preserved zoological specimen from MARGGRAFE's adventures in Dutch Brazil (*fig. 79*).[125]

THE FATE OF A PRESS-READY MANUSCRIPT, LOST IN SPAIN

As explained in the introduction to this book, GOLIUS intended to publish MARGGRAFE's observations. But strangely enough, he planned to do that together with the Latin translation of a ninth-century manuscript by the Persian astronomer AHMAD IBN MUHAMMAD IBN KATHĪR AL-FARGHĀNĪ. The editing of this complex manuscript evidently took far more time than anticipated, but in the meantime GOLIUS – or perhaps KECHEL – had already drafted a neat copy of MARGGRAFE's observations. This edited manuscript had probably been ready for the printing press for years. GOLIUS died in 1667, followed a few months later by KECHEL. Both their libraries were auctioned in 1668,

FIG. 79. Small feather of an unidentified tropical bird (?) from Dutch Brazil, preserved in a manuscript with astronomical tables, composed by Georg Marggrafe (ELO).

but the preserved catalogues of these sales provide no clue what had happened with this press-ready copy of the Brazilian observations.

Nevertheless, this unpublished press-ready manuscript with Marggrafe's astronomical observations emerged in Paris in 1668, then in the hands of Melchisédech Thévenot (1620-1692), a well-known French scholar and a correspondent of Jacob Golius. By coincidence, Thévenot stayed in The Netherlands between 1666 and 1668. Being a renowned book collector, he most likely was present at the auctions of the libraries of both Golius and Kechel. Anyway, in the fall of 1668, Thévenot gave some information on Marggrafe's astronomical observations to Edmond Halley, who transferred this news to the Astronomer Royal James Flamsteed in London.[126] Eventually the Leiden copy of Marggrafe's astronomical manuscript is mentioned in the auction sale of Thévenot's library in 1694: 'Astronomical observations made in Brazil by Marggrafe and their use' (*fig. 80*).[127]

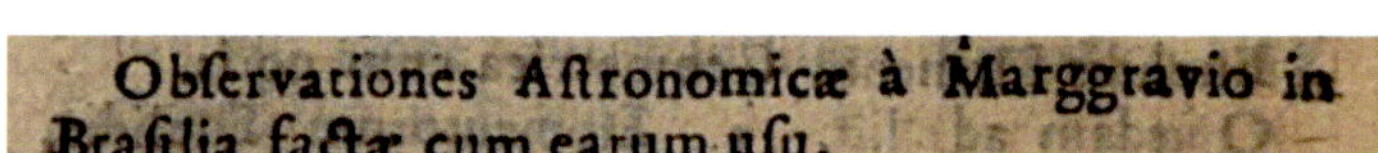

Obſervationes Aſtronomicæ à Marggravio in Braſilia factæ cum earum uſu.

FIG. 80. Detail from the *Bibliotheca Thevenotiana*, section 'manuscrits Latin', heading 'quarto' (Paris 1694), 210. (Google Books, from Public Library Lyon)

According to the French astronomer JÊROME LE FRANÇOIS DE LALANDE, THÉVENOT's manuscript was acquired at the auction by the French mathematician, astronomer and architect PHILIPPE DE LA HIRE (1640-1718).[128] DE LA HIRE was also in the possession of a copy of the observations of TYCHO BRAHE, brought to Paris in 1672 by the astronomer JEAN PICARD.[129] After DE LA HIRE's death these manuscripts came into the possession of his son and heir, the mathematician GABRIEL-PHILIPPE DE LA HIRE (1677-1719), who unfortunately died a year after his father. Whether the manuscripts were then auctioned or sold privately is not known. However, it is certain that the astronomer PIERRE COUPLET DES TORTREAUX (ca. 1670-1743), concierge of the Paris Observatory, became the new owner of the manuscripts. Twenty years before (in 1698) he had made astronomical observations in Paraiba, the region also surveyed by MARGGRAFE, so COUPLET's interest in the Brazilian documents is understandable, all the more where had lost all his own Brazilian souvenirs, when he at his return in 1699 shipwrecked on the coast of Picardie.[130] According to BIGOURDAN, in 1719 the young astronomer JOSEPH-NICOLAS DE L'ISLE (1688-1768) also became involved in the legacy of the young DE LA HIRE.[131] DE L'ISLE (also written as DELISLE) was very eager to collect astronomical records of the past, documents that could help him in the composition of his compilation work of ancient astronomical observations, for which he had projected the title *Traité complet d'astronomie exposée historiquement et demontrée par les observations.*[132] DE L'ISLE convinced COUPLET to grant him permission to make neat copies of both collections of observations.[133] But hiring a careful copyist was a costly undertaking, reason why DE L'ISLE requested financial support from abbé JEAN-PAUL BIGNON, president of the Académie royale des sciences, but also the newly appointed librarian of the French King LOUIS XIV. In a memoir written in late 1719 or early 1720, DELISLE requested 2,500 livres to copy various compilations of astronomical observations, with the provision that these copies would be available to all members of the Académie royale des sciences.[134] This request concerned not only the papers of TYCHO BRAHE and GEORG MARGGRAFE, but also those of JEAN-DOMINIQUE CASSINI (1625-1712), which were still in the possession of his son, the astronomer JACQUES CASSINI. Evidently the money was granted. The reproduction work was carried out by the same copyist, as can be seen today in the Observatoire de Paris, where a set of the copied observations of BRAHE and MARGGRAFE is kept, both in the same handwriting.[135]

In 1743 COUPLET's life also came to an end. The Leiden-made press-ready manuscript with MARGGRAFE's observations then went to the astronomer LOUIS GODIN (1704-1760), one of DE L'ISLE's pupils. In 1735 GODIN had participated in an expedition to Peru, aimed to determine the shape of the earth. So, he too had been in South America. When in 1752 GODIN accepted the post of director of the astronomical observatory in Spanish Cadiz, the Leiden-made manuscript went with him.[136] After his death in 1760, his colleague LALANDE bought from his library in Cadiz a fair copy of the correspondence of JOHANNES HEVELIUS.[137] On that occasion, LALANDE discovered that GODIN had collected the written legacy of more astronomers, including the press-ready manuscript of MARGGRAFE's observations, which at the time remained with GODIN's heirs.[138]

But in a publication of 1772, LALANDE could report that this manuscript now was transferred 'in the hands of admiral ANTONIO DE ULLOA'.[139] This Spanish nobleman had participated in GODIN's expedition to Peru in the 1730s and had played an important role in the founding of the Cadiz Observatory. In January 1758, DE ULLOA had returned to Peru, becoming the local governor, a position he changed in 1766 for governor in then-Spanish Louisiana. After an unsuccessful period in North America, he returned home in December 1768, long after GODIN's death. So, it must have been in the years 1769-1771 that GODIN's heirs transferred the MARGGRAFE manuscript to DE ULLOA. But then the trail is lost. We have not been able to figure out what happened to his library after his death in Cadiz, in 1795.[140] An English traveller, JOSEPH TOWNSEND, has left us the following portrait of the room where the Leiden manuscript must have resided:

> The room was twenty feet long by fourteen wide and less than eight feet high. In this I saw dispersed confusedly chairs, tables, trunks, boxes, *books and papers*, a bed, a press, umbrellas, clothes, carpenters tools, mathematical instruments, a barometer, a clock, guns, pictures, looking-glasses, fossils, minerals and shells, his kettle, basons, broken jugs, American antiquities, money, and a curious mummy from the Canary islands, or at least its trunk, with the head and arms for having been the common plaything for his children, [who] had amused themselves with drawing its teeth and breaking off its limbs.[141]

So, maybe it's still there ... the seventeenth-century manuscript that gave cause to this book. Maybe it is still there…, somewhere in Spain, perhaps in an attic of one of the numerous descendants of this Spanish admiral ANTONIO DE ULLOA. Although the information contained in this lost manuscript has been saved by DE L'ISLE, it would be nice if it surfaced again.[142]

SURVIVING MANUSCRIPTS IN PARIS AND LISBON

Luckily, one of DE L'ISLE's neat copies of the press-ready manuscript drafted under GOLIUS's supervision survived in the *Observatoire de Paris.*[143] JOHN NORTH has already noted that this Paris manuscript in itself has 'the appearance of an edited work'.[144] This manuscript containing MARGGRAFE's astronomical observations in Brazil is partly based on material still extant in Leiden, but also on material that has been lost. Next to this, the library of the *Observatoire* also preserves some extracts from the Brazilian observations. These handwritten notes are made by the astronomer ISMAEL BOULLIAU (1605-1694) and by the commissioner of the copies, JOSEPH-NICOLAS DE L'ISLE. We will return to these notes in Chapter 9.

Another copy of MARGGRAFE's observations, written by the same scribe as the copy in Paris but with a different page formatting, is currently in the *Biblioteca Nacional de Portugal.*[145] Unfortunately this manuscript has been damaged, reason why it now is preserved incompletely. This Lisbon copy was most likely given by DE L'ISLE to GIOVANNI BATTISTA CARBONE (1694-1750), one of DE L'ISLE's astronomical correspondents. This Italian Jesuit travelled in 1722 from Naples to Lisbon, following a request by the Portuguese King JOÃO V, who desired that cartographical work would be done in the state of Maranhão in northern Brazil. However, eventually this journey was cancelled and CARBONE stayed in Lisbon, where he founded an astronomical observatory. From there most of his papers went to the Lisbon national library where they currently are preserved.[146]

PART III

ASTRONOMY

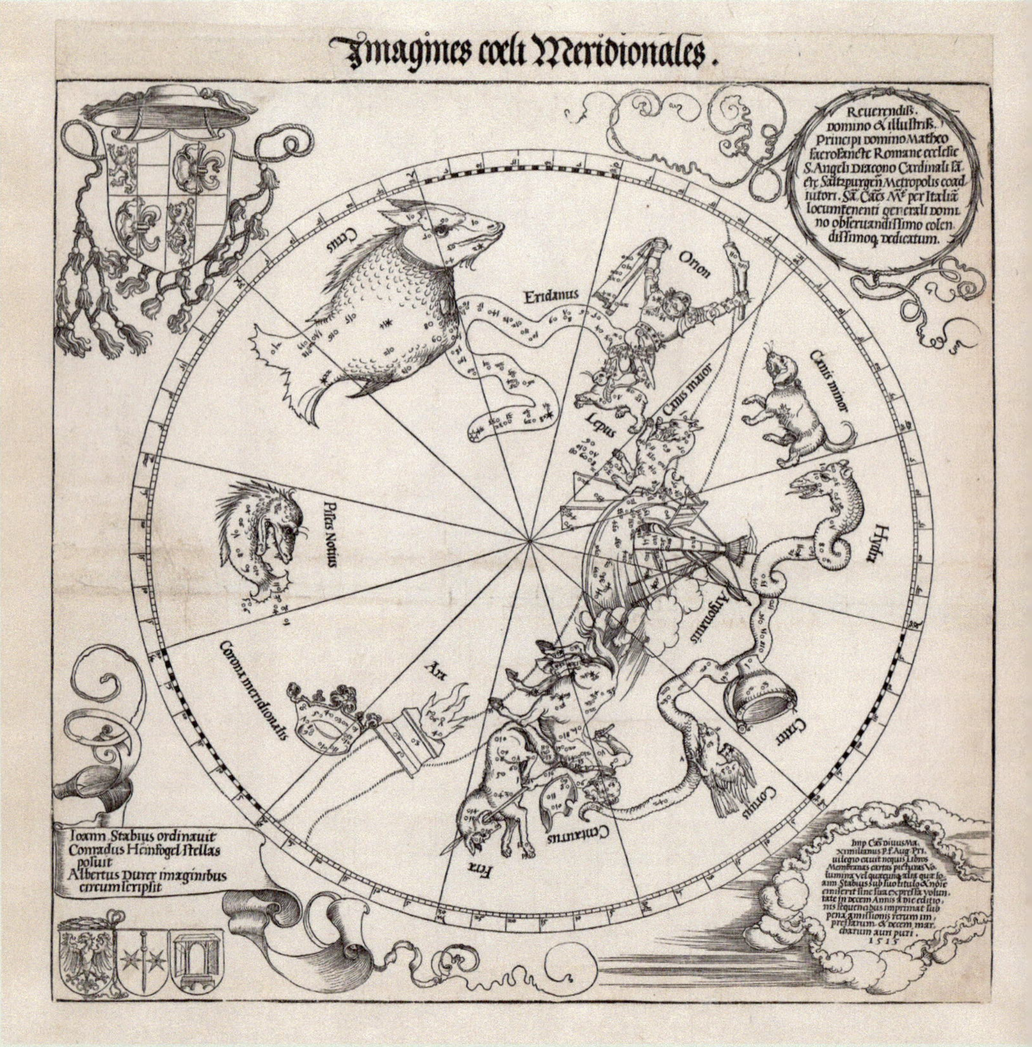

FIG. 81. The earliest celestial map of the southern sky (*Imagines coeli meridionales*), engraved by ALBRECHT DÜRER, in 1515, after a design by CONRAD HEINFOGEL. For the star positions, HEINFOGEL used the *Almagest* of CLAUDIUS PTOLEMY, which he had updated in 1503. (Rijksmuseum, Amsterdam)

CHAPTER 7

ASTRONOMY – BACKGROUND

Early astronomical observations of the southern heavens

Why was Georg Marggrafe so eager to observe the heavens in the Southern Hemisphere? After all, his sibling Christian testified in 1687 that his brother Georg possessed 'a great desire of contemplating the southern stars'.[1] What did he think he could achieve there? Was there so much astronomical knowledge to be gained at that point in time? In short, what was in that time the state of affairs concerning the astronomy of the southern sky?

Many southern stars were already known in antiquity. In the catalogue attributed to Claudius Ptolemy (c.100–c.170) already more than 170 stars belonging to the southern celestial sphere are listed, most of them with positions not far below the equator. A celestial map of the southern sky, engraved by Albrecht Dürer in 1515, shows what was known until the year 1598 (*fig. 81*).

Around the southern pole only a handful of stars were known. Most of the region there was empty space. Understandably, because stars located there are invisible for observers at European latitudes. In the late sixteenth century Tycho Brahe had included these southern stars in his star catalogue, but every astronomer knew that, compared to Tycho's own observations, these data of stars around the south pole had to be considered inaccurate.

This situation changed in 1598 when in Amsterdam a celestial globe was published by the publisher Jodocus Hondius. On his globes information was processed collected during the so-called *Eerste Scheepvaart* ('First nautical reconnaissance'), a voyage of Dutch navigators to the Indonesian archipelago performed during the years 1595-1598. This gathering of information about the southern stars was done at the request of the Amsterdam mathematician Petrus Plancius.[2] He was the real *auctor intelectualis* behind Hondius's globe publishing project. It was Plancius, too, who arranged the 185 southern stars on HONDIUS's globe (among which 13 were non-Ptolemaic) into 12 newly invented southern constellations. Most observations during the *Eerste Scheepvaart* were carried out by the skipper PIETER DIRCKSZ KEYSER, who unfortunately died before his return to the Netherlands. Therefore, it was his companion on the same ship, FREDERICK DE HOUTMAN, who brough the information back to PLANCIUS. DE HOUTMAN himself had collected observations about the variation of the compass needle, which at the time was thought to hold the key to the determination of longitude.

When in 1598 a *Tweede Scheepvaart* ('Second nautical reconnaissance') was set up, again with DE HOUTMAN, a new globe maker, who just had returned in Amsterdam from an apprenticeship at TYCHO BRAHE's island of Hven, requested DE HOUTMAN to bring back new information. This young man, WILLEM JANSZ [who later adopted the surname BLAEU] had the intent to use this new information on globes of his own design. He would produce these in competition with the globes made by HONDIUS (and others). In 1603 DE HOUTMAN returned from the Far East, and the same year he managed to publish a new catalogue of the southern stars, this time with 85 non-Ptolemaic stars. Unfortunately, DE HOUTMAN printed his star catalogue as an appendix to his *Spraeck ende woord-boeck, in de Maleysche ende Madagaskarsche talen*, a dictionary of the Malayan language. This strange publishing strategy resulted in the circumstance that his star catalogue remained virtually unknown to contemporary European astronomers. However, luckily WILLEM JANSZ [BLAEU] used DE HOUTMAN's information on his new celestial globes, which were issued in the same year 1603. Of the 'more than three hundred' stars mentioned in the address to the user, one hundred and seven were Ptolemaic stars with revised positions (*fig. 82).*

As ELLY DEKKER has demonstrated through a careful analysis of the production processes of the two most important and influential stellar publications of the time, namely Bayer's *Uranometria* (1603) and Kepler's *Tabulae Rudolphinae* (1627), the Hondius and Blaeu globes were the source of all information about the southern stars in these two publications. With the understanding that Bayer's atlas only included information from the *Eerste Scheepvaart,* lacking much of the later information.[3] Still, plate number 49 from Bayer's *Uranometria* (1603) provided for a long time the standard cartography of the southern celestial sphere (*fig. 83*).

FIG. 82. Celestial globe by WILLEM JANSZ [BLAEU], first issued in 1603. This is a reissue, improved in 1621. (Christies, London, auction 13 July 2016, lot no. 184).

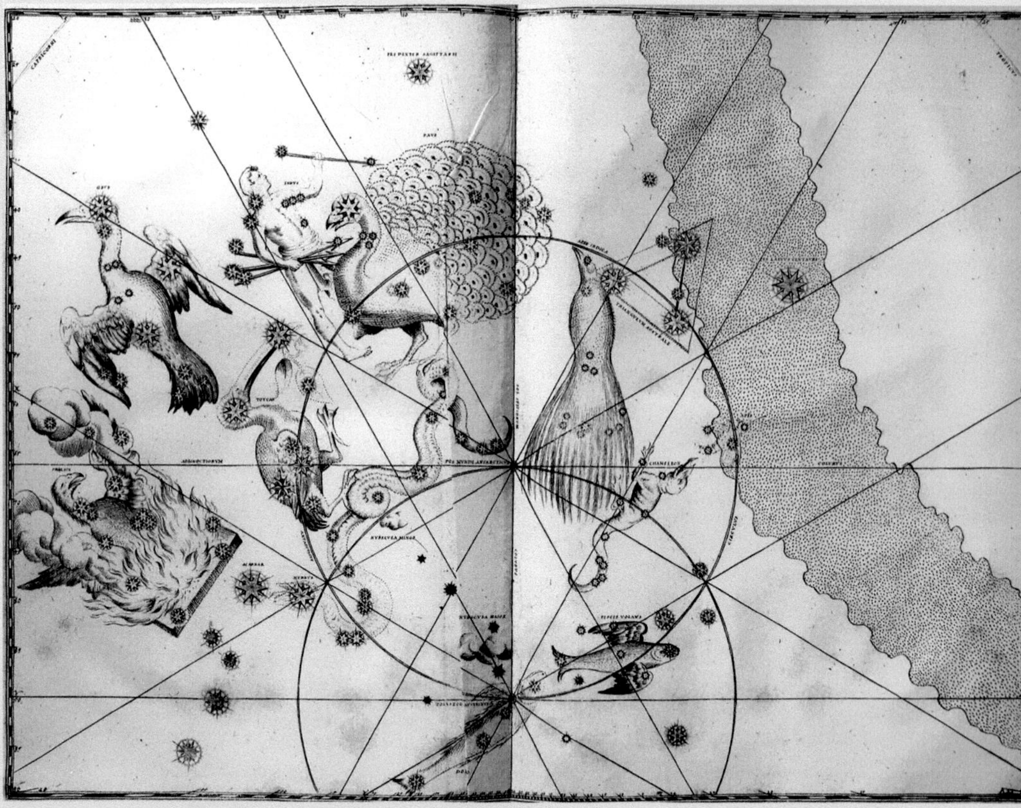

FIG. 83. Copperplate engraving no. 47 from Johann Bayer's *Uranometria* (1661 edition), showing the 12 southern constellations, newly invented by Petrus Plancius. Clockwise from top center: Pavo, Apis, Triangulum Australe, Musca, Chamaeleon, Volans, Dorado, Nubecula Major, Phoenix (left margin), Grus (outside upper left), Indus (left of Pavo), Tucana (below Grus), Hydrus (below Tucana, near center), and Nubecula Minor (just under Hydrus). (University of Oklahoma, History of Science Department).

From the above paragraph it may be evident that in the 1630s a star catalogue with accurate coordinates for the southern stars was indeed an absolute desideratum, not only for the astronomical community, but for navigators as well. For a young astronomer looking for fame, researching the southern sky offered indeed a promising perspective. This is exactly as stated in MARGGRAFE's earliest biography, 'he knew that there was in America a harvest of no small praise'.[4] The question remains, who or what inspired GEORG MARGGRAFE to this mission? In the absence of MARGGRAFE's own elucidation on this question, we can only make an educated guess. But the phrase that 'he turned over every stone, grabbed every opportunity that could lead him to America', seems no exaggeration if we look at the course of his life.

MARGGRAFE probably already encountered astronomy in Strasbourg at the age of 18, following lectures from ISAAC II HABRECHT (1589-1633), the Strasbourg professor of mathematics and astronomy. Just that year 1628, HABRECHT had published his designs of a celestial globe and some planispheres, which book appears on the list of works MARGGRAFE brought with him to Brazil. In Strasbourg, MARGGRAFE could also have met MATTHIAS BERNEGGER (1582-1640), the Strasbourg professor of history and eloquence, but better known as long-time correspondent with both JOANNES KEPLER and the Tübingen astronomer WILHELM SCHICKARD (1592-1635).

MARGGRAFE will certainly have been influenced by the Wittenberg professor CHRISTOPH NOTHNAGEL, whose horoscope is between MARGGRAFE's papers. One may even wonder whether MARGGRAFE's travel to Rostock, in late 1634, was prompted more by a desire to meet TYCHO BRAHE's former apprentice JACOB FABRICIUS, the Rostock professor of higher mathematics, instead of FABRICIUS's almost son-in-law, the botanist SIMON PAULLI.

However, it is plausible that MARGGRAFE was most influenced by the astronomer LORENZ EICHSTÄDT during his almost two-year-long stay as his apprentice in Stettin (*fig. 84*). MARGGRAFE's tendency to follow in BRAHE's footsteps will have emerged particularly during this period. EICHSTÄDT was an industrious physician and astronomer, who shared with MARGGRAFE a keen interest in medicine, alchemy, astronomy, astrology and nature in general. He, too, was an alumnus of Wittenberg's DANIEL SENNERT. In 1634, shortly before MARGGRAFE's arrival in Stettin, EICHSTÄDT had published a medical-alchemical study in line with SENNERT's approach to medicine.[5] His penchant for

FIG. 84. Lorenz Eichstadt (1596-1660), astronomer and calendar maker in Stettin and Danzig. (Österreichische Nationalbibliothek, Wien)

careful observation instead of theoretical reasoning was also based on his practise of medicine. As a physician, EICHSTÄDT held public anatomical dissections and in one of his works he drew a parallel between the careful study of anatomy and the careful exploration of the sky.[6] This is a way of looking similar as those we encounter with MARGGRAFE.

EICHSTÄDT published astronomical almanacs since his settlement in Stettin in 1624, but in the 1630s he also started to create astronomical tables, partly based on work by the Danish astronomer CHRISTEN SØRENSEN LONGOMONTANUS (1562-1647), but building also on the *Tabulae Rudolphinae*, the planetary tables constructed by TYCHO BRAHE and JOHANNES KEPLER published in Ulm in 1627. The result was a series of ephemerides of the stars and planets for the years 1636-1650. These tables were published for the first time in Stettin between 1634 and 1636, together with other tables listing the coordinates of numerous cities in Europe (compared to BRAHE's observatory in Uranienborg) and the logarithms according to the 1614 invention of the Scottish mathematician JOHN NAPIER. In the text of these two volumes EICHSTÄDT also paid attention to the prediction of numerous eclipses, including those that could be observed in places in the Southern Hemisphere. In one case, the lunar eclipse of 14 April 1642, he even mentioned Pernambuco in Brazil.[7] It must also have been EICHSTÄDT who sparked MARGGRAFE's interest in the orbit of the planet Mercury. In the first volume of his *Ephemeridum* he told his readers that he had not adopted the *Rudolphine* values for Mercury, because PIERRE GASSENDI's observation of Mercury in 1631 had forced astronomers to revise this part of planetary theory.[8] At least this explains why CHRISTIAN MARGGRAFE told his visitors that his brother GEORG, 'always observed and speculated about the planet Mercury'.[9]

EICHSTÄDT's work on the two volumes of his *Ephemeridum* was finished in December 1635.[10] This meant that MARGGRAFE's job, too, had come to an end. He had to go on with his life. Perhaps this was the moment on which he moved to Greifswald, not far from Stettin. In MARGGRAFE's biography, the *Vitae*, this small city on the Baltic coast is mentioned as a place that he visited. But unfortunately this sojourn is recorded without a date or other particulars. Had he heard in Stettin of the arrival in Greifswald of the mathematician and theologian JOHANNES BERING (1607-1658)? This young scholar had just returned to his native town in Pomerania, coming from Leiden in the spring of 1636. Although Bering officially had been a student of theology, he had trained himself mostly in mathematics and astronomy, first at Franeker University under the guidance of Adriaen Metius (1571-1635) and after 1632 in Leiden under Jacob Golius.[11] So, Bering had witnessed the construction of the Leiden astronomical observatory with his own eyes and therefore he was able to tell about it firsthand. Was it there, in Greifswald, that Marggrafe learned about the brand new Leiden observatory? It may be, but unfortunately we don't know. A meeting between him and Bering is entirely based on speculation.

Astronomical observations in Leiden before Marggrafe's arrival

In any case, after the summer of 1636, Marggrafe chose Leiden as his new destination. What had happened there shortly before his arrival? What made that city so attractive to him? It may be clear that the Leiden Observatory did not arise *ex nihilo*. As was explained in Chapter 4, its founding was a direct result of the surveying activities of the professor of mathematics Willebrord Snellius (1580-1626). It was only because Snellius left after his death a large Tychonic quadrant that Golius could request the university's trustees to erect a platform for the instrument. And although by this act the Leiden University Observatory was founded, this did not make the place into a location for research. Its purpose was only 'to demonstrate to the students the course of the heavens and the stars'.[12] In the early years the observatory was mainly used in this way, as is evidenced by the payment to an assistant, Pieter Vincentsz van Heemskerck, for his year-long service to Professor Golius and his students 'on the newly made platform for astronomical demonstrations'.[13]

With this emphasis on the educational objective of the observatory, it is not said that no serious observations were made. This happened even before the construction of the observatory. The first recorded astronomical observation at Leiden University dates from 1605. That year the French astronomer Guillaume le Nautonier asked the Leiden professor Josephus Justus Scaliger to witness lunar and solar eclipses. In October 1605 Scaliger indeed tried to observe a solar eclipse: 'There is no one else who took more pains to see the aforementioned [solar] eclipse than I took to observe it', he wrote to Le Nautonier. But it had been too cloudy to see anything. He had not even noticed the announced decrease in brightness. 'Several eyes at this university, which stayed in the same place for this phenomenon', could corroborate his findings.[14] While this observation was a side-step for the philologist SCALIGER, this was different for WILLEBRORD SNELLIUS, who taught astronomy as part of his teaching mandate of mathematics.[15] Unlike his predecessor as professor of mathematics (his father RUDOLPH SNELLIUS (1546-1613) who never made astronomical observations), WILLEBRORD was well versed in practical astronomy. During his *Grand Tour* he had visited TYCHO BRAHE in Prague and had assisted him with his observations, an internship that ended with BRAHE's death in October 1601.[16] However, the first published observations of his own were those of the great comet of 1618, still made without a telescope. In the booklet he published about this comet, SNELLIUS wrote that he could not use his sextant because of troubles with its pivots and because he had no assistant available, needed to handle the instrument.[17] However, these kinds of observations were incidental. Systematic observational research at the university was not yet been conducted.

Remarkably, the founder of the observatory, professor JACOB GOLIUS, was mainly interested in mathematical theory and ancient oriental astronomy. It's telling that he spent more time on the design of a celestial globe with elaborate Arab nomenclature than on practical observations.[18] Before MARGGRAFE's arrival in Leiden, we know of GOLIUS only of two astronomical observations, one of the lunar eclipse of 3 March 1635 (which was witnessed also by EICHSTADT in Stettin, most likely together with MARGGRAFE), and the other of the second lunar eclipse of 27 August of the same year.[19]

Before that more was done by the student MARTINUS VAN HOVE (1605-1639), better known as HORTENSIUS. He had been educated in Dordrecht at the Latin school of the natural philosopher ISAAC BEECKMAN.[20] In March 1628 HORTENSIUS enrolled at Leiden University. However, he soon took a break, leaving for Middelburg in the southern part of the country, where he became an assistant of the astronomer PHILIPPUS LANSBERGEN. After his death, HORTENSIUS assisted his former teacher BEECKMAN at his recently constructed platform on top of the Dordrecht Latin School. In 1630 they witnessed together a solar eclipse, the image of which was projected on a screen by means of a telescope.[21] Returned to Leiden the next year, HORTENSIUS observed a lunar eclipse on in November 1631, measuring the altitude of the Moon with a quadrant, probably the one from SNELLIUS's heritage.[22] In 1632 he also made some observations with a telescope.[23] At the end of 1633 HORTENSIUS moved to Amsterdam, being appointed as professor of mathematics at the recently founded Amsterdam *Athenaeum Illustre.* There, he observed the lunar eclipse of 3 March 1635.

It was only a year later, in September 1636, that MARGGRAFE registered as a student at Leiden University. His first known observation in Leiden dates from 13 January 1637, the first of a long series of observations. His initiative has undoubtedly been applauded by GOLIUS, who, as the founder of the observatory, must have been pleased that – after HORTENSIUS – at least someone was willing to carry out a routine program of observations.

CHAPTER 8

ASTRONOMY – SETTING AND EQUIPMENT

Observatories in Leiden and Recife: the design, organization and instruments compared

In an earlier article, we concluded that in Brazil, Marggrafe used Leiden Observatory as his benchmark. In Recife, he copied to a large degree the site, instruments, and methods that he had become familiar with when he trained himself in observational astronomy in Leiden. To demonstrate this, we will compare below the design, organization and instruments of both locations.

Location and building: 1. Leiden

The Leiden Observatory was housed on top of the Academy Building on the Rapenburg canal, with current geographical coordinates 52° 9' 25" North, 4° 29' 8" East. The building was an old convent that was seized by the civil authorities of Leiden after the Reformation. It was put in use in 1572 and the building still serves today as the place for events with a ceremonial character, such as graduation ceremonies, inaugural lectures and promotions.

In 1633, a rectangular platform was made on one of the the roofs of the building. The preserved specifications give the dimensions as approximately 18 feet long, 13 feet wide, on a tower 19 feet high. The tower had to be clothed with wooden wainscot, with an occasional window to provide light on the spiral staircase leading to the platform.[1] On this platform, a solid vertical pole was placed, to which the quadrant was mounted in such a way that it could rotate around its axis.

The quadrant was originally set up in the open air. To protect the instrument against wind and weather, it was decided in the spring of 1634 to add to the platform a small octagonal turret, with in its centre a sturdy pole, 'motionless and standing straight', so that the quadrant could easily rotate around its centre, in pivots fixed at the bottom and the ceiling of the dome (*fig. 85*).[2] Unfortunately, the drawing of the architect that accompanied the specifications of the dome is lost.

But the text mentions two door frames and four window frames, for 'looking out'. The dome itself had to be clad with 'wood paneling' (*wagenschot*), one inch thick, with a lead roof. On the sides, it had fourteen shutters, eight to turn upward, and six to drop down. These had to be made in such a way that, if opened, the quadrant could be used in a range between 80 and 20 degrees in altitude. Horizontally, the quadrant had to be able to turn easily in all directions.[3]

FIG. 85. Leiden Observatory constructed in 1633-1634 on the attic of the Academy Building on the Leiden Rapenburg. Details from (*left*) a view of Leiden, dated 1669 (ELO, PV_PV1732.3), and (*right*) the frontispiece of Paulus Herman, *Horti academici Lugduno-Batavi catalogus* (Leiden 1687), drawn in 1686 from the adjoining *hortus botanicus* by WILLEM VAN MIERIS. Three years later the platform was enlarged, and a second turret was built for a newly bought sextant. (Leiden University Library).

FIG. 86. The Recife Observatory (marked C). Detail of a view on *Mauritiopolis* by Frans Post, published in Caspar Barlaeus, *Rerum per Octennium in Brasilia* (Amsterdam 1647).

LOCATION AND BUILDING: 2. RECIFE

The Recife astronomical observatory was housed on top of the first residence of the governor of Dutch Brazil, JOHAN MAURITS VAN NASSAU-SIEGEN (*fig. 86*). This was a large Portuguese house on the fluvial island of Antônio Vaz that predated the count's arrival. The location has currently the geographical coordinates 8° 3' 51" South, 34° 52' 37" West.

In 1979, NORTH suggested that MARGGRAFE's observatory had been (re-)located to the *Vrijburg* palace, built in the years 1641-1642. As JORGE POLMAN already has pointed out in a paper published in 1984, there is no evidence whatsoever that this move ever took place.[4] Our analyses of MARGGRAFE's observations (see Chapter 9) also lead to the conclusion that the hypothesis of *Vrijburg* palace being used as astronomical observatory is a hoax.

The present location of site where the observatory in Recife stood is determined by JOSÉ MOTA MENEZES[5], who compared maps of different epochs to images. After the Dutch period, the river bank withdrew considerably, with the result that the location which originally was on the bank of the Capibaribe River is situated today

FIG. 87. *Left*: The location of Marggrafe's observatory (red circle) in today's Recife, at the corner of the streets 'Primeiro de Março' (almost vertical) and 'Imperador D. Pedro II' (almost horizontal). North is to the right. (Google). *Right*: The location of Marggrafe's observatory in 2010. (Photo by the authors).

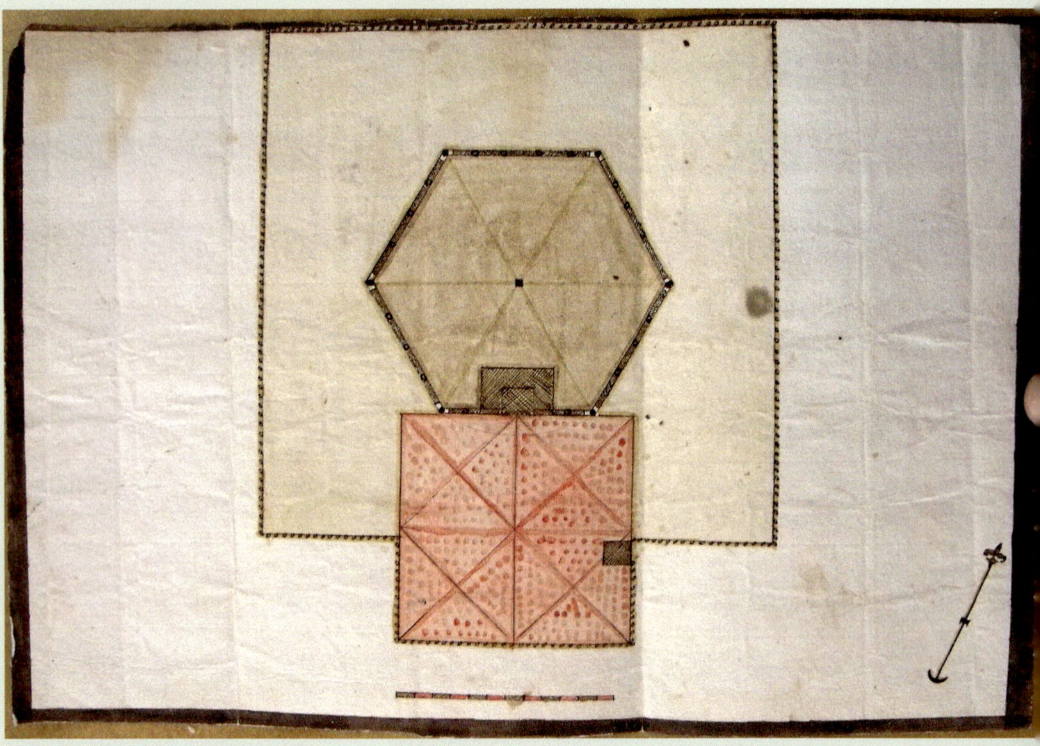

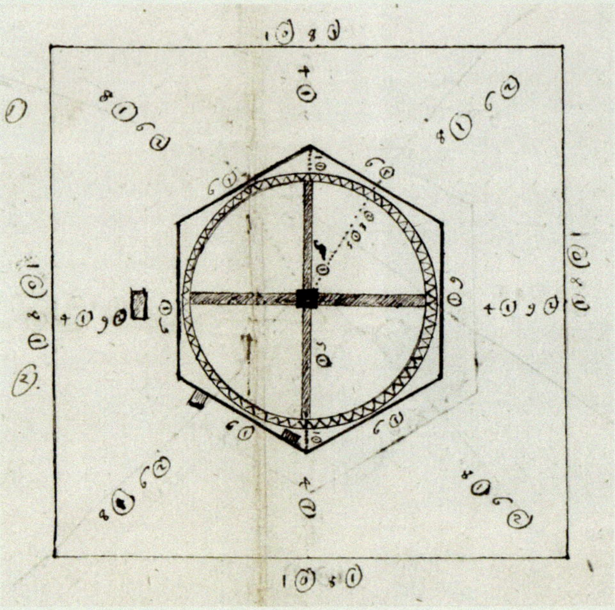

FIG. 88. MARGGRAFE's two (undated) designs of his Recife Observatory, one with an extension on the southern side, the other drawn with the azimuth circle of the installed quadrant. The symbol ① means one Rhineland foot. The symbol ② stands for 1/12 of a foot or one Rhineland thumb (inch). (ELO).

more than a block away from the river. The exact location is the corner of the streets 'Primeiro de Março' and 'Imperador D. Pedro II' in the Santo Antonio quarter (*fig. 87a & 87b*). This corner came into being during the implementation of the urbanization plan designed by surveyors in the service of count Nassau. Today this corner is the last physical remnant of the period, since all original buildings from the Dutch epoch were demolished and replaced by others.

Platform and dome

The Leiden platform had a rectangular shape: a consequence of the configuration of the roof it was constructed on. As mentioned above, it measured 18 by 13 Rhijnland feet (5.7 x 4.1 meters), and was surrounded by a balustrade.[6] In Recife, the platform had a square shape, fitting to the square configuration of the roof it was constructed on. Its sides were about 20 Rhijnland feet (≈ 6.3 m). Here, too, the platform was surrounded by a balustrade.[7] According to one of the designs Marggrafe has left us, the Recife platform had a square extension on the southern side, possibly referring to a staircase (*fig. 88*).

In both observatories, a polygonal turret stood in the middle of the platform. In the centre of both turrets, a solid vertical pole was erected, around which a wooden quadrant could revolve.[8] In Leiden, the octagonal turret had two door- and four window-frames. The turret's roof was covered with lead. In total there were fourteen shutters or hatches: eight on top that could be opened during observations, and six that could be used to close the doors and windows after use. According to the building specifications of 1634, every window should provide the quadrant with an unhindered view on all sides between 20 and 80 degrees, measured from the horizon.[9] The doors in the meridian enabled an undisturbed view to both horizons. Marggrafe's measurements reveal that in Leiden the meridian altitudes were measured from 3°33' up to 70°9' south, but not up to the zenith, which is in accordance with these specifications.

In Recife, the turret or dome was hexagonal, and topped with a copper wind vane bearing the coat of arms of Nassau-Siegen. The vertical walls of this turret had glass windows. These windows were not used during the astronomical observations (when the hatches were opened), but they were useful to monitor ships approaching Recife. That the building also was used in this way is demonstrated by Wagener, who in his drawing depicted someone in the window at the right of his watercolour, watching the horizon with a small telescope (see *fig. 34*).

In both observatories, a lower chamber was constructed, both for storage purposes and as a dark room for solar observations. In Leiden, this space was made below the platform, already in 1633, thus before the construction of the dome a year later. This *solderken* was added on GOLIUS's request, originally only for the storage and use of a pair of large globes from the university's library for which he had received permission to use in the observatory.[10] In Brazil, this was done differently. There a floorboard was made *inside* the turret, five feet above the floor of the platform. It was here that the instruments were placed. In this way MARGGRAFE created an enclosed space that could be used as storage and as a dark room. In this chamber, MARGGRAFE also stored smaller equipment, such as the two pairs of globes he brought with him, as well as Bayer's star atlas *Uranometria* (1603).[11] Here was also the equipment needed for orienting his instruments, such as a bubble level (in Dutch *Waterpas*), a carpenter's square (a tool used to measure out right angles) and three rulers made of a solid wood. Assorted furniture, such as two lanterns, a slate board to write and draw on with chalk, two three-legged stools with three steps for reaching the sighting tubes of the quadrant and sextant, was also stored here (see the reconstruction in *fig. 89*). Both observatories could be reached comfortably from the interior by climbing up a spiral staircase. In Recife we even know the number of stairs, starting at the first floor of the two-story residence: forty-three 'very comfortable steps' from the interior of the house to the platform, and ten more to reach the floor inside the turret, arriving then at a height of in total 18 feet (~ 5.7 meters).

QUADRANT

The main purpose of astronomical observations in the seventeenth century was to constitute ephemerides, or to check their accuracy. An ephemeris is a table of values that gives the positions of astronomical objects in the sky at a given time. The position of such an object is determined by two angles: (1) the elevation above the horizon plane and (2) the azimuth, an angle measured clockwise along a horizontal circle, starting from the local geographic north. For the measurement of these two angles a precision measuring device is needed, such as a quadrant. The larger the radius of such an instrument, the more precise the measurements will be. That's why TYCHO BRAHE designed such large instruments, a description of which he published in his *Astronomiæ instauratæ Mechanica* (1598).[12]

FIG. 89. Reconstitution of the dark room below the 5-foot quadrant (with the walls removed). On the floor, at left, there are several globes, a three-step ladder and, on the bench, there is another globe, a sandglass, a lantern and the portable 20-inch sextant. In the background, three rulers are hanging. In the foreground, there is a square, and behind it the pedestal with the Polish hammer (the tapered end is visible). In the rear is the water level on a tripod. In the middle is a circular screen on which sunspots could be drawn. On the far right, there is a bookcase. Further to the right, outside the dark room, the foot of the 5-foot sextant (in the shape of a cross) can be seen. Nearby, a dark rectangle on the floor is the tilting trap door leading into the house below. Picture made with the 3D-software program *SketchUp*.

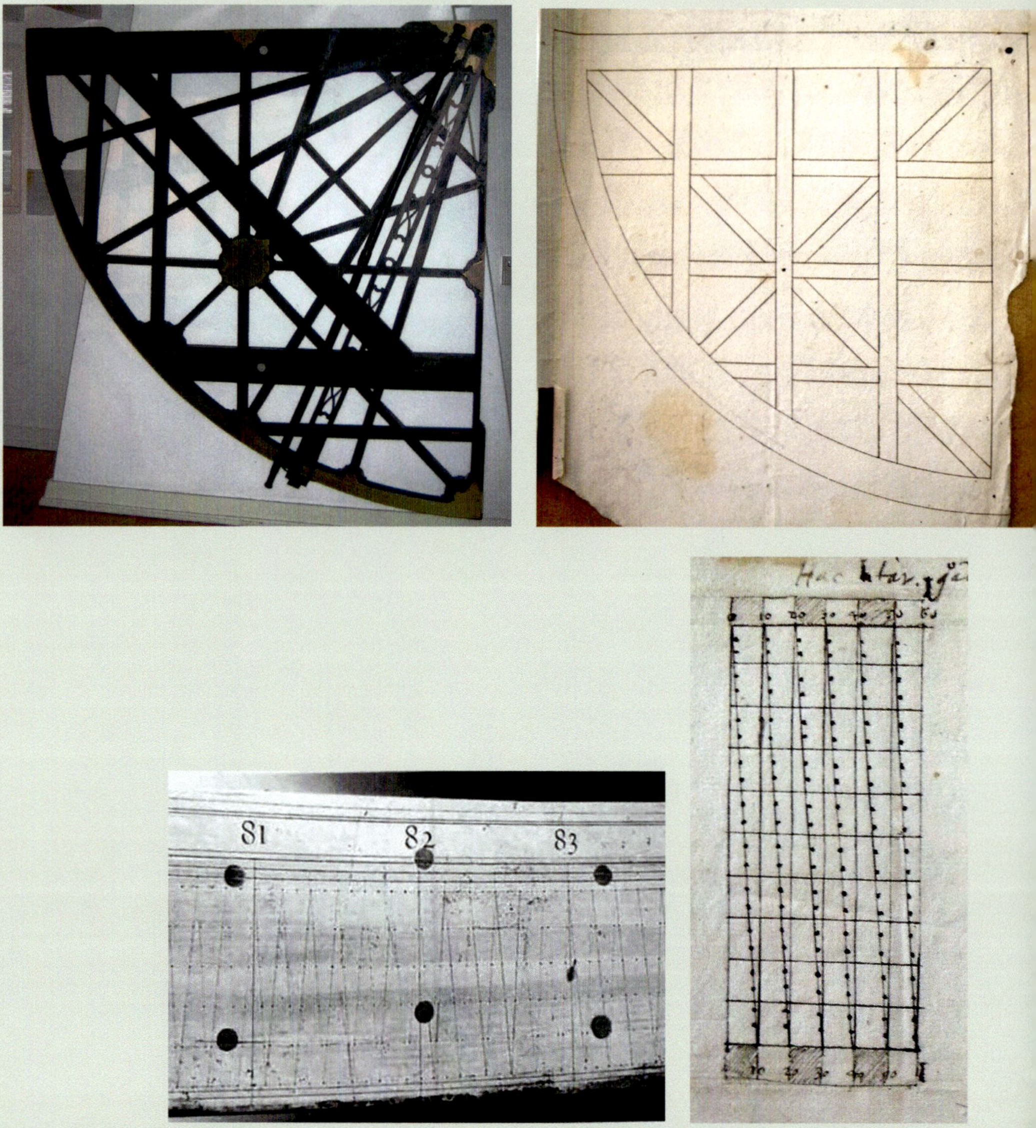

FIG. 90. Top left: (a) Snellius's quadrant built by WILLEM JANSZ [BLAEU] c. 1625 and used by GEORG MARGGRAFE in 1637 (the long telescope tube was added in 1669) (Rijksmuseum Boerhaave, Leiden). Top right: (b) Marggrafe's design for his quadrant made in Brazil. (ELO, North no. 71). Bottom left: (c) Transversals on the limb of the Leiden quadrant. Bottom right: (d) Marggrafe's design for transversals on the Brazil quadrant (ELO, North no. 75)

The Leiden quadrant, made of wood with brass fittings with a radius of six Rhijnland feet and eight inches (≈ 2.1 m), is still extant (*fig. 90a*). Its limb, or graduated scale, is divided into 90 degrees (or a quarter of a circle, hence the name quadrant), subdivided into intervals of ten arc minutes (= 1/6 of a degree). To allow further accuracy, the scale has transversals (*fig. 90c*), also introduced by BRAHE. In Brazil, MARGGRAFE also used this arrangement for a similar quadrant with a radius of 5 feet (≈ 1.57 m), built under his supervision. Except for the metal parts, the bulk was made of hardwood designated *pao sancto* (holy stick) by the Portuguese (*fig. 90b* and *90d*). Before the construction of the Recife observatory, MARGGRAFE also made some observations with a smaller, portable quadrant of one foot (≈ 31.4 cm) that he had brought with him to Brazil.[13]

Both in Leiden and Recife, the quadrant could revolve in all directions around a sturdy vertical wooden pole, to which it was attached by its center of gravity. In Brazil, there was a wrench-like device made of iron to rotate the quadrant. At the base of the quadrant, MARGGRAFE also made a large stationary azimuthal circle of ten Rhijnland feet in diameter (≈ 3.1 m). This circle was supported by 12 pillars at a height of 1½ feet above the floor. A pointer, fixed at the quadrant, indicated the azimuth value on the circle (*fig. 91*). This was in accordance with Brahe's original description of the instrument. However, in Leiden the horizon itself was used as an azimuthal

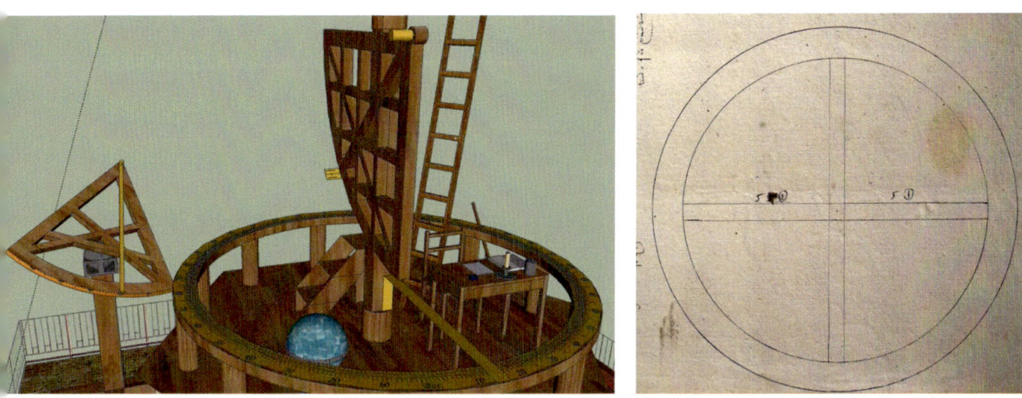

FIG. 9. *Left*: Reconstitution of the interior of the upper room of the observatory. The walls are removed to show the instruments outdoors on the lower platform. Made with the 3D-sofware program *SketchUp*. *Right*: MARGGRAFE's sketch of the azimuth circle in Recife (ELO, North no. 71).

circle.[14] There such a horizontal graduated scaled remained a desideratum until 1656.[15] This explains why in MARGGRAFE's Leiden log all measurements of an azimuth value are absent.

PINACIDIUM TYCHONICUM, OR TYCHONIC ALIGNING SIGHT

Both in Leiden and Recife, the quadrant had a sighting tube that revolved around the right angular point of the quadrant. At the viewer's end, next to the measuring scale or alidade, there was a Tychonic aligning sight, or *Pinacidium Tychonicum* (*fig. 92*). This was a simple but ingenious device with two parallel slits, called *dioptra.* It was invented in 1575 by TYCHO BRAHE, and became standard for such measuring devices across Europe. When an observer aimed at a star or planet, he had to adjust the sighting tube in such a way that the star or planet was seen equally bright in both slits. Only then he could be sure that the correct position was set. Today, on the preserved Leiden quadrant, this sighting tube is missing, because this device was replaced by a telescope tube in 1669.[16]

In 2014, a full-scale reconstitution of the Brazilian quadrant was built in the courtyard of the Museum of Astronomy and Related Sciences (MCTIC) in Rio de Janeiro (*fig. 93*). This reconstruction was made for public exhibition and practical observing sessions for students. One of us (OSCAR MATSUURA), had therefore the opportunity to test in practice the usefulness of the *pinacidium.* In nocturnal observations, the pointing to a bright planet or star was done by making the star just touch one side of the horizontal gap of the *pinacidium.* Then, the star was made to just touch the other side of the gap. Assuming that the most likely position for the null error is in the middle of the extreme positions, it was found that it was relatively easy to guess the 'true' position repeating the oscillation a few times. Several tests with different observers attested that the method worked surprisingly well. For measuring the meridian elevation of the Sun, the quadrant was previously positioned at azimuth equal to 0°. Then, under solar illumination, the shadow of the cylinder projected on the dioptra was observed.

As the meridian passage was approaching, the sighting tube was adjusted so as to keep the shadow of the cylinder within the markings made on the dioptra, indicating the correct pointing of the sighting tube towards the Sun. For checking the performance of the *pinacidium* in daylight, observations supervised by monitors were made by students, twice a month for ten months, in 2014. In each observing session the elevation of the Sun was measured five times

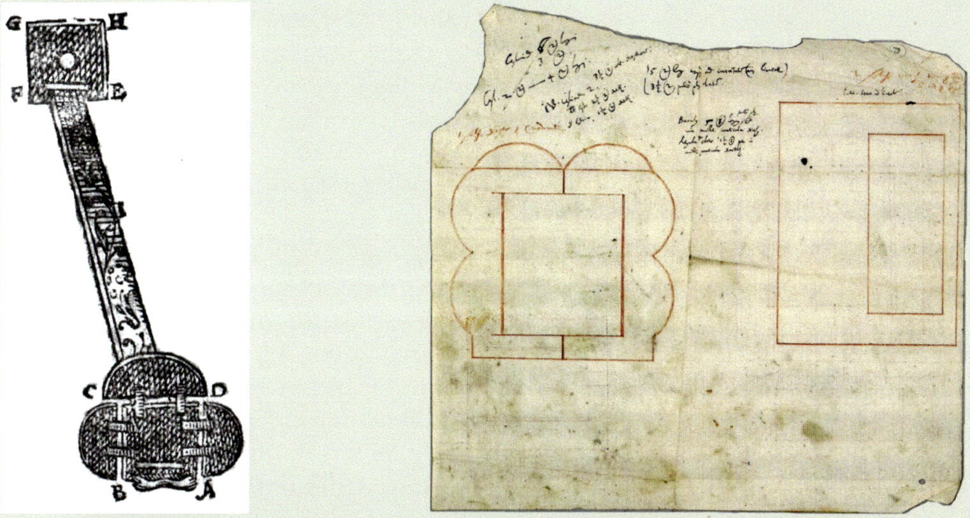

FIG. 92. *Left*: A *Pinacidium Tychonicum*, or Tychonic aligning sight. From: Tycho Brahe, *Astronomiae Instauratae Mechanica* (1598). This sight was attached to MARGGRAFE's quadrant and sextant. *Right*: One of Marggrafe's sketches of his Tychonic aligning sight, with notes in Latin (mentioning a *baculus*, or stick to connect the two parts) and a note in German: ('*1 solche diopter [für] quadrante*', for the left part, and '*2 solche …*', for the right part) (ELO, North no. 73r).

FIG. 93. *Left*: A rotating quadrant as depicted by Bengt Hedraeus, in his booklet on measurements with the Leiden instrument. *See Nova et accurata […] quadrantis astronomici azimuthalis* (Leiden, 1643) (Zentralbibliothek Zürich) *Right*: Full-scale reconstitution of the Brazilian 5-foot quadrant, built in the courtyard of the Museum of Astronomy and Related Sciences (MCTIC) in Rio de Janeiro. See H.M. Gesteira, *Observações do Recife holandês* (MAST, 2014). (Photo by the authors).

around noon, between 11 AM and 1 PM (Local Time). Thus, the local time of the meridian passage could be obtained by calculation, using parabolic approximation. Finally, as a function of the date of the year, 20 points of the local time of the meridian passage, obtained this way, were plotted, along with the calculated curve of the Equation of Time. The results showed to be satisfactory with a standard deviation of 1.5 min.

Other angular measuring instruments

In Leiden, Marggrafe also had a sextant at his disposal, which he used to measure angular distances between a fixed star and a planet. A sextant has an arc of 60 degrees (= 1/6 of a circle, hence the name). Unfortunately, the information about the Leiden sextant is scant. The term 'sextant' appears only in the observation of 25 February 1637. In Recife, before the construction of the observatory, Marggrafe made some observations with a small sextant. Perhaps with one that he brought with him from Leiden.

In Recife he also designed a large wooden sextant (*fig. 94*), with a radius of five Rhijnland feet (≈ 1.6 m), having similar Tychonic sights as the quadrant. In use, this large sextant was hung in a U-shaped fork made of solid wood, attached to a sturdy pole, five feet high (~1.6 m), standing on a cross-shaped base. A ball joint was placed at the top of this pole, to which was attached a square protuberance, 2 inches wide and 2½ inches high, upon which the large sextant could be set for measuring stellar (angular) distances, moving freely in all directions. Marggrafe's sketch of this pedestal can be seen to the left in *fig. 94*. However, among the Brazilian observations, no observation with this instrument is recorded, so very likely this sextant was only part of an initial plan that did not materialize.

Instead, Marggrafe used another, smaller sextant with a radius of 20 Rhijnland inches (~ 52 cm) and a scale graduated in two arc minutes. To this sextant belonged a portable base, a stick made of *Pau Santo* wood, 4½ feet (≈ 1.41 m) high. Its lower part was tapered, so that the stick could be pushed firmly into the ground. This smaller sextant could be easily used during expeditions. At the upper part of the stick, a massive piece of iron was attached in the shape of a so-called 'Polish hammer' (*fig. 96*). This part was about 9 inches (~ 23 cm) long, and had two protrusions of one inch perpendicular to each other. In this way the sextant could be hung vertically for measuring elevations, and set horizontally for measuring the azimuth.

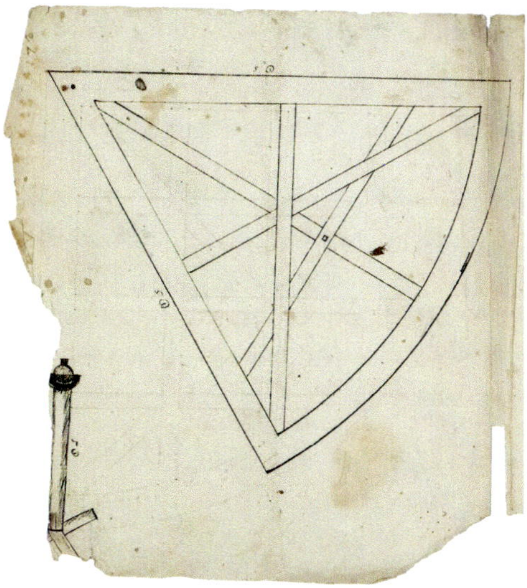

FIG. 94. Marggrafe's design of the triangular structure of a 5-foot sextant. A sketch of the pedestal with a ball joint is shown at left. (ELO, North no. 71).

FIG. 95. A large sextant used simultaneously by two observers at Tycho Brahe's observatory in Uranienborg. Detail from Brahe, *Astronomiæ instauratæ Mechanica* (1598) (Zentralbibliothek Zürich)

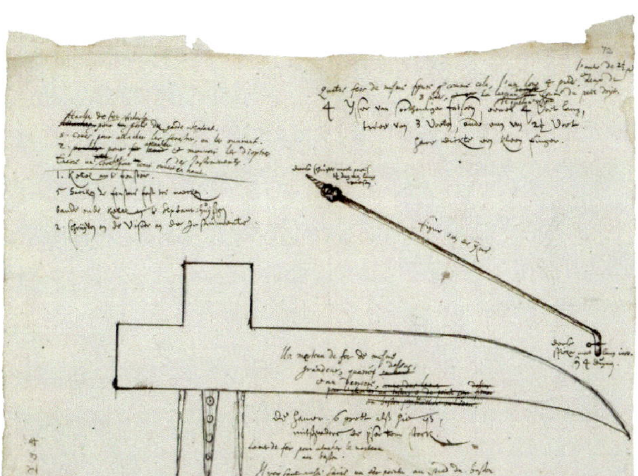

FIG. 96. Drawing for (1) the 'Polish hammer', the iron top of a further wooden base for supporting the 20-inch portable sextant, and (2) the gnomon (or pointer) for a sundial outside the dome. With written instructions in French and Dutch for the smiths who had to make these objects, including the remark, 'It is also necessary to make an iron point at the foot of the stick to easily put it into the ground'. (ELO, North no. 72r).

Telescope

In Marggrafe's days the telescope was still rather a novelty, almost more an object of curiosity than an instrument of research. Moreover, telescopes still lacked a well-defined function in an observatory, whereas the roles of quadrants and sextants in checking the theory of astronomical calculations were well established.[17] At that time, only the Dutch (or Galilean) and the Kepler configurations were known. The first configuration with a convex (converging) objective lens and a smaller concave (diverging) lens, had the disadvantage of a rather low magnification. When a magnification of more than about 20 times was applied, this restricted the field of view so much that the instrument became virtually useless. On the other hand, the Kepler configuration with two convex lenses suffered from such serious spherical and chromatic aberration that the images were useless for astronomical observers.[18] Moreover, preserved lenses of the period show that many suffer from small air bubbles, striae and inhomogeneities. As a rule, the best ground part of the objective lens was the centre, reason why the apertures of the telescopes were stopped down to about 1.5 – 2.0 cm. This practice had, of course, as consequence that the light-gathering power of the instruments was limited.[19]

At both observatories, Marggrafe used a refracting telescope with an upright image for diverse observations. The Leiden telescope was most likely Golius's property. It was customary until the early eighteenth century that an observer at Leiden Observatory used his own equipment.[20] Golius nevertheless stored this telescope at the observatory, so that it could be used by everyone. A French tourist, who in July 1636 climbed 'the high tower of the observatory' to look over the city, did not only witness there 'the unspeakably precise instruments, with whose help one observes the sky', but he also could use the telescope in order to view 'the whole province, and also into the individual houses of this vast city'.[21]

Marggrafe brought one telescope to Brazil, as he explained in one of his letters to Golius.[22] He had bought the instrument – 'just like the one we used in Leiden' – shortly before he left the Netherlands. Regrettably he did not note where he acquired them, or which instrument maker made these telescopes.[23] He only stated that the telescope possessed 'a good glass to detect and display [the sun] in a closed dark room'.[24] To enable comfortable observations of celestial objects, Marggrafe installed a pedestal with a hollow trough or cradle on which his *tubus opticus* could be placed. This gutter was

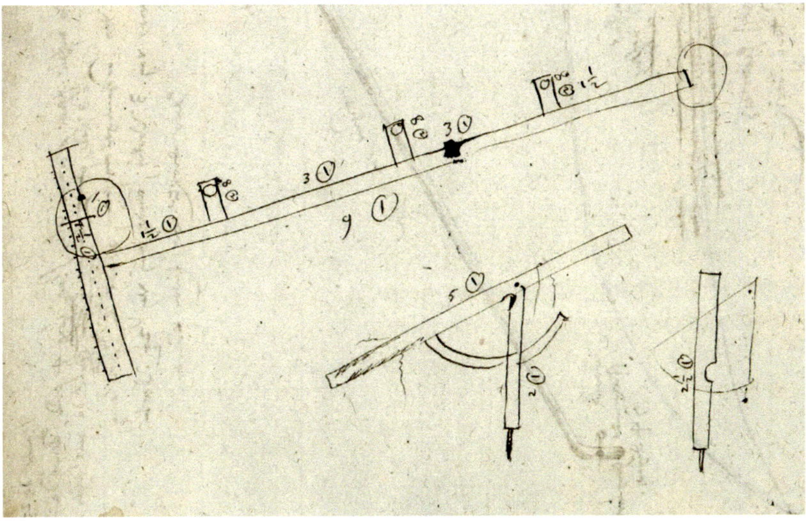

FIG. 97. *Top*: MARGGRAFE's design for an adjustable support for his telescope, to be used when projecting solar images onto a screen, for the drawing of sunspots or solar eclipses. *Below*: The smaller sketches refer to other pedestals for MARGGRAFE's telescope. (ELO, North no. 72vs).

supported by two metal semicircles to ensure its stability. The whole was vertically adjustable to a proper height by some screws. In use, the pedestal could rotate horizontally by inserting its tip at the bottom into a matching hole in the platform's balustrade (See *fig. 97*).

HELIOSCOPE

The dark room on the ground floor of both observatories was designed for observing sunspots, solar eclipses and optical experiments. According to MARGGRAFE's description, the interior of the room could receive the solar rays through circular apertures, one in the east side of the turret and another in the west. Inside the room was again a long wooden cradle as support of the telescope, as well as a clipboard for drawing sunspots. The entire structure, which could be named a *helioscope*, could be dragged laterally on the ground to allow adjustments for a changed position of the Sun.

In Leiden, MARGGRAFE indeed used such an optical assembly for the drawing of sunspots, but – as will discussed in the next chapter – no sunspot observation was recorded in Recife. Although MARGGRAFE, in his description of the Recife Observatory, presents an outline of the dark room and the setting installed there, it seems that this dark room was never put into use.[25]

Clock, sandglass and pendulum

Time accuracy was very difficult to achieve in Marggrafe's time. The invention of the pendulum clock by Christiaan Huygens was still several decades in the future. To establish the local solar time, Marggrafe designed a suitable sundial for each side of the platform, so that in various ways the time could always be registered from the shadow of the sun, cast by a pointer. But measuring time intervals was more difficult. For that Marggrafe used two timepieces, both in Recife and in Leiden.

First there was the sandglass, playing the role of the ordinary *clepsydrae* or water clock. In this device, the time is measured by the regulated flow of liquid (or sand) from one vessel into another. The amount transferred is then a measure for the time elapsed. In Leiden he possessed at least one sandglass and in Brazil, two. Marggrafe also used the pendulum (*perpendiculum mobilis*), counting the oscillations (*pulsus*) of a swinging cord. According to his description, Marggrafe brought to Brazil a pendulum made of a turned metal cylinder of 41¾ ounces, suspended by a chord with a length of 29 Rhijnland inches (~76 cm). According to the simple pendulum formula, for the full length of the string, the oscillation period would be 1.75 seconds. Details of the Leiden pendulum are not known.

Marggrafe's concise library

During his stay in Leiden, Marggrafe had the university library at his disposal. Such a fortunate situation was completely lacking in Recife. So, in order to be able to consult essential astronomical literature, he had to collect a small library before his departure to Brazil. A small note among Marggrafe's Leiden manuscripts, written in shorthand, reveals the small number of books that he took with him.[26] Was it an order list for a Leiden bookseller? It looks like it. In one of his letters to Golius, Marggrafe wrote that he had bought a copy of the *Tabulae Rudolphinae* (1627) before he left Holland.[27] This reference work is among the books listed on the above-mentioned note. It contained a crucial set of planetary tables compiled by Tycho Brahe and continued during 26 years by his assistant Johannes Kepler. In Stettin, Marggrafe had assisted Eichstädt in his work to supplement exactly these tables, so he must have been familiar with this book since that time. Another book on this concise inventory is JOHANNES BAYER's *Uranometria* (1603), an illustrated star catalogue important for stellar nomenclature. The note further lists KEPLER's *Epitome astronomiae Copernicanae*, a popular textbook about the Copernican system, in which KEPLER had revealed his famous third law of planetary motion. JOHN NORTH, who discussed MARGGRAFE's list in 1979, caracterized this book as 'perhaps the most widely read astronomical book of the time'.[28] Other works listed were by KEPLER's son-in-law, JACOB BARTSCH and the Strasbourg professor of astronomy ISAAK II HABRECHT, the constructor of a celestial globe based on TYCHO BRAHE's observations. So, MARGGRAFE was evidently impressed by BRAHE and KEPLER and their achievements in the field of celestial cartography.

Indirectly related to astronomy were two books on navigation. One was written by the mathematician GUILLAUME LE NAUTONIER DE CASTELFRANC (1560-1620), in which he (unsuccesfully) proposed a method for the determination of longitude by magnetic observations. The other was by the Leiden professor WILLEBRORD SNELLIUS, in which a detailed study was presented of the *loxodrome* (the curved line that intersects every meridian at the same angle). Finaly, two other – not clearly defined – books seem to have been general works on nature. One of these seems to point to a undefined edition of ARISTOTLE, and the other is listed as a book by a certain 'De Castor'. For this title, NORTH suggested a medical work of the Italian scholar GIROLAMO FRACASTORO (c. 1476-1553). In our view, the Spanish Jesuit JOSÉ DE ACOSTA (1540-1600) is a better candidate. In the Paris manuscript, MARGGRAFE refers to a chapter in ACOSTA's book *Historia natural y moral de las Indias* (1590; Dutch edition 1598). ACOSTA had observed the southern sky in Peru. So, at least this testifies to the fact that in Recife MARGGRAFE had access to one of the editions of this rare book.

It should be clear, however, that these were not all the books MARGGRAFE owned. From other sources it is evident that in Recife he also used EICHSTADT's *Ephemeridum*, understandable, because he had contributed himself to those tables. And, as already discussed, he gave JOHAN MAURITS the book *Cyclometrica Elementa Duo* by the Leiden humanist JOSEPHUS JUSTUS SCALIGER, a book that he had purchased in 1636. Finally, between the Leiden manuscripts is also one leaf from a language manual, published in Leiden in 1637. This booklet presented various expressions in Latin, French, Italian and German.[29]

CHAPTER 9

ASTRONOMY – OBSERVATIONS IN LEIDEN AND RECIFE

1. Training in Leiden

An eager student in practical astronomy

Marggrafe came to Leiden with the firm intention to train himself in practical astronomy. But in the first few months, he had to adjust to normal life at the university, following lectures from Jacob Golius and other Leiden professors. After all, he needed to win Golius's trust before he would be allowed to work at the observatory on his own. From these first months in Leiden two documents are preserved, one with mathematical notes from Golius's lectures,[1] and another with mostly illegible notes (caused by crabbed handwriting) concerning astronomical problems and related calculations, in which various times the expression *Lugdunum Batavorum* (Latin for Leiden) can be distinguished, with dates ranging from 20 May to 24 December 1636.[2] Despite the scratchy handwriting, it is clear that the notes relate mainly to positions of the planets in the zodiac, mostly theoretical in nature, but with a single entry on 17 November 1636 suggesting an observation of the Moon and the planet Jupiter.[3] These notes may have to do with Marggrafe's activities in preparing horoscopes for some of his fellow students.

Shortly after new year 1637, Marggrafe received the green light to use the Leiden Observatory and its instruments on his own. His Leiden observation notes reveal that he started his regular astronomical observations in early January 1637.[4] It was a week in which it was so cold that all classes at the university had to be cancelled.[5] But as a consequence of the frost, it were also days with a very bright sky. Marggrafe started in the early morning, observing some obvious phenomena. With the 6-foot Blaeu quadrant and a portable sextant he measured the altitude of a few stars, looked with the telescope at the Moon, witnessed the gibbous shape of Venus, and tried to find the satellites around Jupiter, one of which he indeed was able to discern.[6] After sunrise, he used a telescope to look at the Sun, and probably to his surprise, on 18 January he found that there was a sunspot visible on the Sun's surface. It seemed a good start, because Marggrafe returned that very evening for more measurements of stellar positions with the sextant and the large quadrant.

Witnessing a learning curve: Marggrafe's observations of sunspots

Before we discuss Marggrafe's observations with the quadrant and other instruments in more detail, it's revealing to look more closely at Marggrafe's sunspot observations, because they demonstrate his learning curve at the observatory. When he recognized a sunspot on the very first day, he recorded this phenomenon in a small freehand sketch of 5 mm wide. In the sketch the spot is to the left of the picture, at – so to speak – 10 o'clock. But another note, made on the same day, shows a more careful drawing with a 9 mm disk, now with the spot below right (*fig. 98*). So, the orientation of the solar disk of the freehand sketch can be transformed into the orientation of the larger drawing through a 180° rotation and vice-versa. He repeated this procedure the following day. This reversal in direction seems to indicate that the first sketch represents what Marggrafe saw when he looked directly through the telescope (via a darkened or coloured protecting glass), whereas the second reversed, 9 mm drawing – which is essentially the same image – represents the actual trace of this projection on a screen. Marggrafe probably used the objective lens of the telescope (with the then-common Galilean or Dutch configuration[7]) to obtain this

FIG. 98. Marggrafe's 5 mm freehand sketches and the 9 mm solar disks depicting sunspots seen on 18 and 19 January 1638. (ELO, North no. 22 and 52)

image. If this interpretation is correct, the focal length of this lens would be around 94 cm.[8] Moreover, the comment that the sunspot disappeared on 22 January indicates that in the freehand sketch west is to the left, while in the 9 mm solar disk it is to the right.

Later that year, on 9 June 1637, MARGGRAFE applied a new method for outlining the solar image. From then on the drawings of the solar spots were recorded in a much larger pre-drawn circle with a diameter of about 16 cm (*fig. 99*). The projection of such an image with a single lens would require a focal length of about 16.5 meters, an impossible distance for the observatory's platform.[9] This means that MARGGRAFE must have been instructed how to use the entire telescope for his projection. This method of telescopic projection was introduced in 1612 by BENEDETTO CASTELLI, a pupil of GALILEO GALILEI. The latter described the method in a publication, later that year:

> Direct the telescope upon the Sun [...] Having focused and steadied it, expose a flat white sheet of paper about a foot from the concave lens; upon this will fall a circular image of the sun's disk [...] In order to picture them [*the sun spots*] accurately, I first describe on the paper a circle of the size that best suits me, and then by moving the paper towards or away from the tube, I find the exact place where the image of the Sun is enlarged to the measure of the circle I have drawn. [...] But one must work dextrously, following the movement of the Sun and frequently moving the telescope, which must be kept directly on the Sun. [...] Next, one must note that the spots come from the telescope inverted, and reversed from their positions on the sun; that is, from left to right and from top to bottom; for the rays intersect one another inside the tube before coming through the concave lens. But since we draw them on the side of the paper facing the sun, we have the picture opposite to our sight, so that the right-to-left reversal is already effected. [...] If we merely turn the paper upside down [...] we have then only to look through the transparency of the paper against the light, and the spots will be seen precisely as if we were looking directly at the sun.[10]

This technique was used for many years, for instance in Rome by CHRISTOPH SCHEINER, who published his results in 1630 (*fig. 100*).[11] The first known use in the Netherlands of such a solar projection dates from June 1630. On the platform of the 'Latin School' in Dordrecht,[12] the rector ISAAC BEECKMAN observed an impressive

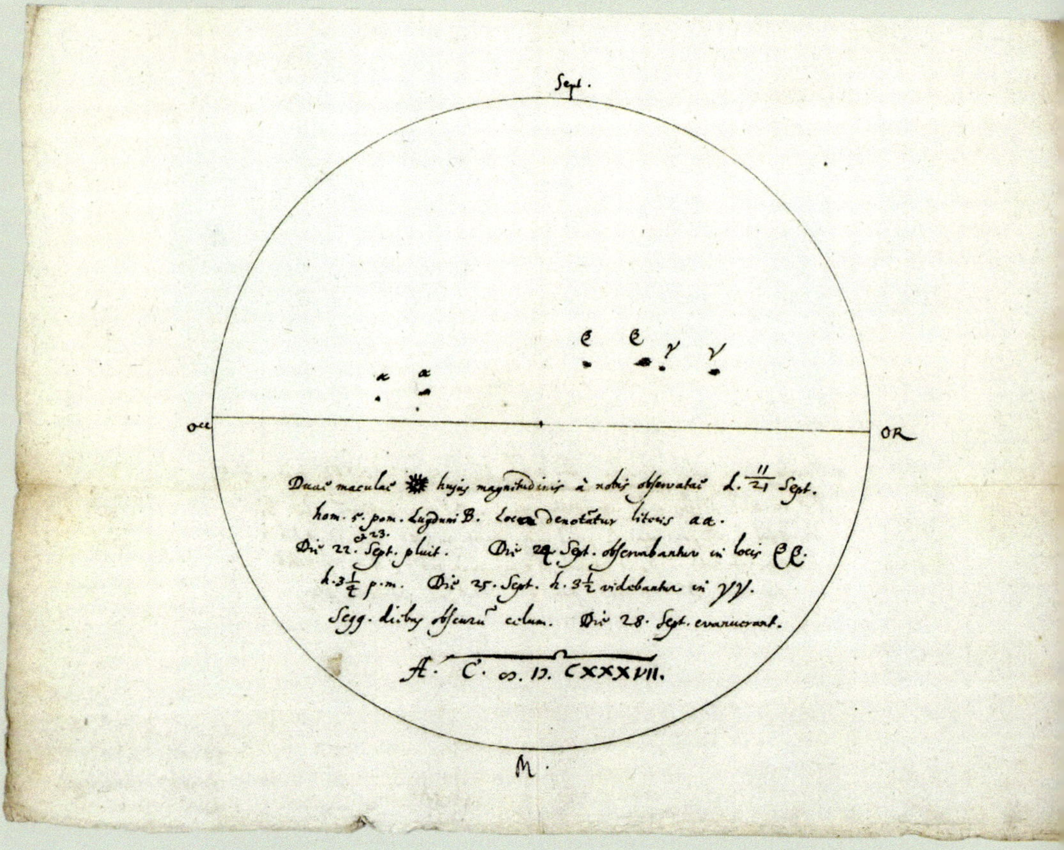

FIG. 99. Drawing of a sunspot observed from 21 to 28 September 1637. The successive positions of the sunspot are indicated by Greek letters. MARGGRAFE's sunspot observations are the only known from the year 1637. (ELO, North no. 24c)

solar eclipse using this method, being assisted by his former pupil MARTINUS HORTENSIUS.[13] In 1634, HORTENSIUS, then recently being appointed professor at the Amsterdam Illustrious School, also used a telescope in a dark room to determine the Sun's angular diameter.[14] MARGGRAFE evidently followed in their footsteps. It is interesting to note that his original projection sheets show that he first marked the position of the sunspots by piercing a small hole in the paper. Later these pinpricks were highlighted with black ink. In the first projection of 12 June 1637, he oriented the solar disk not yet according to the cardinal direction, but in September he evidently had learned to place the disk with the Sun's rotation axis in a vertical direction. In short, by following these documents one almost witnesses how MARGGRAFE step-by-step improved his observation skills. MARGGRAFE's sunspot observations are world-wide the only known from the year 1637.[15]

CHARACTERISTICS OF THE LEIDEN TELESCOPE

MARGGRAFE's projections of the Sun enable us to make some comments about the characteristics of the telescope he used. First, knowing that the Sun represents an angular diameter of 33 arc minutes, it may

FIG. 100. CHRISTOPHER SCHEINER's method of observing sunspots. From his *Rosa Ursina* (1630), 150. (Leiden University Library; Photo by the authors).

be concluded from the mutual relationships between the diameter of smallest sunspot and those of the Sun that this sunspot was about 11 arc seconds in diameter. The detection of such a small arc requires an objective lens with a free aperture of at least one centimetre in diameter.[16] This is in line with other telescopes of the period which usually have a diaphragm in front of the objective lens up to two centimetres in diameter.[17] Above we concluded that the focal length of his telescope was around 94 cm. Assuming further that MARGGRAFE's telescope had a magnification of 20 times (a usual value in those days), and that at night his eye pupil was 8 mm wide, with the eye 1.0 cm removed from the ocular, then it is possible to calculate the theoretical field of view of the telescope, using the appropriate formulae for a Galilean lens configuration.[18] This resulted in a field of view at night of almost 22 arc minutes. This result is in good agreement with the largest field of view we have found among MARGGRAFE's Leiden observations, namely 18 arc minutes.[19]

But at daylight, when the eye pupil is smaller (ca. 5 mm), this calculated field of view is only some 13 arc minutes. Compared to the angular diameter of the sun (33 arc minutes), this means that in direct view MARGGRAFE was only able to witness less than half of the total solar diameter. So, apart from the danger to his retina, it is unlikely he used a direct view. Such an observation was much easier when the Sun was projected on a screen. When used in projection, the aperture of the eyepiece acts as the field stop. Assuming this to be 2 cm in diameter, we derive a projected field of view of around a degree, so more than enough to project the total solar disk on the screen. To obtain this projection of 16 cm in diameter, using a telescope with the above specifications, MARGGRAFE had to pull out the drawtube of his ocular some 3 mm, receiving the projected image 88 cm behind the ocular lens.

EARLY OBSERVATIONS OF JUPITER'S SATELLITES

It is unknown whether MARGGRAFE had already observed through a telescope before his arrival in Leiden. His former employer EICHSTÄDT does not mention this instrument in his ephemerides. But if MARGGRAFE didn't know the instrument yet, he more than made up for that deficiency. In Leiden, he used the *tubus opticus*, as he called the telescope, very intensively. He applied the instrument to the observation of planetary phases, satellites, conjunctions and occultations. Among these recordings, his telescopic observations of the satellites of the planet Jupiter are especially noteworthy. Already during his first days at the observatory, he noted Jupiter's moons, and he repeated this observation on several later occasions.[20] This is the more remarkable because around the same time, the Amsterdam astronomer Martinus Hortensius wrote to the famous Italian astronomer Galileo Galilei that no telescope for sale in Holland could show Jupiter's four moons in a proper way: 'Even the best show Jupiter's disk bushy and poorly defined, so that the satellites in its vicinity are not properly seen'.[21] This was a serious deficiency that at the time was regarded as a substantial problem. In December 1635, Galilei had submitted to the States General a new proposal for finding the longitude at sea, in the hope of receiving the prize of 15,000 guilders awarded by the Dutch for solving this crucial navigational problem.[22] In this proposal, Galilei used the configuration of Jupiter's satellites as a clockwork with which the local time could be established. A committee was formed to judge about the practicality of the method, among whom were Hortensius, Jacob Golius, and Isaac Beeckman. But without their knowledge, and against Galilei's wishes, the secret of his method was revealed by the chairman of the committee, Laurens Reaal, *bewindhebber* (governor) of both the VOC and WIC.[23] So, in the spring of 1637 several others were aware of the content of Galilei's proposal. To Hortensius's surprise, even Golius had told the importance of observing the satellites of Jupiter to some of his students. We can be sure that Marggrafe was one of them. No wonder he was eager to try this observation.

Contrary to what Hortensius had asserted, Marggrafe had no problem at all in distinguishing Jupiter's satellites. Apparently, the telescope he used was of a good quality. Some sketches still testify about what he saw. A present-day reconstruction of the configuration of Jupiter's satellites corroborates one of his observations (*fig. 101*). The sketch proves also that the field-of-view of his telescope was at least 9' 35", at that moment the angular distance between Jupiter's moons Ganymede and Callisto.

Marggrafe and Mercury

Marggrafe had a special interest in Mercury. He expressed that several times. Why was that planet so intriguing for him? What made that celestial body so special, not only for him, but for the whole astronomical community? There were two reasons. Firstly, there were the stunning findings of Pierre Gassendi made during the transit of Mercury across the surface of the Sun in 1631.[24] A transit occurs when an inner planet (one circling the Sun in an orbit closer to the

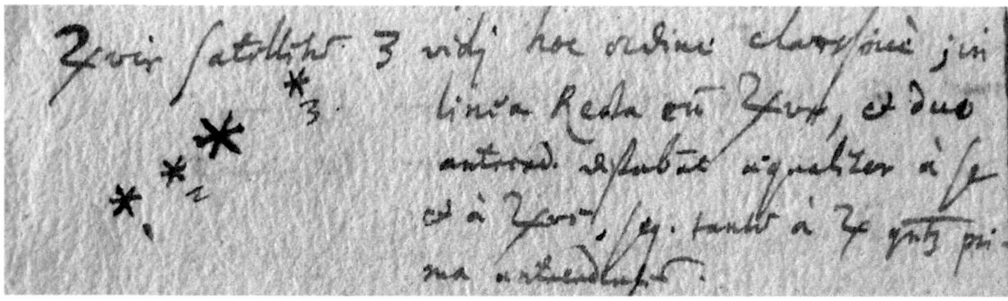

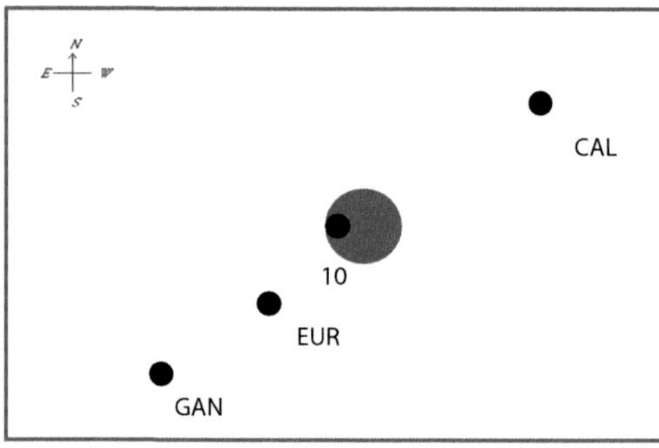

FIG. 101. MARGGRAFE's drawing of Jupiter satellites on the night of 14/15 March 1637, compared with the calculated configuration.

Sun than the Earth) aligns with the line of sight between the Sun and the Earth. For Venus that is a very rare occasion, not yet seen when MARGGRAFE started observing, but for Mercury this happens roughly once every decade. This occurrence was predicted in 1629 on mathematical grounds by JOHANNES KEPLER and published by his son-in-law JACOB BARTSCH.[25] So, in 1631 many astronomers prepared themselves to witness this event for the first time. However, at the time it was generally accepted that Mercury was much larger than the planet actually is. So, most witnesses followed KEPLER's advice to use a *camera obscura* (without a lens) in the supposition that Mercury's shadow would be seen distinctly on the projected surface of the Sun. Only GASSENDI used a projection lens. He, too, thought initially to have missed the event, until the moment he discovered that a very tiny spot, much smaller than a usual sunspot, moved too fast over the Sun's surface to be a sunspot. Therefore, in his report about the event, GASSENDI concluded that this tiny spot actually must be identified as the planet Mercury.[26] In the end, all other astronomers had missed the transit because the planet was far too small to distinguish it in a *camera obscura*. Although it should be noted here that some astronomers – such as HORTENSIUS – could not conduct an observation due to bad weather.

This was the first mystery, which raised questions such as, why was Mercury that small? What did this size say about the dimension and distance of the other planets, and as a consequence, the solar system? It sparked an intense discussion between several authors, such as in Holland by the earlier-mentioned Hortensius, who in turn was attacked for his views by the Danish scholars Petrus Bartholin and Longomontanus.[27] But secondly, Gassendi's transit observation also demonstrated that the orbital data on Mercury, as published in all planetary tables of the time, fell short. Improving these data was difficult because the planet moved so close to the Sun. After all, the planet was never farther away from the Sun than about 23°. For that reason Mercury could only be sighted – let alone its coordinates measured – very rarely. To make things worse, its orbit was dramatically elliptical. To revise this part of planetary theory was therefore one of Marggrafe's ultimate dreams. It was his key problem. No wonder that he tried to observe Mercury whenever he could. In Leiden, he observed the planet together with the Danish astronomer Jørgen From, who reported about this to his tutor Longomontanus.[28]

Among the Leiden papers, a drawing of Mercury has been preserved, made on 28 September 1637 (*fig. 102*). Marggrafe reports that the upper or oriental part of Mercury was getting dark. According to modern calculations, the planet was visible for only a quarter of an hour, because the Sun set at 18:26 and Mercury at 18:45.

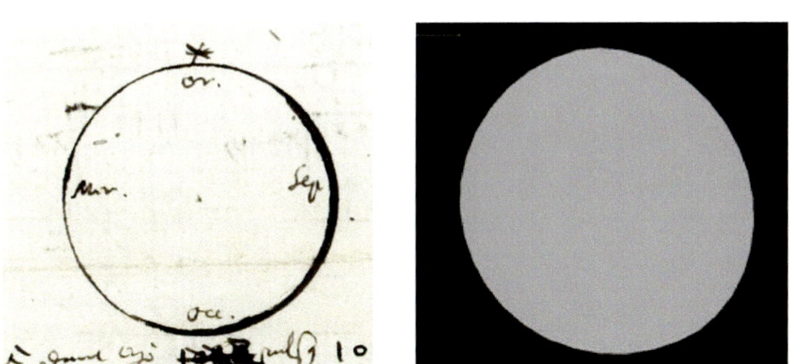

FIG. 102. *Left*: Marggrafe's drawing of Mercury in the evening of 28 September 1637 (ELO, North no. 51vs). *Right*: Mercury at 18:36:00 of the same date, according to SkyMap Pro11. The phase of Mercury was 0.924.

Measurements with the large Tychonic quadrant made by Willem Jansz Blaeu

The main asset of the Leiden Observatory was the large Tychonic quadrant made by Willem Jansz [Blaeu]. It was an instrument that Marggrafe had never used before. He still had to learn how to use such a large measuring device. So what exactly did he measure in Leiden during the 10 months that he worked there as an observer? What did he learn, and how accurate was he in his measurements?

The Leiden observation register contains a total number of 338 altitudes of celestial bodies measured with Snellius's quadrant. From these, 86% were observations made during a crossing of the meridian.[29] Without exception, all meridian altitudes were southern. In that case the azimuth was fixed on 180 degrees, so that it was sufficient to measure only one coordinate. In the remaining 14% of extra-meridian observations two coordinates had to be noted.

To evaluate Marggrafe's accuracy as an observer, we analysed the observational error of his quadrant observations. For that purpose, we calculated all meridian altitudes with astronomical software, corrected these for the atmospheric refraction and compared them with those actually observed.[30] When we plot the observational error O–C (= Observed minus Calculated position) versus the date of the observation, is it possible to see any improvements as a result of Marggrafe becoming more experienced? The answer on the last question is a firm 'no'. There is no sign of Marggrafe's getting more experienced (*fig. 103*). If any trend can be established, it is a seasonal modulation of the observational error, which tends to reach a minimum in May and June, for all meridian observations. Wesley found a comparable result for some of Brahe's instruments, which he attributed to temperature effects on the brass scale of the instrument.[31] A variation of the humidity of the quadrant's wood (or the platform) could also contribute to such an instrumental error.

An instrumental error seems also to be present when we look at the observational error plotted against the measured altitude (*fig. 104*). In 1968, Haasbroek argued that the reading error of Snellius's quadrant should be directly proportional to the measured angle.[32] His theoretical analysis was based on the assumption that the reading of the quadrant was similar to the reading of a cross-staff. However, this assumption is wrong. When a *pinacidium Tychonicum* is used, theoretically the error should be the same at each measured angle. Theoretically a graph of (O – C), plotted against the altitude, would show a horizontal line along the axis of zero error with an error distribution symmetrical around it. As *fig. 104* shows this is obviously not the case.

The mean error tends to decrease at higher meridian altitudes. This effect is probably attributable to a small bending of the sighting tube on the quadrant caused by gravity; the bending becoming larger towards the horizontal position.[33] Nevertheless, the overall mean error of the Leiden quadrant for meridian observations appeared to be rather small (49' for all meridian altitudes), but with a large scattering in the readings (resulting in a standard deviation of 1' 34"). This result is in the same order of magnitude as the mean error of 32' that Walter Wesley has calculated for Brahe's meridian observations with his 5-foot quadrant, also with a broad range of scattering.[34]

Measurements with the sextant

Before an observation of the angular distance of a planet to a fixed star, Marggrafe often pre-calculated this angular distance. Sometimes he also measured the distance between two fixed stars, maybe for checking the accuracy of his instrument. For this class of observations the quadrant was unsuitable, so he must have used a sextant or a mathematical instrument with similar capacities. In one instance (25 February 1637), Marggrafe stated explicitly that he used a sextant.

We have collected from the Leiden observation register a total of 223 measurements of angular distances. Except one case of an angular distance larger than 60°, all distances were less than 60° (in fact less than 55° 20'). The smallest angular distance measured was 1°10'. The mean angular distance was 31° 34'. The largest measured angular distance was 72° 18' (between Venus and Jupiter on the evening of 19 January 1637). Marggrafe must have resorted to some particular procedure to measure an angle that exceeded the 60 degrees range of the sextant, but nothing about this is reported.

For Marggrafe's measurements with the sextant, the observational error is also evaluated. The mean error of all angular distances is –7' 34" (with a standard deviation of 7' 32"). This mean error is about ten times larger than with the quadrant, indicating that the sextant was small and portable. The negative sign of the error indicates a systematic tendency by the sextant to produce angular distances smaller than the calculated ones. This tendency, which was almost negligible for small angular distances, increased with the angular distance. This leads us to the conclusion that the Leiden sextant was not equipped with a Tychonic aligning sight, because theoretically the mean error would be independent of the measured angular distance,

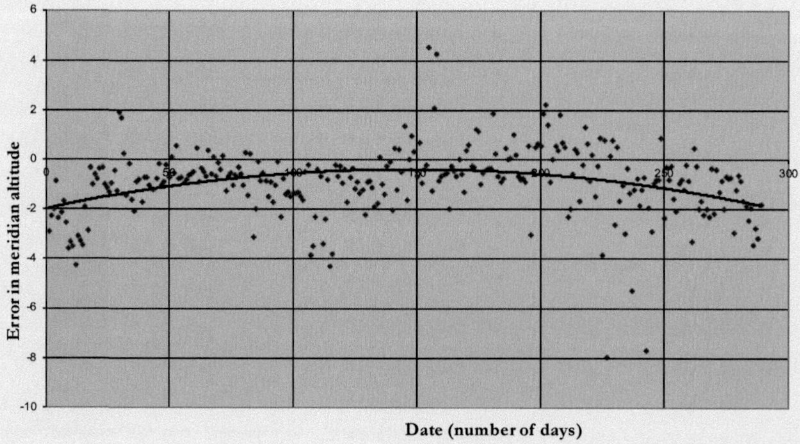

FIG. 103. Observational error O-C (in arc minutes) for all measured meridian altitudes versus the date of the observation. Number of days plotted from 19 January until 6 November 1637.

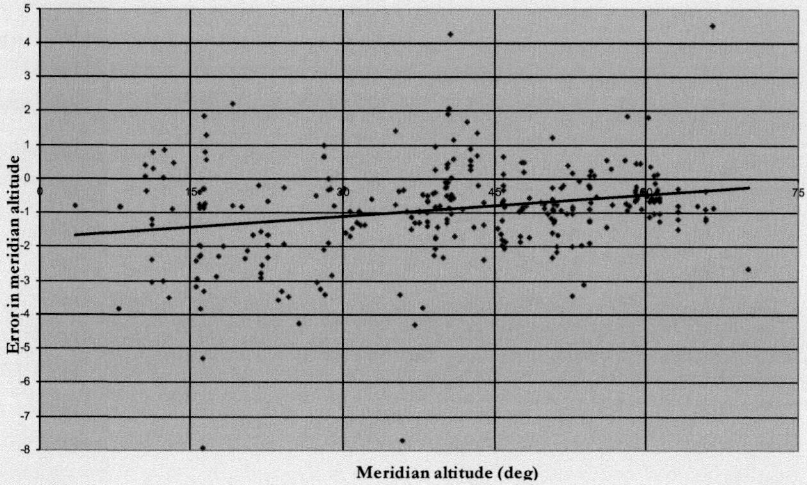

FIG. 104. Observational error O-C (in arc minutes) of the meridian altitude plotted versus the meridian altitude for all meridian observations made with SNELLIUS's quadrant between January and November 1637.

just as (theoretically) in the quadrant's case. Instead, this tendency to increase with the measured angular distance is in line with the theoretical hypothesis formulated by HAASBROEK (1968) that with this type of instruments the systematic errors are directly proportional to the measured angle.[35]

MEASURING TIME

One of the very important, but also problematic aspects of astronomy is the measurement of time. For example, when observing a fixed star, the moment of observation is important because its position in time is needed to calculate from the observations of elevation (or altitude) and azimuth the time-independent coordinates of right ascension and declination. This process of conversion is called the reduction of the observations. At every location, a crossing of the Sun through the meridian provides the local time of 12:00 h. If during the day the sky is clear, time can also be recorded via a sundial. But to know the local time during the night it is necessary to 'preserve the time' with a timepiece. In the 1630s, mechanical clocks could only register the time with an accuracy never better than a quarter of hour (15 minutes). A better method was to use a sandglass, but that method required a very careful procedure of handling this device. For the measurement of relatively small time differences, it was best to use a loose pendulum.

When MARGGRAFE started his observations on a given date, he usually wrote down the *Hora Urbica* [the local time] obtained mostly from a mechanical timepiece with a bell. In Leiden, such a clock was installed at the academy building in 1636.[36] In Recife, there was 'below the platform' also a 'large iron clock that by the stroke of the bell announced the time to all inhabitants of the city'.[37] These clocks might already have been endowed with a verge escapement as a regulator. Reinforcing the assumption that the *Hora Urbica* was obtained from such a clock, there is one record for 28 September 1637 mentioning the clock bell ringing at 7 AM: *Et horologium sonabat 7 mat.*[38]

For a measurement of the elapsed time between two observations, MARGGRAFE rarely used the sandglass. On most occasions, he preferred to count the oscillations (*pulsus*) of a pendulum (*libella*). For instance, in Leiden on the night of 12 April 1637, MARGGRAFE explicitly mentioned his use of a *clepsydra* (= sandglass) for counting the 'minutes'. An accompanying note states, however, that each minute of the clepsydra corresponded to 72 oscillations of the pendulum. So it is clear that the *pulsus* were not identical to seconds, but were an arbitrary subdivision of the minute. As NORTH already explained, such an arbitrary subdivision causes no problem when these measurements were used during the entire time span of an observation. In that case, the time could be found by interpolation.[39] Nevertheless, sometimes it is amazing how many *pulsus* MARGGRAFE had to count during one observation. A personal record is noted on 8 October 1637, when he counted 2,220 *pulsus* from the pendulum.

THE ACCURACY OF MARGGRAFE'S TIMING

For an evaluation of the accuracy of MARGGRAFE's timing, we compared his measurements of the time lapse between a meridian crossing of two stars or planets, with those calculated with astronomical software (SkyMap Pro11). For instance, on the evening of 12 April 1637, MARGGRAFE measured a time difference of 4 minutes between the crossings of Jupiter and the Heart of Lion (a Leo), while our calculated time was 5 minutes and 22 seconds. We collected from the Leiden observation register in total 62 cases in which MARGGRAFE measured the time interval between two such astronomical events. The observed time difference ranged from 0 to 70 minutes, but the smallest non-null interval was 1 minute and 20 seconds.

Our analysis reveals that astronomical events which according to MARGGRAFE's notes happened simultaneously, could in fact be separated by a substantial time interval. From the set of 62 data points, 17 events are reported by MARGGRAFE as being simultaneous (by definition, with a time difference of 0 seconds). However, the calculated mean time space for such 'simultaneous' events was 6 minutes and 48 seconds, with a large scattering of the error (the standard deviation was 15 min 24 sec.).[40] This clearly reflects the difficulty MARGGRAFE experienced of measuring time intervals without a reliable timepiece.

THE PERSONAL EQUATION: MARGGRAFE'S ACCURACY AS CALCULATOR

During his training as a practical astronomer, MARGGRAFE made astronomical observations of various types. Of these, the determination of planetary positions was the most complex, because it required a coordinated chain of actions. On several of these occasions, MARGGRAFE first tried to predict the position of a planet by calculating the positions from the printed ephemerides. In those cases, MARGGRAFE often wrote *Observationes expectanda* at the top of a record. Then the actual observation followed. Clearly, MARGGRAFE wanted to compare his predictions with the actual observations, in the hope to improve the planetary ephemerides. He started with the measurement of the elevation of a fixed star (often meridian, but not always). Then, in general, he measured the angular distances of the observed planet to three neighbouring fixed stars. At the end of this, a new measurement of the elevation of a fixed star was made. Our analysis shows that MARGGRAFE's prediction of the angular distance of a planet to fixed stars was affected by a mean error of +2' 17" (standard deviation 6'). This relatively small error is explicable if we consider that the distance of a planet to a star is a differential quantity, in which some systematic errors cancel out.

WHAT DID MARGGRAFE ACHIEVE DURING HIS TRAINING AT LEIDEN OBSERVATORY?

To summarize: In Leiden MARGGRAFE learned to measure the coordinates of several celestial bodies with a large Tychonic quadrant and with a smaller sextant. He used time measuring devices such as the sand glass (*clepsidra*) and the pendulum, and learned to handle a telescope with a Galilean or Dutch configuration. The observations we analysed show that the overall accuracy of his work with the quadrant was a bit less than achieved by TYCHO BRAHE, but his observational errors were still the same order of magnitude as those of BRAHE. His time measurements were sometimes problematic. They clearly reflect the difficulty of measuring time intervals without a reliable timepiece. His observations of sunspots, however, demonstrate his learning capacity. His observation of Jupiter's satellites and his registration of the planet Mercury attest also to his ability as a skilled observer.

2. Observing in Brazil

A leap forward at the very start

Strengthened by his Leiden experiences, Marggrafe longed to observe the Brazilian evening sky. To his disappointment, this was possible only after his return from the military campaign to São Salvador, where he was sent immediately upon his arrival. Nevertheless, even before the Dutch fleet returned to Recife, he made tremendous progress in fulfilling his dreams. Shortly after the lost battle, while still being at the Bay of All Saints, he was able to arouse Count Johan Maurits's interest in his astronomical project in such a way that the count promised his actual support. As it turned out, this promise involved financing the construction of a copy of the Leiden Observatory, including a large quadrant, on top of the count's own residence. Full of enthusiasm, Marggrafe informed Golius in a letter sent at the end of June 1638, only a few weeks after his return from the campaign to São Salvador.[41] He told the Leiden professor how eager he was to study the southern sky. Until then he had been able to make only a few meteorological observations, probably also because of the monsoon, which caused heavy rain during the fall and winter season, meaning from March till September.[42]

The aspired Progymnastica Mathematica Americana

What do we know about Marggrafe's first observations in Brazil? What were the aims he targeted? From his notes, we learn that he started to observe long before the observatory was ready. But these documents also reveal that his observations were only incidental. Apart from elaborate notes concerning the eclipse of December 1638, discussed below, we know of only two sets of observations performed before the observatory was ready. The first concerned two successive days in September 1638 and the other set a few days in May 1639. Both sets relate to conjunctions in which the planet Mercury was involved. The sightings were evidently intended to capture an accurate position of Mercury at these moments. But in the absence of a suitable measuring instrument, he had to visually estimate the angular distance between two celestial bodies, for example between Mercury and the star Spica or between Mars and – what he called – 'the left shoulder of Sagittarius'. Interesting is Marggrafe's remark that during the last event, in the absence of the moon, the light from the planets Venus and Jupiter did cast a shadow on the floor. He also noted that, when observed with the naked eye, the Magellanic

Clouds were devoid of stars, and commented insightfully that in his view these clouds consisted of the same substance that formed the Milky Way.

Still, this harvest seems rather disappointing for a person who claimed to have come to Brazil to chart the southern sky. Indeed, in a letter to his friend SAMUEL KECHEL, written in May 1639, he confessed: 'I haven't been able to observe much in astronomy'.[43] Since his arrival, his daily work had prevented any devotion to astronomy. A factor that probably contributed to this circumstance was that, after the death of his companion KRALITZ, he had to do the work of two. After all, he now was the only assistant left to his employer PISO. His hope was pinned on the observatory, but that superstructure on the count's residence became operational only in September 1639, more than one year after his arrival in the country.

MERCURY REVISITED

What actions were taken when the observatory became operational? What changed in his approach? First of all, MARGGRAFE started a real observation register, partly preserved among the Leiden manuscripts (*fig. 105*). This register (copied neatly in the Paris and Lisbon manuscripts) clearly shows that MARGGRAFE devoted his first nights again to the observation of Mercury's position. Overall, Mercury is mentioned 129 times in the Paris manuscript; Venus only 28 times, Mars 26, Jupiter 27, and Saturn 31 times. So, it's clear that Mercury definitely was MARGGRAFE's darling. He really wanted to fulfil the third objective formulated in his design of the *Progymnastica Mathematica Americana,* namely to find a 'new and true theory' of this planet. The improvement of Mercury's orbital elements was possible only by a series of careful observations, something that coincided with his ambition to observe 'all the planets'.[44]

Among Mercury's numerous sightings, there is one that really deserves our special attention. On the evening of 28 September 1639, just after sunset, an occultation (covering) would occur of Mercury by the Moon. In itself this would not be so special, but it happened to be New Moon, so the Moon was actually in its invisible phase. Only a very trained observer could see its very thin crescent. MARGGRAFE did! As dusk fell, he saw the planet just above the horizon, very close to the moon's thin crescent. Only moments later MARGGRAFE saw through the telescope how the planet plunged into the dark Moon and instantly disappeared from the sky. It was a spectacle that he really enjoyed witnessing (*fig. 106*).

FIG. 105. First page of MARGGRAFE's observation register started on 15 September 1639 at the newly completed Recife Observatory. (ELO, North no. 53, fol. 2r).

MARGGRAFE was as excited when, in September 1640, he saw Mercury's crescent. For such a tiny planet, that close to the Sun, and therefore always to be observed close before sunrise or after sunset, this is not an obvious event. In any case, it is much more difficult to observe than with Venus. Mercury's phases can be seen only when the planet is in (or near) its greatest elongation. During this event, MARGGRAFE again enjoyed what he saw: 'Immediately after it emerged from the twilight, I inspected Mercury carefully with the telescope. And as long as I looked, the planet appeared to me clearly hollow [...], which means that Mercury is horned like Venus ...'.[45] With this sighting was MARGGRAFE actually the first astronomer in the world to have actually witnessed Mercury's phases! In current historical literature, this achievement has so far been attributed to JOHANNES HEVELIUS who observed the planet's cycle between November 1644 and May 1645.[46] Still, as JOHN NORTH already has concluded, all these observations yielded little:

> It's sad to say, there is not the slightest evidence that MARGGRAFE ever reached the stage of integrating his calculations of planetary positions with a comprehensive theory, new or old. [...] The few ecliptic longitudes for Mercury that I have reworked from his observations are, as it happens, generally no more accurate than those he could calculate from the Rudolphine Tables, and only the most generous of optimists can see in MARGGRAFE's work the seeds of a reformation of the Mercury theory.[47]

AN EXCELLENT OBSERVATION WITH A DUBIOUS RESULT: THE DETERMINATION OF RECIFE'S LONGITUDE

One of the prime concerns for any astronomer is to know where he is. Therefore, one of the primary goals in MARGGRAFE's program for the *Progymnastica Mathematica Americana* was to present 'A theory of the longitude of the earth and a manner of computing the same'.[48] For that purpose, he needed two crucial coordinates: the latitude and longitude of Recife. Latitude is the angle between the equatorial plane and the vertical direction to a point on the Earth's surface when viewed from the centre of the Earth. This coordinate specifies the north–south position on the Earth's globe. The angle ranges from 0° at the Equator to +90° (North) or –90° (South) at the poles. Longitude is the angle measured from a prime meridian (a great circle running across both poles) to the location in question, ranging from -180° to +180°; positive longitudes are east of the prime

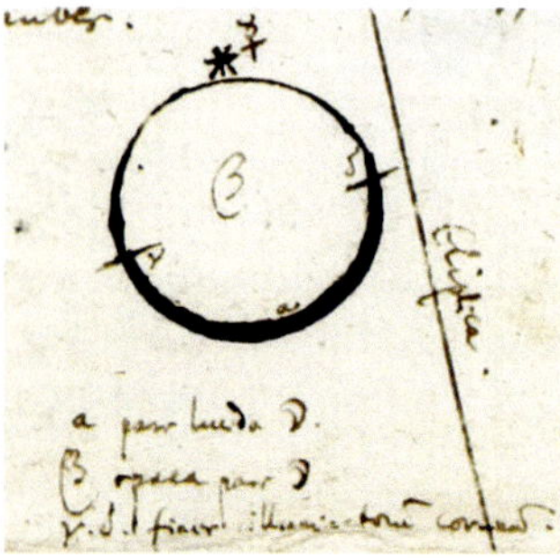

FIG. 106. Marggrafe's observation of the occultation of planet Mercury (☿) by the Moon on 28 September 1639, during 'New Moon', when the Moon is in its (almost) invisible phase. (ELO, North no. 53, fol. 4r).

meridian, and negative ones are west. Today, the Prime Meridian passes through the Royal Observatory in Greenwich, England.

In the Northern Hemisphere the determination of the latitude is solved rather easily by measuring the elevation of the Polar star, but as such a star is missing at the southern sky, such a procedure was not applicable in Recife. Another, more accurate method is measuring the elevation of the Sun during its passage through the meridian. From his observations in the years 1639 and 1640, Marggrafe extracted a result for Recife's latitude of 8° 8', later improved to 8° 7' 30" (see *fig. 112*).[49] Compared to the latitude accepted today (8° 3' 51"), this is a small difference of only 3' 39", or in length, 6.7 km too far south.

The determination of the longitude is quite a different problem. Because the Earth rotates in 24 hours, the solution can be found if the time difference with another location of known longitude can be found. In Marggrafe's case, he aspired to seek the time difference with *Uranienborg*, the observatory used by Tycho Brahe. In the seventeenth century, this location was often used as a reference point.

Because a reliable portable chronometer was not available before the middle of the eighteenth century, the best way to find a time difference between two locations was through the observation of a solar or lunar eclipse. That's why his observation program had listed the goal to observe 'all solar and lunar eclipses, worked out in an original way'.[50] Celestial tables, such as Kepler's *Tabulae Rudolphinae* (1627) or Eichstädt's tables (1634), offered the calculated times for when an eclipse would be seen at *Uranienborg*. A measurement of the times of ingress and egress of the passing celestial body during an eclipse would therefore bring Marggrafe the desired knowledge of Recife's longitude. This result would become even better if he could exchange results with an observer elsewhere in the world.

However, when the lunar eclipse of 20 December 1638 approached, the plans for the observatory and its instruments were still on the drawing table. The desired platform with its large measuring instruments was not yet available. Nevertheless, on the very day, the sky was clear, so MARGGRAFE did not hesitate for a moment.[51] After all, the observation required mainly a telescope and a time measuring instrument. He had brought these instruments with him when he went to Brazil, next to a small quadrant with a radius of one foot, required to take some reading of relevant elevations. So, he was well prepared to collect the relevant data. MARGGRAFE was lucky because the sky was cloudless that evening. All five crucial moments: the time of ingress, inner touching, the middle, inner touching before egress and exit, could be measured. The data gave him, by straightforward calculation, a time difference with *Uranienborg* of three hours and thirty-four minutes, or expressed in degrees 53° 30'. This led to a longitude for Recife, recalculated to the current Prime Meridian belonging to the so-called *Insulis Fortunatis*, of 343° 15' (*fig. 107*).[52] Converted to today's Prime Meridian, this gives a longitude of 40° 48' west of Greenwich, which compared to the current longitude of 34° 53' is in fact almost six degrees too far west![53] A later 'improvement' of his calculation made this result even worse. In the *Historia Naturalis* of 1648, a longitude of 340° 30' is presented, an error of almost nine degrees. This means that MARGGRAFE had determined his location to be *almost a thousand kilometres inland in Brazil!*

How was this result possible? And why was this value not corrected on a later date, when the observatory was operational? The main reason is that the 1638 observation had been executed under very favourable conditions. MARGGRAFE had measured a duration of the transit that differed only 3% from the expected duration in *Uranienborg*, taken from the Rudolphine tables, which seemed to be an excellent result.[54] The determination of the duration of the eclipse went so well, probably because MARGGRAFE had used a pendulum. At any rate, MARGGRAFE was very convinced of the reliability of his measurements. For the finding of the local time, he measured the elevation of two stars with known coordinates during their crossing through the meridian, first of Procyon and then of Regulus. After all, at the moment of the eclipse, in December 1638, the mechanical clockwork at the count's house was still not available, and at night evidently a sundial does not work.

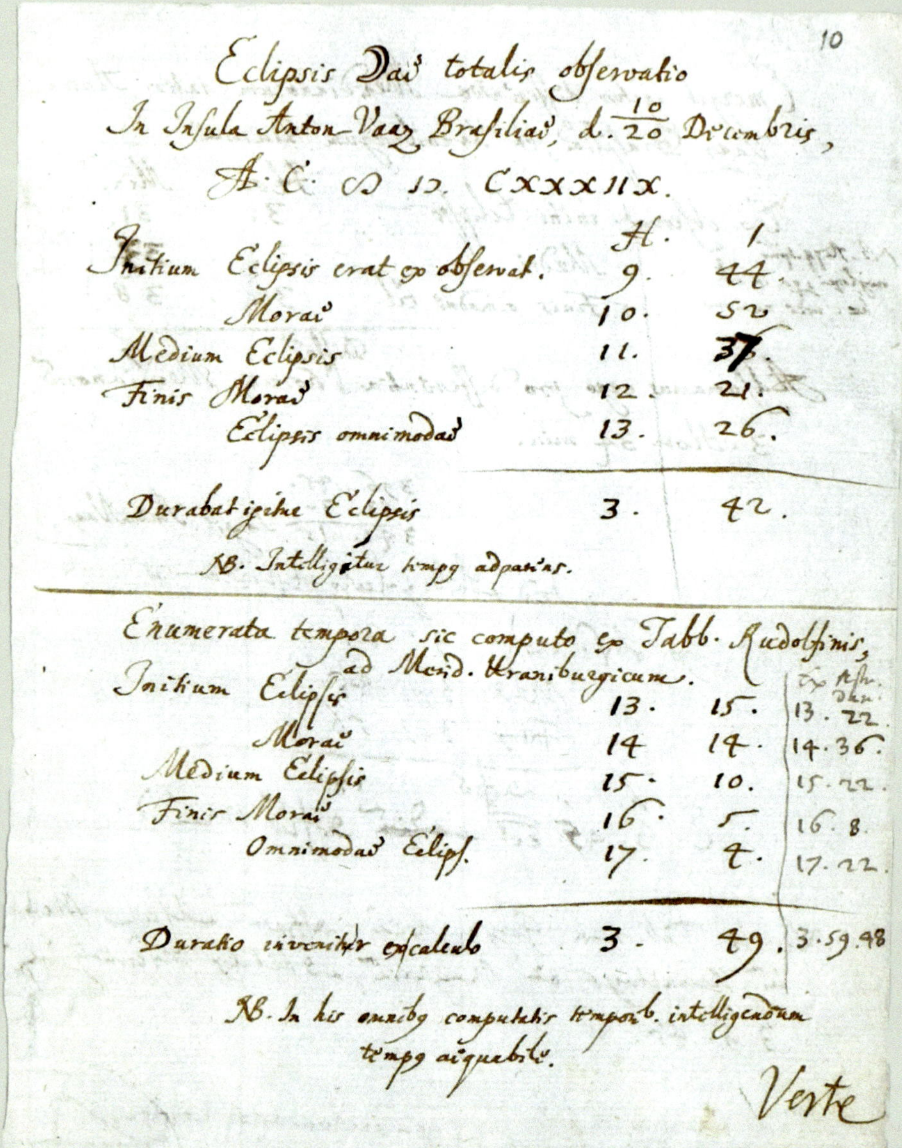

10

Eclipsis Lunae totalis observatio
In Insula Anton-Vaaz Brasiliae, d. 10/20 Decembris,
A. C. CIↃ IↃ CXXXIIX.

	H.	'
Initium Eclipsis erat ex observat.	9.	44.
Morae	10.	52
Medium Eclipsis	11.	37.
Finis Morae	12.	21.
Eclipsis omnimodae	13.	26.
Durabat igitur Eclipsis	3.	42.

NB. Intelligitur tempus adparens.

Enumerata tempora sic computo ex Tabb. Rudolphinis, ad Merid. Uraniburgicum.

			ex Ast. Dan.
Initium Eclipsis	13.	15.	13. 22.
Morae	14	14.	14. 36.
Medium Eclipsis	15.	10.	15. 22.
Finis Morae	16.	5.	16. 8.
Omnimodae Eclips.	17.	4.	17. 22.
Duratio invenitur ex calculo	3.	49.	3. 59. 48

NB. In his omnibus computatis temporib. intelligendum tempus aequabile.

Verte

FIG. 107. MARGGRAFE's determination of Recife's longitude from his observations of the lunar eclipse of 20 December 1638. He first compared the duration of the eclipse in Recife (3 h 42 m), compared with those calculated for *Uranienborg* (3 h 47). Then he took three time differences of ingress, middle and egress (which differed by only seven minutes).

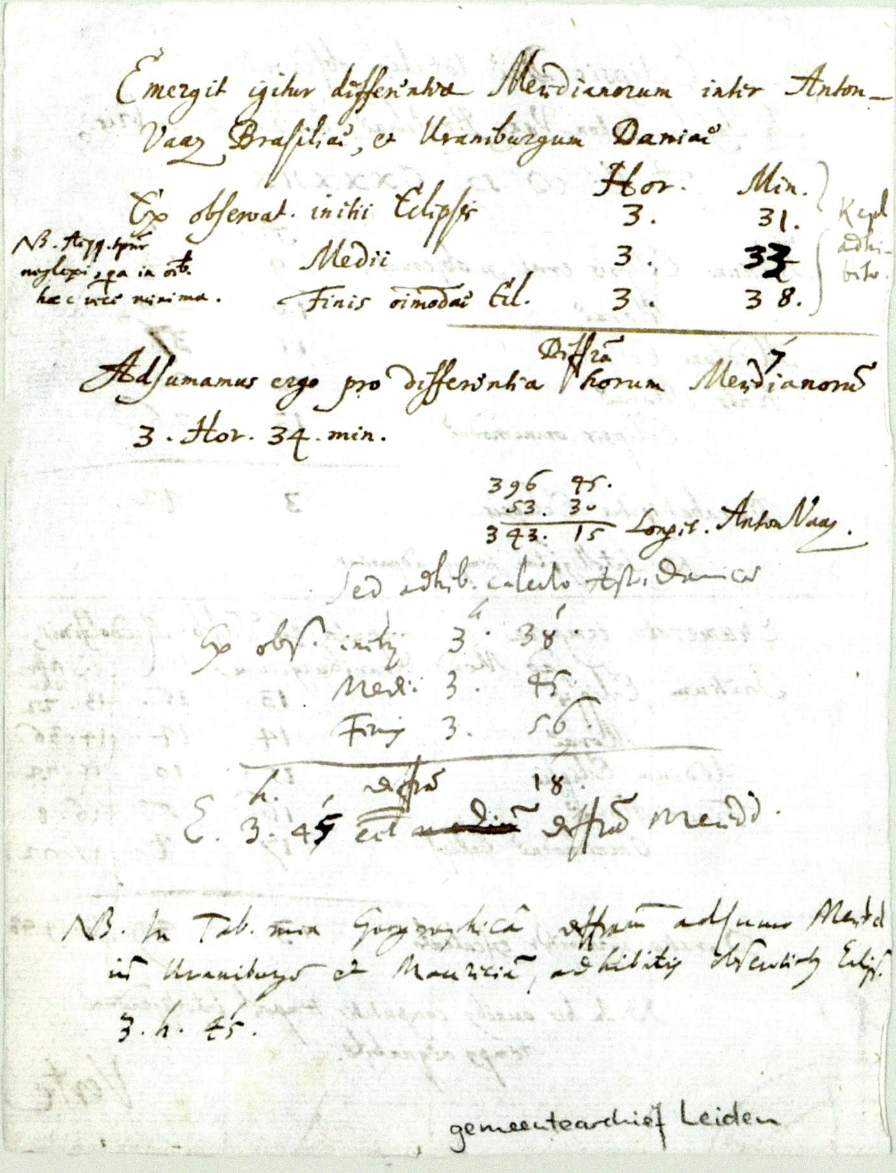

The mean value of these three time differences (3 h 34 m) corresponds to a difference in longitude of 53° 30'. The Rudolphine tables gives as longitude for *Uranienborg* 36° 45' (relative to so-called 'Fortuna Islands'). The difference with 360° delivered MARGGRAFE's first result, 343° 15'. *ELO, North no. 10.*

Why then was there, seen in retrospect, such a deviant result? An analysis of all known observations of the lunar eclipse of 20 December 1638 reveals what went wrong (*See Table 1*). The error lay in the Rudolphine tables. If we calculate the difference in longitude that follows from the comparison of MARGGRAFE's measurements with those of other observers in Europe and compare these results with the present-day coordinates, we may conclude that given the possibilities of the time, MARGGRAFE did a very good job. In all cases, the differences between the measured and actual longitude differences are less than two degrees. When we realize that not only MARGGRAFE contributed to these differences, but the other parties too, this means that we may conclude that the uncertainty in his work stayed in the order of magnitude of only one degree.

Blame, however, must be pointed at JACOB GOLIUS, who in Leiden also measured some of the essential times of the lunar eclipse, but evidently failed to share his readings with MARGGRAFE, who did send his own![55] This in spite of the fact that MARGGRAFE, in his letter to

TABLE 1.

Differences in longitude, calculated from time measurements of the Lunar Eclipse of 20 Dec. 1638

Observer	**Location**	**Longitude (degrees) relative to Recife**		**difference measured & real**	**number of measure-ments**
		in reality	**measured in 1638**		
Rudolph. Tables	Uranienburg	47.6	53.4	5.9	4
Gassendi	Digne	41.1	43.1	2.0	1
Holwarda	Franeker	40.4	41.8	1.4	2
Wendelin	Herck	40.1	41.3	1.3	4
Hortensius	Amsterdam	39.8	41.1	1.3	2
Golius	Leiden	39.4	40.3	1.0	2
Langren	Brussels	39.2	40.3	1.1	3
Gasconge	Leeds	33.3	37.6	4.2	4
Crabtree	Broughtown	32.6	33.5	0.9	1
Horrox	Toxteth	31.9	31.7	-0.2	4
Marggrafe	Recife	0.0	0.0	0.0	4

GOLIUS of 16 February 1640, explicitly requested to send him all available observations from Leiden over the years 1638 and 1639.[56] Although GOLIUS had been able to measure only the moments of immersion (inner point of intersection at ingress) and emersion (inner point of intersection at egress), had he shared his results with MARGGRAFE, the resulting difference in longitude between Leiden and Recife would have been only one degree (or in length, 112 km).

This demonstrates the excellence of MARGGRAFE's measurements. Had this difference been applied to the then-specified longitude of Leiden, taken from the Rudolphine or EICHSTÄDT's tables (which by coincidence was 0.5 degrees too far East),[57] the difference between MARGGRAFE's final longitude and the one known today would have been reduced to only 0.5 degree. This result is almost 18 times (!) better than the one published in the *Historia Naturalis* (1648) and the *Indiae utriusque re naturali et medica* (1658).

Nevertheless, had MARGGRAFE been taught by HORTENSIUS, he surely had heard HORTENSIUS's opinion of building a result on only one incident. To his French correspondent PIERRE GASSENDI, he had issued this warning in 1636, 'Be careful. After all, who would dare to draw conclusions from a single eclipse about the true difference of the longitudes?'[58] So, the question remains, why didn't MARGGRAFE use more data from other eclipses?

OTHER ECLIPSES

The fact is, he did try to do so! But in those attempts, MARGGRAFE was not fortunate. During his stay in Brazil there were 13 lunar and two solar eclipses. However, of the lunar ones, seven were penumbral (stayed in the half shadow of the Sun) and were therefore unusable for performing measurements. MARGGRAFE did not waste any time on these events. In the case of almost all other eclipses, he attempted to take observations. However, as Table 2 shows, none of these were seen as clearly as the lunar eclipse of 20 December 1638. Most of the time he found the sky covered with clouds, obscuring the eclipse from his view.

At the time of the first lunar eclipse, 25 June 1638, MARGGRAFE was troubled by bad weather, reason why he had seen hardly anything. In the two letters he sent the next day (one to GOLIUS and the other to FROM), he did not even mention the event. From another lunar eclipse, the partial one of 18 October 1641, MARGGRAFE left no record at all, probably because he was on a long expedition to inland Brazil. And although MARGGRAFE mentioned the solar eclipse of 1 June 1639

TABLE 2.

Solar and Lunar Eclipses in the years 1638-1643, listed by Pingré				
No.	**Date**	**Type**	**Specifics**	**Observed?**
1	25 June 1638	Lunar eclipse	Total. Partially visible from Recife	No measurements, because it rained all night with strong winds
2	20 December 1638	Lunar eclipse	Total. Entirely visible from Recife	Successful measurements of all essential times.
3	1 June 1639	Solar eclipse	Annular. Not visible from Recife	Expected, but invisible.
4	13 November 1640	Solar eclipse	Total. Partially visible from Recife	Beginning and end were roughly observed. Marggrafe's elaborate prediction is published in Barlaeus (1647)
5	18 October 1641	Lunar eclipse	Partial. Partially visible from Recife	No observation possible during expedition to Rio São Francisco
6	14 April 1642	Lunar eclipse	Total. Entirely visible from Fortress Ceulen	Partly observed in Fortress Ceulen, because of clouds
7	7 October 1642	Lunar eclipse	Total. Entirely visible from Recife	Partly observed in Recife, because of clouds
8	4 April 1643	Lunar eclipse	Partial. Partially visible from Recife	Partly observed in Recife, because of clouds
9	27 September 1643	Lunar eclipse	Partial. visible from Africa	No observation due to illness in Angola?

as an upcoming event in a letter to KECHEL,[59] this phenomenon would not be visible in Recife, so obviously no preparations for observation were made. During the solar eclipse of 13 November 1640 (the elaborate prediction of which is published in BARLAEUS's *Rerum per octennium in Brasilia*),[60] MARGGRAFE again had bad luck. Hindered by an overcast sky, he could only roughly pinpoint the start (10:30) and end (13:30) of the event. But even then, in Europe this eclipse was not visible, so in this case a comparison with other data was impossible anyway. The remaining lunar eclipses didn't yield anything useful either. During the lunar eclipse of 18 October 1641, he was on expedition in the area around the Rio São Francisco; the one of 14 April 1642 he observed from Fort Ceulen in Natal, but only partly, because of bad weather conditions. A clouded sky also disturbed a full observation of the lunar eclipses of October 1642 and April 1643, when he was prepared to witness these events at the Recife observatory. In all cases, it was impossible to measure the crucial times. So, in the end MARGGRAFE had no other choice than to trust the values of 1638, which he had obtained under the most favourable conditions.

In summary, we may conclude that MARGGRAFE indeed tried to collect more relevant data during the next eclipses, but that he failed to obtain them due to unfavourable weather conditions. In this respect, it is telling for his perseverance that one of MARGGRAFE's last notes from Brazil concerns a document in which he, probably in preparation for his trip to Angola, noted the dates of two eclipses he planned to observe in Africa.[61]

THE DETERMINATION OF THE SUN'S ANGULAR DIAMETER

In the morning of 28 June 1640, not long after the roof of the count's residence had been restored and the observatory could be used again, MARGGRAFE settled in its darkroom (below the turret with the quadrant) to determine the angular diameter of the Sun (*fig. 108*). Why he desired to do so is not recorded. This goal is not formulated in his programmatic *Progymnastica Mathematica Americana*. Perhaps he saw it as a preparation for the upcoming solar eclipse in the fall. But given the month of June in which he made this observation, it seems more likely that he had heard of a very topical astronomical discussion.

The value of the Sun's diameter was one of the subjects intensely discussed in the correspondence between MARTINUS HORTENSIUS and the French astronomer PIERRE GASSENDI in the years after 1634. Both scholars worked on a publication on the subject. HORTENSIUS's

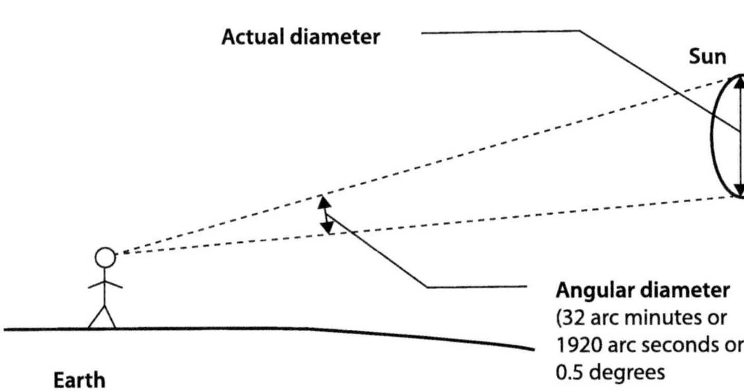

FIG. 108. The angular or apparent diameter of the Sun (From: Wikipedia).

book was halted by his sudden death in 1639, but in 1642 GASSENDI indeed published his *De apparente magnitudine solis humilis et sublimis* ('The apparent size of the Sun in low and high altitudes'). The problem was that the astronomer CHRISTOPH SCHEINER, in his famous *Rosa Ursina* of 1630, had argued that in the month of June, the Sun's angular diameter was at least 46', this in contrast to his mean value, for which SCHEINER assumed a diameter of 32', the same as TYCHO BRAHE had done.[62] HORTENSIUS did not understand how SCHEINER had been able to find such a divergent value: 'This man turns the diameter of the Sun into 42, 47, 52 or 56, etc. minutes, which is not true and also does not fit with what is found during the eclipses. And yet no one doubts that he perceived it, because of the honesty he shows in all his experiments', HORTENSIUS wrote to GASSENDI.[63] So, after 1634, HORTENSIUS submerged himself in various experiments to check this value. In October 1636, he could inform GASSENDI that his frequent examinations had led to the conclusion that 'the diameter of the Sun does not change even a tiny bit, at whatever height above the horizon this body is observed', except for some phenomena at very low elevations, related to the atmospheric refraction.[64] In February 1637, GASSENDI reported his final conclusion, namely that the Sun's angular diameter varied between 0° 30' 12" (when the Sun was in its apogee), and 0° 31' 06" (for the perigee position).[65] Only at the very low elevation of 3 degrees above the horizon he had found a value of 0° 28' 52", which confirmed HORTENSIUS's conclusion about the influence of atmospheric refraction.

It is against this background that MARGGRAFE performed his experiment to determine the Sun's angular diameter. To achieve this, he used the method of the pinhole camera in the same way as SCHEINER had done (thus projecting the Sun's image without a lens; see *fig. 109*).

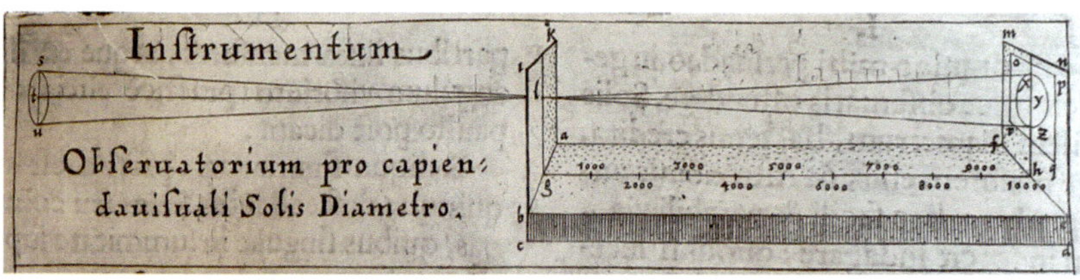

FIG. 109. CHRISTOPH SCHEINER's method to determine the solar angular diameter using a *camera obscura* with a pinhole. From: Scheiner, *Rosa Ursina* (1630), 575. (Leiden University Library; Photo by the authors).

FIG. 110. The pinhole in a blackened piece of vellum used by MARGGRAFE on 28 June 1640 to measure the angular diameter of the Sun. The note *Dist. 7340 tabell* / Diam. ....10' refers to the relative distance between the screen and the opening of the pinhole (ELO, North no. 79).

His further approach was as follows. On a piece of parchment, blackened on both sides (to reduce the scattering of light), a small hole was punched with an arbitrary measure of 10 divisions (the surviving parchment shows an opening of 3.1 mm; see *fig. 110*).[66] This 'pinhole' was placed in front of a small opening at the sunny side of the darkroom. Beforehand, a strip of paper was attached to a screen on which three pre-prepared small circles were drawn, with slightly different diameters (*fig. 111*).[67] By moving the screen in the direction of the pinhole, at some moment the Sun's image on the screen matched with one of the pre-drawn circles. At that time the distance to the hole was determined as 7,340 divisions. If we apply the dimension of the hole's opening (3.1 mm), this makes a distance of 2.28 meters, which fits well within the diameter of the darkroom. After all, this space had to be larger than the diameter of the azimuthal circle

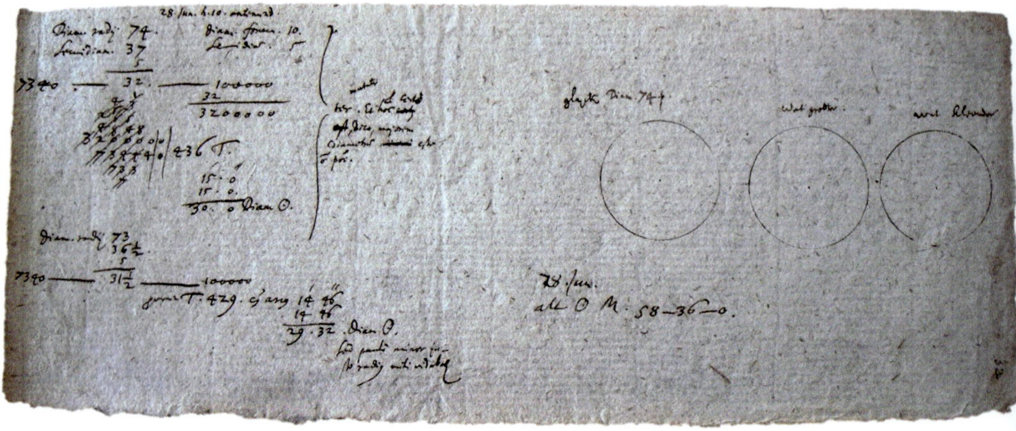

FIG. 111. Sheet used by Marggrafe on 28 June 1640 to measure the angular diameter of the Sun. The calculations to the left of the circles present Marggrafe's two results for the angular diameter of the Sun, 30' 00" and 29' 32". (ELO, North no. 32).

in the room above, which was 10 Rhijnland feet, or 3.14 meters. This procedure was repeated in the afternoon. By a straightforward calculation from the data of the two measurements, Marggrafe produced a value for the Sun's angular diameter of respectively, 29' 32" and 30' 00", which, in fact, is a bit too low; the currently accepted value of the Sun's diameter at the end of June 1640 was 31' 27".[68]

A noteworthy detail in one of the notes about this experiment is the fact that above the circles in *fig. 111* three words can be distinguished, which are clearly written in old Dutch: *glijck 74* [a mark for 'division']*; wat grooter; wat kleender* ('equal to 74 [divisions]; somewhat larger; somewhat smaller'). Although these words are put down in the same ink as the text written by Marggrafe, these three words may be written by a Dutch companion to this observation.[69] At least, not a single scrap of paper is known on which Marggrafe has written anything in Dutch. It is even questionable whether he ever used that language. Neither in the academic circles of Leiden University, nor in Dutch Brazil was a command of the Dutch language really necessary. The use of German or Latin must have been sufficient. These three Dutch words are therefore perhaps the only indication that, during his observations in Recife, Marggrafe may have been assisted by a Dutch companion. Perhaps this assistant was the unnamed *jongen* ('boy'), marked as such in the 1643 list of the retinue of JOHAN MAURITS.[70] But we are by no means sure of that.

CHARTING THE SOUTHERN STARS WITH THE TYCHONIC QUADRANT

The first objective for MARGGRAFE's intended *Progymnastica Mathematica Americana* was to compose an overview of all southern stars.[71] How far did he come in this ambition? The preserved manuscripts reveal that MARGGRAFE measured the elevation of 241 different stars during a meridian passage, observed in 780 different sightings. The overall mean error of the Recife quadrant for these meridian observations appeared to be much larger than for the Leiden quadrant (approximately 3.5 arc minutes for all meridian altitudes, compared to 49", or 0.82 arc minutes in Leiden), with a large scattering in the readings (resulting in a standard deviation of 9 arc minutes, compared to of 1' 34", or 1.57 arcminutes in Leiden). The mean error of the Recife quadrant was therefore approximately five times those of Leiden. This is an understandable difference. First, in Recife, MARGGRAFE was a victim of the poor quality of the woodwork on the mansion's roof. Furthermore, the quadrant from Recife will not have had the same quality as the Leiden quadrant, made in Amsterdam by a professional craftsman as WILLEM BLAEU.

In his notation of the stars, MARGGRAFE followed in general the current nomenclature, but for less conspicuous stars he used provisional designations, many of which he invented himself, which makes it more difficult to identify them. This all the more because MARGGRAFE died before he had reduced the recorded changeable observations of 'elevation' and 'azimuth' into the steady coordinates of 'declination' and 'right ascension'. In other words, he left us no catalogue of the observed stars. So, it is difficult to tell how many non-Ptolemaic stars he saw. During his observations MARGGRAFE noted on several occasions that a star was 'not on the globe', suggesting that he witnessed such a non-Ptolemaic star. But of the 18 different stars that received such a comment, our identification revealed that seven were already in HOUTMAN's star catalogue, leaving an independent observation of eleven stars that had not been seen before.[72]

All in all, this means that even a beginning of his aspired *Tabulae Mauritii Astronomicae* was not achieved.[73] This table obviously was intended to parallel KEPLER's *Tabulae Rudolphinae* for the Southern Hemisphere, evidently paying homage to Count JOHAN MAURITS. This is unfortunate, because MARGGRAFE had the necessary skills for these calculations. After all, making astronomical tables was something that he learned very well during his two years in Stettin helping Lorentz Eichstädt to construct his acclaimed astronomical ephemeris.

The obliquity of the ecliptic

Another aim formulated in the program of Marggrafe's *Progymnastica Mathematica Americana* was to determine 'a new figure for the obliquity of the ecliptic'.[74] This is the crucial angle between the plane of the earth's orbit (ecliptic) and that of the earth's equator. It is the same as the tilt of the Earth's axis of rotation and the perpendicular to the ecliptic. Marggrafe indeed achieved this goal at some moment in 1641. In the years 1639 and 1640, he had performed many observations, several during the day of the Sun's elevation at noon, and others at night of the Moon, bright stars and outer planets. In total, 603 meridian elevations were measured, 448 of the Sun and 155 of other bright nocturnal objects. Although Marggrafe did not have much spare time, next to his other work, to analyse his observations – this evidently was something he planned to do at a later date – he did find the time to extract from these systematic meridian observations his own number for the obliquity of the ecliptic. The result of his calculations is among the Leiden manuscripts. He found an angle of 23 degrees and 28.5 arc minutes (*fig. 112*).[75] This result is *not* included in the Paris and Lisbon manuscripts, which provide only his observations and do not include any calculations. Therefore, Marggrafe's number for the obliquity of the ecliptic was not known to the English Astronomer Royal John Flamsteed, who in 1668 received an extract of Marggrafe's observations. From these notes, Flamsteed also

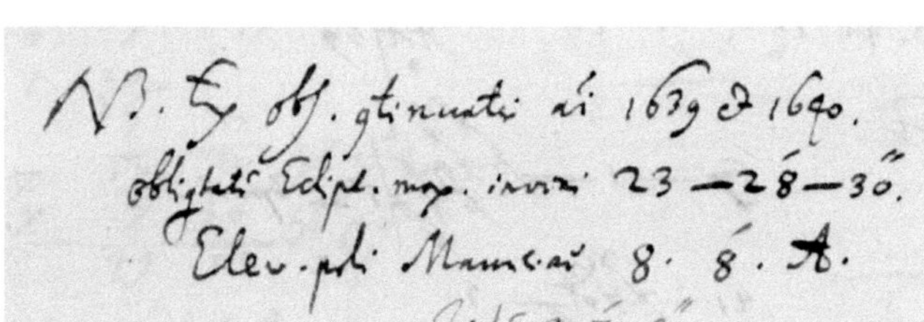

FIG. 112. The result of Marggrafe's calculations for the obliquity of the ecliptic, as well as the latitude of Recife, determined from his meridian observations over the years 1639 and 1640. (ELO, North no. 54, last page).

calculated the mean angle for the obliquity, using three groups of observations. His result? FLAMSTEED obtained the number of 23° 29', almost the same as MARGGRAFE's, and also very close to what is known today (23° 27').[76]

The determination of the Earth's radius

The second part of MARGGRAFE's intended observation programme would be 'geographic and geodetic in nature'. However, of all objectives formulated in this section, only one has been achieved, 'a demonstration of the true dimensions of the Earth', again 'based on his own observations'. He indeed delivered a number for the Earth's radius. How he collected the required data has not been passed on to us. He probably made measurements in the period between 9 February 1641 and 16 June 1642, when he was away 'for the sake of Geography, Chorography and Natural History'.[77] His result for the size of the Earth is published in the last chapter of the *Historia Naturalis*, a part added by JOHANNES DE LAET from several loose notes, which are since lost. DE LAET evidently did not grasp very much of what was in these scattered notes. His text prompted JOHN NORTH to the following sigh: 'Seldom can a more confused piece of writing have been committed to print'.[78] Still, as NORTH concluded: 'Without going into details, it is reasonably certain that embedded in the thinking behind the text there is a figure of 28,175 Rhineland rods per degree on the Earth's surface. This gives a radius of 6,081.1 km for the Earth'.[79] Compared with the modern figures of 6,378.4 km (equatorial) and 6,356.9 km (polar), this is a deviation of less than 5%, not bad for the middle of the seventeenth century.

'Accidental data and other astronomical rarities'

One of the goals formulated in the *Progymnastica Mathematica Americana* was to record 'accidental data' and 'other astronomical rarities'. MARGGRAFE indeed noted down a number of special phenomena that caught his eye. He left, for example, several notes relating to atmospheric dispersion. In his observations near the horizon, he assessed the limiting brightness of the background stars and wondered under which conditions the brightness of the sky's background caused the impossibility of seeing Mercury. For that reason, he also evaluated his ability to read a text without the aid of a candle, during dawn or dusk. At the moment he could not read anymore, he measured the position of a star to calculate *a posteriori* the dusk's duration. According to our calculations, this happened about 50 minutes after sunset, when the Sun was already 11° below the horizon.[80] By making these observations, MARGGRAFE showed an interest in phenomena that future astronomers would attribute to the scattering of light by small particles.

Intriguing, too, is the note made on Christmas, 25 December 1638, when MARGGRAFE observed one of the brightest lights in the Southern sky.[81] This concerned the star named *Fomalhaut* (currently α *Piscis Austrini*), a star in the constellation 'The Southern Fish', already described in PTOLEMY's *Almagest.* Its name came from the Arabic word 'Fom al-Haut' (فم الحوت), literally, 'mouth of the whale'. Given the date, it could hardly be otherwise than that he was inspired to this observation by the biblical story of the Star of Bethlehem.[82] Was he demonstrating this bright star to others? Perhaps even to JOHAN MAURITS himself? Such a gesture would certainly fit the support the count gave to MARGGRAFE's project by ordering for him the construction of an astronomical observatory as a superstructure on top of his own house.

WHAT PART OF THE *PROGYMNASTICA MATHEMATICA AMERICANA* WAS ACHIEVED?

MARGGRAFE's observational program remained, understandably, very incomplete. A lot of his promises remained wanting. There is, as we saw, not even a beginning of his *Tabulae Mauritii Astronomicae.* Neither can anything be found of a 'theory of astronomical refraction and parallax'. And from his promised 'new and true theory' for the orbits of the planets Venus and Mercury, 'based on own observations', he got stuck in the observation phase. So the question is relevant: which part of the programme for the *Progymnastica Mathematica Americana* was achieved? To our surprise, the harvest is not disappointing. Given MARGGRAFE's many other obligations and duties, added to his untimely death, it is amazing how much he achieved. He had formulated ten goals for his *Progymnastica Mathematica Americana.* Let's check, point by point, what he desired to achieve and what he actually achieved.

MARGGRAFE's goals were to obtain:

I-i *An overview of all the southern stars.*
- ✓ Only the observational section is partly achieved.
- o The reduction of the observed 241 different stars had not even started.

I-ii. *Observations of all the planets and of eclipses of the sun and moon, 'worked out in an original way'.*

✓ Only the observational section is partly achieved. All eclipses MARGGRAFE could observe, are observed. Planetary observations have been made only sparsely. The Sun's angular diameter is 'worked out', as well as the calculation for the prediction of the solar eclipse of 1640.

○ For this subject there was still work to be done.

I-iii. *New and true theories of the inferior planets Venus and Mercury, based on own observations.*

○ This goal is NOT achieved. New elements for the planets are NOT calculated.

I-iv. *A theory of astronomical refraction and parallax.*

○ This goal is NOT achieved.

I-v. *A new figure for the obliquity of the ecliptic.*

✓ This goal is completely achieved.

I-vi. *Accidental data, not only on sunspots, but also on other astronomical rarities.*

✓ This goal is reasonably achieved.

II-i. *A theory of the longitude of the earth and a manner of computing the same.*

✓ This goal is completely achieved. The calculation of Recife's longitude, based on a lunar eclipse, was methodically correct.

II-ii. *A demonstration of the true dimensions of the earth, based on his own observations.*

✓ This goal is achieved.

II-ii. *A discussion of previous geographical errors.*

○ This goal is NOT achieved, although some minor fragments were published in the 'Historiae Rerum Naturalium Brasiliae'.

III. *The composition of the Tabulae Mauritii Astronomicae.*

○ This goal is NOT achieved.

Overall, we may conclude that MARGGRAFE, despite the fact that he had far less time available for astronomy than he had hoped, still managed to fulfil <u>*six*</u> out of the <u>*ten*</u> goals he had formulated for himself. In this assessment we have to realize that, as NORTH already remarked, 'MARGGRAFE simply did not live long enough to accumulate or assemble the materials for the promised book'.[83] Elsewhere NORTH noted, 'there is so little theoretical reduction of the observations as to make the material no more than a basis for further work'.[84] Apart from the determination of Recife's coordinates, the obliquity of the ecliptic, and the Earth's radius, there is not even the beginning of a theoretical elaboration of his collected observations. This although 'all the evidence suggests that he planned to produce either a whole new theory, or at least new sets of parameters, for Mercury, Venus, and the Sun'. Nevertheless, it is clear that MARGGRAFE was engaged in the hot astronomical topics of his time and took charge of serious and promising observational demands to clarify them. Had he survived Africa, it is legitimate to assume that he would have worked out the theoretical requirements. His project for the *Progymnastica Mathematica Americana* was well thought out and entirely feasible.

MISSING OBSERVATIONS

THE FAMOUS 1639 TRANSIT OF VENUS

What phenomena did MARGGRAFE not observe? This is an interesting question, with in one case a very regrettable answer. Like so many others, MARGGRAFE did not observe the first known transit of Venus in December 1639, but by sheer coincidence he almost did! The story is as follows. The possibility of a Venus transit was suggested by KEPLER in the same pamphlet in which the 1631 Transit of Mercury was announced. In his small tract *De raris mirisque Anni 1631,* KEPLER had included a 'warning to all astronomers', announcing not only the Mercury transit, but also the transits of Venus, occurring in 1631 and 1761. However, caused by a flaw in the Rudolphine tables, he did not predict the Venus transit of 4 December 1639. Only a few weeks before the transit would actually occur, this mistake was discovered by the English astronomical enthusiast JEREMIAH HORROCKS. Working rather in isolation, HORROCKS only communicated this finding to his friend WILLIAM CRABTREE, living nearby. Both men indeed observed the Venus transit, each from his home. They were the first and only persons in the world to witness this event.[85]

However, by sheer coincidence, MARGGRAFE came very close to becoming the third observer. In November and December 1639, he measured *every day* the elevation of the Sun at noon. He also intended to do so on Sunday 4 December 1639, the very day of the – for him unexpected – transit. According to our calculations, seen from Recife, Venus would enter the solar disk at 11:57, when the elevation of the Sun was 71.8°.[86] So, during MARGGRAFE's measurement of the Sun's elevation, Venus would actually have entered the Sun's disc. Contrary to Mercury, the planet Venus, with an angular diameter of just over one arc minute, would be clearly visible with the naked eye; projected on a screen this result would even have been unavoidable. We therefore may conclude that in case of a bright sky, MARGGRAFE had a good chance of discovering the dark spot on the Sun. From the speed of the spot across the Sun's disc, he also would have realized that this was no regular sunspot, but the planet Venus. But, for the umpteenth time, he was affected by bad luck: it was stormy weather and the sky was filled with clouds. So, the planned measurement was never carried out. Luckily, MARGGRAFE never knew how close he came to this unique experience.

Nevertheless, this bad weather condition on 4 December 1639 was really unfortunate for the whole astronomical community. Had MARGGRAFE succeeded in this observation and had measured the times of egress, it would have enabled later astronomers to combine his measurements with those of HORROCKS and CRABTREE. From these parallel sightings of the transit, it would have been possible to calculate the first real distance to Venus. With the knowledge of that value, the third law of Kepler would have provided the distance between the Sun and the Earth. Now, the establishment of this crucial distance in the Universe, also known as the Astronomical Unit or Solar Parallax, had to wait until the next transits of Venus in the years 1761 and 1769.[87]

SUNSPOTS

Striking, too, is the lack of sunspot observations, a subject he intensely studied in Leiden. At first sight, it seems a mystery why he omitted these observations in Recife. After all, he did have a darkroom in the observatory, and he had the expertise to perceive and record them. Moreover, had he not said in one of his letters to GOLIUS that his telescope possessed 'a good glass to detect and display [the Sun] in a closed dark room'?[88] However, we have to realize that the dark room became available only after September 1639. In that year, JEREMIAH HORROCKS in England was the only astronomer in the world who reported one single sighting of some minor sunspots, during his famous observation of the Venus transit when MARGGRAFE's observation register states it was stormy weather in Recife for many days.[89]

In the next year, 1640, sunspots were seen in Europe only between 21 August and 6 September. This was indeed in a period in which MARGGRAFE was very engaged in observing ... but mostly at night! In day time, he probably was busy in performing the tasks for his – then still – employer, WILLEM PISO, who was one of the attendees of the Legislative Assembly. This first form of parliamentary consultation in the history of Southern America was a meeting between governor JOHAN MAURITS and his High Council with representatives from all walks of life in Dutch Brazil. The meeting was held in Recife between

27 August 1640 and 4 September, the very days on which in Europe the sunspots were sighted.

And even then, we have to remember most of the time sunspots cannot be seen with the naked eye. One has to check their existence by projecting the sun image on a screen. This takes time. MARGGRAFE's observation register reveals that even on these busy days of the Legislative Assembly, he tried to measure the Sun's elevation at noon. He succeeded in taking this single measurement five times between 20 August and 10 September 1640, but complained in this same period *twelve* times that it rained during the day, or that storm and clouds were making his observations difficult. So, we may forgive him for not having checked by projection whether or not there were sunspots visible at the time.

Further, no European astronomer reported any sunspot in the year 1641, and when, finally, some sunspots were visible again in June 1642, *nota bene* the very moment at which MARGGRAFE returned from an expedition to inland Brazil, he simply had no time anymore for regular astronomical observations. Then, more pressing duties had to prevail, such as the construction of his maps of Dutch Brazil and redeeming his promise to JOHANNES DE LAET that he would write down everything he had seen of Brazilian nature in his manuscript *Historiae Rerum Naturalium Brasiliae,* later becoming part of the *Historia Naturalis Brasiliae.*

Some planetary riddles

Relating to the planets, there are in this section about missed observations two comments to be made. Firstly, it's remarkable that in the Paris manuscript only one observation of Jupiter's satellites can be found.[90] Given MARGGRAFE's knowledge of GALILEI's method to use these moons as a heavenly clockwork, one would have expected a series of such observations, just to check whether or not this method would work.[91] This raises the question was his telescope perhaps of a lesser quality than the one he used in Leiden?

This assumption aligns with our second observation, namely that MARGGRAFE never made any comment on the shape of Saturn. Since the existence of the telescope, this planet was seen by GALILEI and many others as 'having two handles'. In 1638, the year when MARGGRAFE started his observations in Brazil, this kind of appearance was very clearly drawn by the Italian astronomer FRANCESCO FONTANA in Naples.[92] In 1642, however, to the surprise of many astronomers, Saturn suddenly seemed to have lost its extremities. This 'solitary' appearance of Saturn was first observed by GASSENDI, but after that confirmed by men all over Europe. Today this phenomenon is understood as the coinciding of the line of sight from Earth with the plane of Saturn's ring. But before CHRISTIAAN HUYGENS's understanding that Saturn possessed a ring, this 'disappearance' was a sheer mystery. It is an intriguing question: Did MARGGRAFE miss this phenomenon for lack of a suitable telescope, or because he just didn't pay enough attention to Saturn anymore?[93]

COMETS AND SUPERNOVAE

Other phenomena that MARGGRAFE does not mention are sightings of comets and supernovae, but that lacuna is easily explained: there weren't any! The only comet spotted in Recife appeared in December 1652, when MARGGRAFE was long gone. Unfortunately it is not known who made the surviving drawing of the comet (and where), but it is remarkable that for that observation a telescope was used, which instruments were not yet that common at the time (*fig. 113*).[94] This comet was also observed by HENDRIK HAECX, starting on 15 December, a day

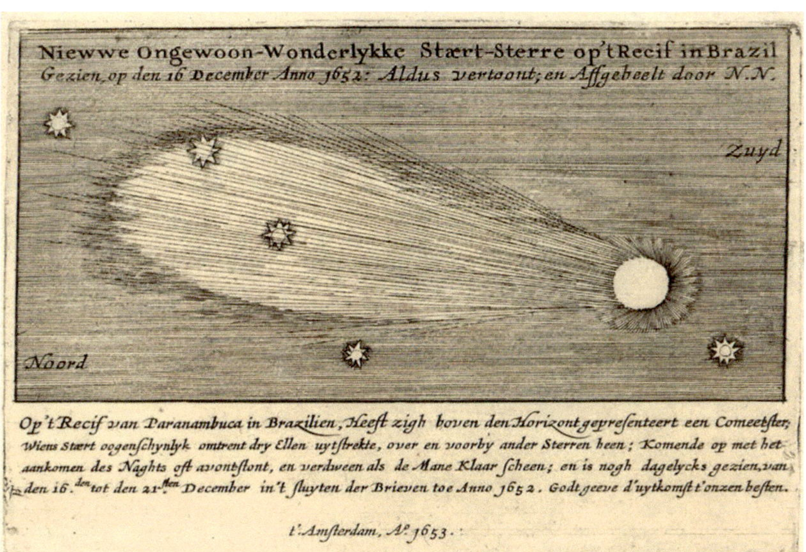

FIG. 113. First known sighting of the bright comet of 1652, observed through a telescope (!) between 16 and 21 December by an anonymous observer in Recife. The comet was also seen from the Northern Hemisphere and in South Africa. (Koninklijke Bibliotheek, Pamphlet 7353)

before the date on the engraving, so he appears not to be the person who made the drawing.[95] HAECX was a member of the High Council since 1646, being appointed by the *Heeren XIX* when he resided in Amsterdam. But he lived in Recife as a 'free merchant' in the early 1640s.[96] History doesn't reveal whether HAECX ever saw MARGGRAFE at work, so it's unclear whether MARGGRAFE eventually gathered him as a follower. In that regard it's also noteworthy that years later, in 1664, MARGGRAFE's former acquaintance, ZACHARIAS WAGENER, also is noted as the observer of a comet, then in Dutch-governed South Africa.[97] Here, too, the question arises whether WAGENER's interest in astronomical matters went back to his old days in Recife.

A MEETING WITH CASPER SCHMALKALDEN?

With respect to other possible astronomical observers in Dutch Brazil, it is intriguing to look more closely to the case of CASPAR SCHMALKALDEN (1616-1673) from Thüringen, who served as a soldier in Brazil between 1642 and 1645. His biographer and first editor of his travel journal, WOLFGANG JOOST, claims that SCHMALKALDEN came to Dutch Brazil after studying astronomy at Groningen University.[98] Later in life, when he was employed by the Dutch East India Company (VOC), SCHMALKALDEN also served as a surveyor, just as MARGGRAFE did for the Dutch West India Company (WIC). So the interesting question arises: did these two kindred souls meet?

SCHMALKALDEN arrived in Recife on 11 December 1642, along with 36 other soldiers and four women, aboard the merchant ship *Den Oliphant* ('The Elephant'). But because he joined the WIC fleet to Chile under the command of HENDRIK BROUWER that left Recife on 12 Januari 1643, he stayed in the capital of Dutch Brazil for only one month.[99] But fortunately, this was a month we know for sure that MARGGRAFE was also in Recife, busy preparing his map of Dutch Brazil. As the observatory was a prominent landmark in Recife, we can be sure that SCHMALKALDEN has noticed the building. Whether or not he actually met his compatriot MARGGRAFE over there, that, unfortunately, remains hidden in the mists of history. In any case, the single isolated observation of a solar altitude recorded on 20 December 1642 is striking. As if this was a one-time demonstration. With a little imagination we can indeed read this event as the meeting of two astronomically interested men. Anyway, when SCHMALKALDEN returned to Recife in December 1643, the observatory was deserted, and MARGGRAFE had already perished in the deluge of African fevers.

The French astronomer ALEXANDRE GUY PINGRÉ (1711-1796) was the first who really studied the Paris manuscript with MARGGRAFE's observations. After having made a partial analysis of the meridian observations (150 data points), he realized that there was a periodization in the observations.[100] He noticed what seemed to be a stepwise change in the latitude of the observatory. In this way, he distinguished the following three periods:

Period I From the opening of the observatory on 15 September 1639 until its collapse on the night of 18 March 1640.
Period II From when the observatory resumed its operation on 11 June 1640 until MARGGRAFE's departure on 7 February 1641.
Period III From the return of expeditions on 3 October 1642 until the last observation on 22 June 1643.

Based on the current geographic coordinates of the observatory, we have checked PINGRÉ's hypothesis by calculating the difference between the observed and the calculated elevations for all meridian elevations observed by MARGGRAFE.[101] For our analysis, the elevations were corrected for the atmospheric refraction. Then, we made the statistical analysis of the errors for all these measurements.[102] The graphic result of the statistical analysis is presented in *Fig. 114*.

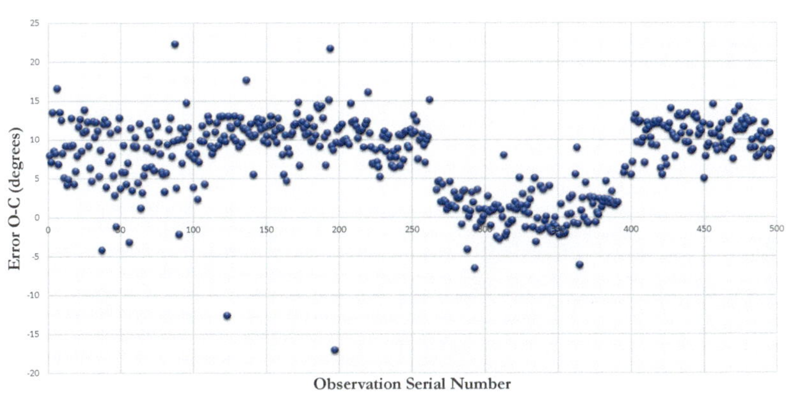

FIG. 114. The vertical axis displays O-C (difference between observed and calculated values) of the meridian elevation in arc minutes. The horizontal axis displays the serial number of the successive measurements. The stepwise behaviour of the error appears indeed correlated to PINGRÉ's periodization: Period I (September 1639-March 1640); Period II (June 1640-February 1641) and Period III (October 1642-June 1643).

The stepwise behaviour of the difference (O – C) indeed correlates to PINGRÉ's periodization. While the uncertainty did not vary much from period to period, the systematic difference does. Obviously, the larger difference in period I may be associated with a structural problem in the roof of the count's house, on top of which the observatory was built.[103] The partial collapse, at the end of period I, is enough proof of the roof's instability. However, period II, after the repair, shows a significant reduction of the systematic difference. But over time, in Period III, which started in October 1642, when MARGGRAFE had not used the observatory for more than one and a half years, the larger differences returned. This is in line with the archival information that over time the condition of the count's first residence had deteriorated again. When in January 1643 the city council of Mauritsstad requested the usage of the abandoned residence of the count to hold their meetings, it appeared that 'his Excellency's old house' was so dilapidated that without repair the building could not be used as their town hall (*Raethuijs*).[104] This information corroborates our hypothesis that the instability of the roof jeopardized the quality of MARGGRAFE's observations.

JOHN NORTH, sparked by PINGRÉ's periodization, suggested in 1979 that between Period II and Period III, MARGGRAFE's observatory was relocated from the count's first residence to one of the towers of the new residence *Vrijburg*.[105] This palace was put into service in June 1642, when MARGGRAFE had just returned from a lengthy expedition.[106] In doing so, NORTH connected his knowledge that MARGGRAFE started to observe before *Vrijburg* was built, with an assertion that had been around for almost a century, namely that *Vrijburg* eventually became the location of MARGGRAFE's observatory. But where did this idea come from? *Vrijburg* palace was already described by BARLAEUS in 1647, who said that the two towers 'served as a landmark to sailors', especially because one of the towers bore a light in the top and helped sailors in this way to find a safe entrance to the harbour.[107] The other tower was used as a lookout to 'see the surrounding stretch of land, and any ships that might be approaching'. During the day, BARLAEUS continued, the towers also provided a convenient way for the guards to keep an eye out for pirates. This account is corroborated by eye-witness CASPAR SCHMALKALDEN, who as a person interested in astronomy, would certainly have reported if one of the towers had been used as an astronomical observatory.[108] And finally NIEUHOF, another eye-witness, confirmed that the two towers 'could be seen five or six miles from the sea' and were used as 'a beacon to the sailor'.[109]

The idea that one of *Vrijburg*'s towers would have accommodated an astronomical observatory is not mentioned until 1899, when FRANCISCO AUGUSTO PEREIRA DA COSTA (1851-1923), a Brazilian lawyer, journalist, historian and politician made this suggestion in an article about the meteorological observations MARGGRAFE made during his time in Dutch Brazil.[110] A few years later this suggestion was adopted as a reality by ALFREDO DE CARVALHO (1870-1916), an engineer and historian from Recife. After mentioning that JOHAN MAURITS built *Vrijburg* palace, CARVALHO continues to say that the count established an observatory in it, the first in the Southern Hemisphere, 'where MARGGRAFE spent the clear tropical nights contemplating the wonders of the meridional skies'.[111] In 1912, GUDGER followed in his footsteps when he wrote: 'This building [*Vrijburg*] had two towers which were visible six to seven leagues at sea and which served as beacons to the mariners. It was probably one of these which Margrave used as an observatory'.[112]

Despite the cautionary 'probably', it seems that by then the myth of the observatory located in one of the towers of the *Vrijburg* palace was already accepted as a historic reality. From then on, this myth was systematically repeated and thus widespread, for example, in 1914 by the Brazilian biologist RODOLPHO VON IHERING (1883-1939);[113] in 1917 by the Afro-Brazilian physician and psychiatrist JULIANO MOREIRA (1872-1933);[114] in 1920 by the historian and diplomat MANOEL DE OLIVEIRA LIMA (1867-1928);[115] in 1921 by the German historian HERMANN WÄTJEN (1876-1944);[116] in 1942 by the Brazilian historian AFFONSO TAUNAY (1876-1958);[117] in 1955 by the astronomer ABRAHÃO DE MORAES (1916-1970),[118] and finally in 1979 by JOHN NORTH, followed by PETER WHITEHEAD.[119] Since then, many others have followed them.

All in all, the story of an astronomical observatory in a *Vrijburg* tower became so ingrained that this point of view is virtually ineradicable today. It is now even mentioned as an established fact in the official *Biographical Encyclopedia of Astronomers* (2014).[120] This all in spite of the fact that already in 1979 the Dutch-Brazilian priest and astronomer JORGE POLMAN (1927-1986),[121] during a discussion with NORTH at a conference in Recife, argued that NORTH's assertion has no basis in the available evidence.[122] NORTH, however, insisted on his position, and in the Portuguese excerpt of his talk he added even a third location when he wrote:

> After the collapse of the house where Markgraf made his first observations [1], it appears that he had successively built two observatories. The first, on top of an old Portuguese house in which the prince resided [2]; the second, a magnificent building, on top of one of the towers of the Vrijburg Palace [3], a beautiful residence erected on the island of Antonio Vaz, surrounded by a beautiful garden park ...[123]

So, for each period distinguished by PINGRÉ, NORTH attributed at that moment its own location for MARGGRAFE's point of observation. However, in 1989, in an addition to a reprint of his 1979-article, NORTH finally admitted that 'Mr POLMAN's suggestion offered distinct possibilities', although he still did not correct his earlier point of view. All things considered, however, there is no reason whatsoever to conclude that MARGGRAFE ever used *Vrijburg* as a point of observation. Firstly, in the available written documents, there is no mention at all of a relocation of the observatory. On the contrary, the Leiden, Paris and Lisbon manuscripts present a chronologically uninterrupted narrative. Secondly, a displacement of some 500 meters, corresponding to a variation of 0.27' in latitude, is absolutely unnoticeable within the noise of the much larger errors (see *fig. 114*). PINGRÉ's periodization provides no proof at all for such a hypothesis. Moreover, such a laborious relocation of the observatory would make no sense. In the summer of 1642, JOHAN MAURITS already had manifested his dissatisfaction to his superiors, the *Heeren XIX*, and had requested his return to Europe. It is difficult to imagine that MARGGRAFE, whose work in Brazil fully depended on the personal support of the count, would undertake such laborious task of moving the observatory to the palace. Rather, he would have desired to accelerate the observations, even in precarious conditions, in order to fulfil his promise, written down in his observation register in December 1641, that he – the Lord willing – would 'finish all that still remains to be done in this part of the world'.[124] Therefore it is our conclusion that the hypothesis of *Vrijburg* palace being used as astronomical observatory is completely false.

THE FATE OF THE GOVERNOR'S OLD RESIDENCE

After the relocation of JOHAN MAURITS's court to *Vrijburg* in June 1642, the old residence had to be renovated for a second time. The repairs made after the partial collapse of the building in March 1640 had proved to be inadequate. Over time, the edifice deteriorated further. We already discussed that when MARGGRAFE returned to use the observatory again in October 1642, after an interval of more than a year and a half, our analysis showed that the large differences between the measured and actual positions of the celestial bodies returned. In January 1643 it became clear that the abandoned residence still needed a makeover before it could be used as the town hall (*Raethuijs*) for Mauritsstad.[125] So this renovation was indeed carried out. The view of Recife on the 1648 wall map by CORNELIS GOLIATH (in the lower left corner of *fig. 55b*) shows the building renamed as *'t Raethuys ofte out Hof* ('Town hall or old residence'). The old residence kept that function until the conquest of the colony by the Portuguese. When Dutch Brazil finally capitulated in January 1654, it was precisely in this building with its symbolic significance as former centre of government where the surrender ceremony took place. The triumphant Portuguese army commander FRANCISCO BARRETO DE MENESES, who arrived with a retinue of 70 horsemen and four thousand infantrymen, took up residence in this former court, next to the strategic bridge to Recife. One can imagine how he, accompanied by the defeated Dutch commander SIGISMUND VON SCHOPPE, watched his conquest from the platform on the further dismantled observatory. There was his catch, the city of Recife, returned to Portuguese rule after 24 years.[126]

The history of the observatory after 1654 is shrouded in mist. In the inventory of all the houses in Recife, drawn up in 1654, it is clear that the ownership of the complex on Antonio Vaz (including several houses built by the Dutch) was claimed by the Portuguese captain BELCHIOR ALVEZ CAMELLO, as being the heir of PEDRO ALVARES, who owned the property before 1630.[127] When the observatory on top of the building is dismantled is unknown. The building itself seems to have survived well into the nineteenth century. Today, the property is divided into many houses, mostly with late nineteenth-century facades (see Chapter 8, *fig. 87b*). According to the Brazilian architect JOSÉ LUIZ MOTA MENEZES, who made a reconstruction of 'the House of His Excellency' (Chapter 5, *fig. 35*), it is plausible that some of its inner walls are still present in the buildings that now stand on this site.[128] Only future archaeological research may reveal more about this.

THE QUALITY OF THE INSTRUMENTS USED IN BRAZIL

THE 5-FOOT QUADRANT

At the end of this review of the observations made in Dutch Brazil, we have to consider for a moment the performance of the instruments used in Recife. As *fig. 114* shows, the observational difference (O – C) between the observed and calculated values of the meridian elevations measured with MARGGRAFE's 5-foot quadrant were optimal only in Period II, when the governor's residence was reinforced after its partial collapse in March 1640. The mean error in that period was 8.9 arc minutes (or 533 arc seconds), which compared with the performance of the Leiden quadrant (49 arc seconds for all meridian altitudes) means that the Leiden instrument performed almost 11 times better than the one in Brazil. But, of course, we have to realize that the Leiden quadrant was constructed by WILLEM JANSZ BLAEU, a very skilled instrument maker, who had learned to build instruments in Uraniborg from TYCHO BRAHE himself. In his workshop in Amsterdam, BLAEU had materials and skilled labour available that MARGGRAFE by no means could get in Recife.

MARGGRAFE'S TELESCOPE

Little is known about the telescope MARGGRAFE used in Recife. It probably was one with drawtubes made of tin-plated iron, with removable lenses. That was the usual design at that time for telescopes made in the Dutch Republic (*fig. 115*).[129] We have seen that the Leiden telescope had a theoretical field of view of almost 22 arc minutes, and an actually measured field of view of 18 arc minutes. From Recife, there is only one observation from which we may conclude that the field of view of MARGGRAFE's own *tubus opticus* was approximately 11.25 arc minutes.[130] This is somewhat smaller than the one from GOLIUS, but perhaps MARGGRAFE's telescope had a larger magnification. However, in the discussion of the sightings of Saturn we already have expressed the assumption that MARGGRAFE's telescope (in terms of the quality of the imaging) was of a lesser quality than the one in Leiden.

FIG. 115. Telescope with Dutch or Galilean lens configuration, most likely made in Holland around 1650, with drawtubes made of tin-plated iron and removable lenses (Skokloste Castle, Sweden; Photo by the authors, 2012).

THE PENDULUM

When the observatory was inaugurated, MARGGRAFE started using the sandglass (*clepsydra*).[131] There are notes of time intervals given in minutes. Our analysis showed that the errors were rather large. Apparently for that reason, after three days he abandoned the sandglass and used only the pendulum. But this device – described by MARGGRAFE as a chord with a length of 29 Rhijnland inches (= 75.7 cm), weighted with a turned metal cylinder of 41.75 ounces (=1.29 kg) – was also problematic.[132] PINGRÉ was the first to note that the duration of the swings of MARGGRAFE's pendulum varied over time, leading him to the conclusion that he started with a longer chord and shortened the length along the way.[133] As NORTH has explained, this has no influence on the determination of the moment of measurement of – for example – the position of Mercury, when before and after that observation the elevations of a star passing through the meridian are measured, keeping the same pace of the pendulum.[134] If the coordinates of these stars were known from the Rudolphine tables, MARGGRAFE could calculate *a posteriori* the actual time of observation.

However, when he desired to connect these measurements with the *Hora Urbica*, the local time, the length of the pendulum did matter. In Recife, this time was provided by an iron clock below the platform that 'by the stroke of the bell announced the time to all inhabitants of the city'.[135] This local time had to be checked in daytime with a sundial. This is exactly what MARGGRAFE did on 20 and 21 June 1640, when he measured several times the length and the direction of the shadow of a gnomon (vertical rod) cast by the Sun on the flat ground. To check his readings on the sundial, he made extra-meridional observations of the Sun with the 5-foot quadrant on the same days (and also on 7 October), measuring the elevation and azimuth of the Sun several times, supplying sometimes the local solar time. But such measurements were scarce, and consequently, the statistics of his time measurements are rather poor.[136]

At the end of this survey of MARGGRAFE's Brazilian observations, one final remark about the pendulum. In his survey of the recorded observations, PINGRÉ dismissed those made with the pendulum on 24 September 1639. In his view, those measurements led to incomprehensible results. Our recalculation, however, shows that these data make sense when one supposes that at that date a swing of the pendulum, back and forth, was counted as one pulse. MARGGRAFE himself would never have changed the way of counting. Therefore, this result provides a second indication that in Recife MARGGRAFE sometimes used an assistant who counted the swings.

3. Is Marggrafe's astronomical legacy ever applied?

It is almost unbelievable that – as far as we could establish – only one of the previous owners of Marggrafe's manuscript (Philippe de la Hire) has ever used them. The other owners, namely Golius, Thévenot, Couplet, Godin and Ulloa, never mentioned Marggrafe's observations in their publications. The first astronomer who consulted the Brazilian observations was the French scholar Ismaël Boulliau (1605-1694), who made a handwritten copy of the first three years of Marggrafe's notes.[137] When he did this is not clear, perhaps as early as 1662, when he visited Jacob Golius and Samuel Kechel in Leiden.[138] Another opportunity presented itself in 1668, when Boulliau paid a visit to Thévenot at his mansion in Issey.[139] We know that in that same year at Thévenot's residence, Edmond Halley also copied some of Marggrafe's observations for the English astronomer Royal, John Flamsteed.[140] But, just as the others who had access to Marggrafe's observations, neither Boulliau, nor Halley seem to have used them.

In Halley's case this is the more remarkable, because he of all people would be the astronomer who finally would reach the goal that Marggrafe had aspired to accomplish: publishing a genuine catalogue of the southern stars. In 1677 Flamsteed would send Halley to island of St Helena, in the middle of the Atlantic, with the assignment to chart the fixed stars of the Southern Hemisphere. Halley's *Catalogus Stellarum Australium* was published in 1679, as an addition to Tycho Brahe's star catalogue, just as Marggrafe had wanted to do.[141] But in the introduction Marggrafe's name is missing among the few remarks Halley devoted to earlier attempts to chart the Southern Hemisphere. Halley fully credited Frederick Houtman and Willem Blaeu for their contributions, but he mentioned only in passing that he had heard from a certain 'Lord Maurice', who as Jean Richer in the 1670s, had attempted to find some new southern stars.

Ironically, it was also Halley who during his two expeditions in the years 1699-1700 literally followed in Marggrafe's footsteps. In March 1699 he visited *João Pessoa*, the capital of Paraiba (formerly known as *Frederikstad*), and in May 1700 he paid a visit to Recife, without realizing that Georg Marggrafe had observed the southern stars from that very spot. In Recife Halley even was received by the local Portuguese governor, in his residence, the *Pallacio das duas tores*, the Portuguese name given to 'Lord Maurice's' former palace *Vrijburg* (*fig. 116*).[142]

FIG. 116. The *Pallacio das duas tores*, or the remains of *Vrijburg* Palace, visited by Edmond Halley in May 1700, when he was received by the Portuguese governor of Recife. Here the depiction of the largely dismantled building, as drawn by James Henderson and published in his *A History of the Brazil* (1821), 392. The two towers and the wings of the building were already demolished by the Dutch in 1646, to create a larger firing range for their cannons. (Biblioteca nacional de Portugal).

In Paraiba, HALLEY observed the satellites of Jupiter with a – then current – long compound telescope, which had a far better resolution than those with a Galilean or Dutch configuration as used by MARGGRAFE. In Recife HALLEY also determined the coordinates of the city. For the latitude HALLEY obtained 8° 10' south, remarkably (given the improved equipment) a value worse than the one established by MARGGRAFE sixty years earlier. His value, then, was 8° 7' 30", while the presently accepted value is 8° 3' 51".[143] For the longitude HALLEY obtained 35° 30' west from London (= Greenwich) a difference of 0.62 degrees with the current value of 34° 52' 37". In December 1638, MARGGRAFE had obtained a difference with the two most accurate English observers of respectively 0.9 and 0.2 degrees (*see table 1*), a result thus comparable to HALLEY's.

According to LALANDE, it was PHILIPPE DE LA HIRE who, for the first time, would really apply MARGGRAFE's observations, when he constructed his famous astronomical tables, dedicated to the French King Louis XIV.[144] Indeed, in the first edition of his *Tabularum Astronomicarum* (1687), there is a note stating that THÉVENOT provided him with MARGGRAFE's observations made in 'la nouvelle ville Maurice dans l'Isle d'Antoine Vaas' (*fig. 117*).[145]

From these observations, DE LA HIRE deduced the meridian height, both during the summer solstice and during the winter solstice. This led him to a value for the tilt of the Earth's axis (its obliquity) of 23° 29' 12". A similar result was obtained earlier at the other side of the Strait of Dover, by the English astronomer JOHN FLAMSTEED. But this value was only revealed in 1725, in FLAMSTEED's *Prolegomena*, the posthumously published third volume of his famous *Historiae Coelestis Britannicae.* Using three different methods, FLAMSTEED had derived from MARGGRAFE's observations a mean angle for the Earth's obliquity of 23° 29' 0", a value indeed corresponding with what is known today.[146] So, with these two publications at least some results of MARGGRAFE's observations found their way to the wider world. However, other results stayed indoors. The archives of the *Observatoire de Paris* reveal that the commissioner of the copies of MARGGRAFE's observations, JOSEPH-NICOLAS DE L'ISLE, made several extracts, showing his interest in subjects such as atmospheric refraction and the apparent mean height of the summer and winter solstices, leading to DE L'ISLE's own result for the Earth's obliquity of 23° 28' 46".[147]

In other words, a value for the obliquity that MARGGRAFE had calculated personally, but which his Leiden editors, due to inattentiveness, had *not* included in their press-ready manuscript of

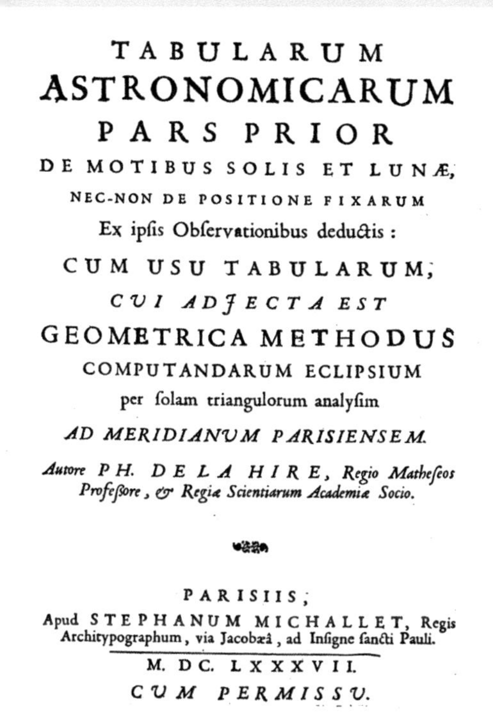

TABULARUM
ASTRONOMICARUM
PARS PRIOR
DE MOTIBUS SOLIS ET LUNÆ,
NEC-NON DE POSITIONE FIXARUM
Ex ipſis Obſervationibus deductis :
CUM USU TABULARUM,
CUI ADJECTA EST
GEOMETRICA METHODUS
COMPUTANDARUM ECLIPSIUM
per ſolam triangulorum analyſim
AD MERIDIANUM PARISIENSEM.
Autore PH. *DE LA HIRE*, *Regio Matheſeos Profeſſore*, *&* *Regiæ Scientiarum Academiæ Socio.*

PARISIIS,
Apud STEPHANUM MICHALLET, Regis Architypographum, via Jacobæâ, ad Inſigne ſancti Pauli.
M. DC. LXXXVII.
CUM PERMISSU.

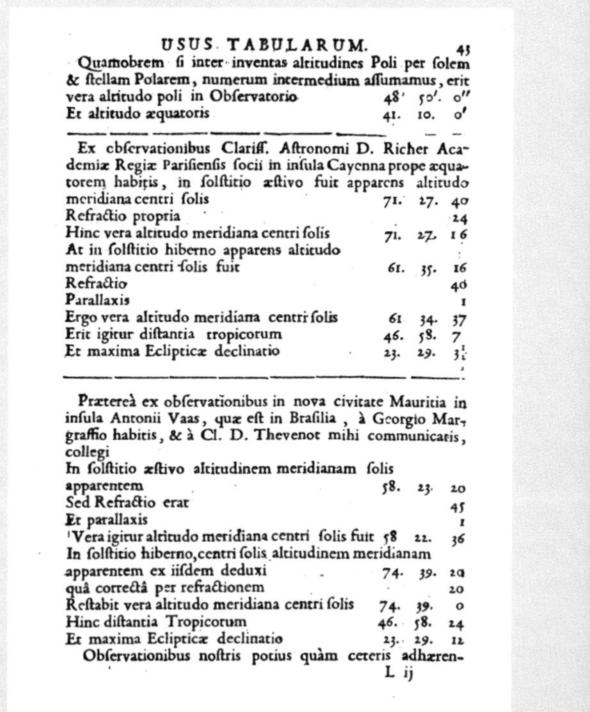

USUS TABULARUM. 43

Quamobrem ſi inter inventas altitudines Poli per ſolem & ſtellam Polarem, numerum intermedium aſſumamus, erit

vera altitudo poli in Obſervatorio	48'	50'.	0''
Et altitudo æquatoris	41.	10.	0'

Ex obſervationibus Clariſſ. Aſtronomi D. Richer Academiæ Regiæ Pariſienſis ſocii in inſula Cayenna prope æquatorem habitis, in ſolſtitio æſtivo fuit apparens altitudo

meridiana centri ſolis	71.	27.	40
Refractio propria			24
Hinc vera altitudo meridiana centri ſolis	71.	27.	16
At in ſolſtitio hiberno apparens altitudo			
meridiana centri ſolis fuit	61.	35.	16
Refractio			40
Parallaxis			1
Ergo vera altitudo meridiana centri ſolis	61	34.	37
Erit igitur diſtantia tropicorum	46.	58.	7
Et maxima Eclipticæ declinatio	23.	29.	3½

Prætereà ex obſervationibus in nova civitate Mauritia in inſula Antonii Vaas, quæ eſt in Braſilia, à Georgio Marggraffio habitis, & à Cl. D. Thevenot mihi communicatis, collegi

In ſolſtitio æſtivo altitudinem meridianam ſolis			
apparentem	58.	23.	20
Sed Refractio erat			45
Et parallaxis			1
Vera igitur altitudo meridiana centri ſolis fuit	58	22.	36
In ſolſtitio hiberno, centri ſolis altitudinem meridianam			
apparentem ex iiſdem deduxi	74.	39.	20
quâ correctâ per refractionem			20
Reſtabit vera altitudo meridiana centri ſolis	74.	39.	0
Hinc diſtantia Tropicorum	46.	58.	24
Et maxima Eclipticæ declinatio	23.	29.	12

Obſervationibus noſtris potius quàm ceteris adhæren-

L ij

FIG. 117. The first published use of MARGGRAFE's Paris Manuscript by PHILIPPE DE LA HIRE in 1687. (Google Books)

MARGGRAFE's observations, was considered so important by both DE LA HIRE, FLAMSTEED and DE L'ISLE, that they all calculated this result independently. The results (*Table 3*) show that the most recent value achieved by DE L'ISLE, after painstaking calculations and corrections, is almost identical with the value already obtained by MARGGRAFE in 1641.

TABLE 3.

The Earth's obliquity				
Calculated from the observations on Antonio Vaz				
Calculator		*deg*	*min*	*sec*
MARGGRAFE	1641	**23**	**28**	**30**
DE LA HIRE	1687	23	29	12
FLAMSTEED	1725	23	29	00
DE L'ISLE	c. 1725	23	28	46
DE L'ISLE	(corrected)	**23**	**28**	**32,5**

Then, in 1785, MARGGRAFE's name suddenly appeared among the very few successful observers of the planet Mercury. An anonymous writer in *The Edinburg Magazine* wrote that successful observations of Mercury always had been very scarce. Or, to quote him literally:

> PTOLEMY in his *Almagest* has only fifteen [observations], two of which are erroneous. After these, until TYCHO's time I find none that have been of any use, except a few made by GAULTERE towards the end of the fifteenth century. COPERNICUS complained that he never could get a good observation. TYCHO BRAHE, LONGOMONTANUS, *MARGRAFF* [our italics], and RICCIOLI were more successful and made several. But most of them are of little use for correcting the tables, as the planet then was not in the requisite part of its orbit.[148]

In this quote MARGGRAFE is finally mentioned in the same breath as TYCHO BRAHE and other famous astronomers. However, at that time MARGGRAFE's observations, including those of Mercury, were not yet published. Therefore, this British author must have had access to handwritten notes of these observations. In England only FLAMSTEED had possessed these, which points to NEVIL MASKELYNE (1732-1811) as the author. After all, as the then-current *Astronomer Royal* he certainly had access to the papers of his predecessor FLAMSTEED.

MARGGRAFE's observations of Mercury were for the first time published in a book composed by the French astronomer ALEXANDRE GUY PINGRÉ (1711-1796). But ... this book was only issued in 1901! How on earth was this possible? The explanation is as follows: Around 1756, visual limitations caused PINGRÉ to abstain from astronomical observations. As in the years 1756-1758 DE L'ISLE was completely absorbed by his search for the expected return of comet Halley, he had abandoned his intention to compose a compilation work of ancient astronomical observations.[149] PINGRÉ was thus free to join this project to further collect and analyze all the observations that DE L'ISLE had collected since 1714. In a way, PINGRÉ became DE L'ISLE's successor. One of his results was a compilation of all comets ever observed. This two-volume *Cométographie, ou traité historique et théorique des comètes* was published in 1783. A second result was PINGRÉ's *Annales célestes du dix-septième siècle,* an overview of all known seventeenth-century observations. This manuscript was prepared for the press in the 1790s. But guided by bad luck, the French Revolution prevented its publication. PINGRÉ's work was only published thanks to GUILLAUME BIGOURDAN (1851-1932), then director of the Paris Observatory, who had found PINGRÉ's manuscript in his archives.[150] BIGOURDAN realized the vital importance of PINGRÉ's *Annales célestes* for the history of seventeenth-century astronomy.

For this reason, the astronomical community had to wait more than two and a half centuries before one could learn about the ambitions and large scale of MARGGRAFE's Brazilian observations. Only then could MARGGRAFE gain some appreciation for the work in which he had desired to excel. But, of course, 1901 was far too late to be of any meaning for the current astronomy. Yet, the *Annales célestes* still missed one part. PINGRÉ had only access to DE L'ISLE's manuscript of the Brazilian observations; unfortunately, he was unaware of the existence of MARGGRAFE's Leiden register. Therefore, MARGGRAFE's Leiden documents remained unstudied until JOHN NORTH's publication in 1979. It is therefore wonderful to see that a real contribution to current astronomy was made as recently as 2011, when MARGGRAFE's Leiden sunspot data gave rise to a new scenario for the onset of the well-known Maunder minimum, the period between 1645 and 1715 during which sunspots became exceedingly rare.[151]

PART IV

CONCLUSION AND EPILOGUE

CHAPTER 10

RETROSPECTIVE

MARGGRAFE'S IMPORTANCE AS AN EARLY MODERN SCHOLAR

Our account of MARGGRAFE's life and work reveals that he was a remarkable and talented scholar with an inquisitive mind. Born in Saxony into a family of Lutheran pastors, he escaped the atrocities of the Thirty Years' War by making a *Peregrinatio Academica* through a large number of important scholarly sites in continental Europe. In this ten-year journey, he acquired a broad spectrum of knowledge and skills. At home in Liebstadt, he mastered the classical languages and developed his skills, both in drawing and music; elsewhere in Germany, he learnt alchemic medicine, mathematics and theoretical astronomy. But to understand his success as a naturalist in Brazil, MARGGRAFE's practical training at Leiden University seems crucial. His stay at Leiden University perfected his knowledge and skills. It was in the Leiden *Hortus Botanicus* where MARGGRAFE learned the basics of hands-on botany; it was in the Leiden *Theatrum Anatomicum* that he was taught how to dissect, and it was at the brand-new Leiden Observatory – the *Theatrum Astronomicum* – that he learned to use mathematical and optical instruments. There he trained himself to systematically observe and carefully record everything he had seen in a journal.

With this set of skills, MARGGRAFE embarked to Brazil with the ambition to chart the Southern Hemisphere. Besides that, his 1640-letter to DE LAET demonstrates that he had promised this *bewindhebber* to investigate the rarities of Brazilian nature. This as repayment for the support that DE LAET had provided him to facilitate his journey to Brazil. So, when the circumstances prevented him from working on his astronomical ambitions, MARGGRAFE used his capacities to make careful observations of the novelties he came across in the non-European nature of this South American continent. Already in August 1638, only a few months after his arrival in Dutch Brazil, he was fascinated by Brazilian snakes. It is telling that he not only caught a *boa constrictor* and dissected the animal to investigate its structure, but also started taking notes about his observations.[1] These notes were supplemented over time by many other observations of particulars in Brazilian nature. Further, he was the first to record daily meteorological observations on the South American continent; the same can be said about his founding of the first European-style astronomical observatory on this continent, where he made observations of the heavens on a regular basis, according to a pre-designed observation program. These impressive achievements of a passionate, unmarried and highly educated man, who did not live past the age of 33, are already more than enough to secure him a lasting place in the history of science.

Marggrafe's importance for natural history

But Marggrafe's pioneering work went further. He was also the first naturalist in the Dutch realm to use a simple microscope (or *megascopium* as he called it) for the study of insects and seeds. His observations about the Brazilian nature that were structured into the *Historiae Rerum Naturalium Brasiliae*, demonstrated not only – as Peter Whitehead remarked – his 'keen eye for detail and a flair for seizing on taxonomically significant features'.[2] But more importantly, as recognized by a contemporary scholar as Øle Worm, Marggrafe was innovative because he reported only what he himself had witnessed and avoided in his text all rhetoric and adornments. This made him an example for a future generation of naturalists. As Brienen pointed out, Marggrafe's objective way of describing nature would become the standard in the eighteenth century.[3] That this approach followed really Marggrafe's own idea is confirmed by the fact that his editor De Laet carefully classified his own *Annotatio* under separate headings, thus guaranteeing the integrity of Marggrafe's text. All this made the *Historia Naturalis Brasilae*, or the combination of Marggrafe's natural history observations with Piso's medico-botanical studies (supplemented by De Laet's comments), the reference work for the knowledge of the flora and fauna of South America for at least two centuries.

Marggrafe's importance for cartography

The empirical approach of 'don't fill in what you haven't seen yourself' also characterized Marggrafe's maps. He did not furnish his maps with imaginary mountains, rivers, or fancy animals. Instead, he designed an innovative system of symbols, explained in a key, which provided a wealth of information about the economic and social value of the surveyed areas, without cluttering his maps. PIMENTEL and PEREIRA's establishing that in this way MARGGRAFE succeeded to present about 1,800 geographic entities is in itself proof enough of the success of this clever method. Moreover, the fact that MARGGRAFE in all his works provides information from mixed Luso-Portuguese, indigenous and Dutch sources attests to an astonishing level of intercultural collaboration. Further, MARGGRAFE's maps of Brazil belong not only to the most elegant examples of seventeenth-century Dutch cartography, they are also among the most accurate made at the time. An obvious reason for the accuracy of MARGGRAFE's maps is that he supplemented the available surveying techniques with astronomical measurements.

MARGGRAFE'S IMPORTANCE FOR EARLY MODERN ASTRONOMY

MARGGRAFE's biography shows that being an astronomer was his first ambition. More precisely, he strongly desired to map the heavenly bodies of the Southern Hemisphere in a similar way as TYCHO BRAHE had done for the northern sky. So, what can be said about MARGGRAFE's role and importance as an astronomer? May we regard him as the New World's TYCHO? Or, to rephrase the question, had he lived, would he indeed have become the southern BRAHE? An unequivocal answer to this question can, of course, never be given. But taking into account MARGGRAFE's background and circumstances, it is reasonable to assume that, as an observer, he could have matched BRAHE. But it's hard to imagine how a man of relatively low social status and utter lack of wealth, like GEORG MARGGRAFE, could ever have matched the immense fame of the wealthy nobleman TYCHO BRAHE, who – supported both by a king and an emperor – was able to found three observatories (two in Denmark and one in Prague), all lavishly equipped with costly mathematical instruments, where he could observe and theorize for decades.

But it is obvious that to achieve TYCHO's fame wasn't MARGGRAFE's ambition. The story of his life shows that he foremost followed his eager curiosity. Once he had found a subject that challenged him to be researched – presumably as a result of discussions with his Stettin employer EICHSTÄDT – he simply desired to explore this gap in the current astronomical knowledge. TYCHO had charted the northern sky, leaving the Southern Hemisphere with many questions to answer. MARGGRAFE took up this challenge. The tasks MARGGRAFE formulated for himself in his *Progymnastica Mathematica Americana* constituted, all in all, a feasible research program. It is plausible that MARGGRAFE, given enough time and money, would have succeeded in completing this project.

In fact, as shown in Chapter nine, MARGGRAFE succeeded in fulfilling *six* out of the *ten* goals he had set for himself in the relatively short time he could devote to astronomy in Brazil. Had his Brazilian observations been published in full in 1648, like the *Historia Naturalis Brasiliae,* they would have been acclaimed as a valuable contribution to the knowledge of astronomy. This all the more so, because a comparison of MARGGRAFE's observational data with the results of modern calculations showed that in Leiden his observational errors were of the same order of magnitude as TYCHO BRAHE's. Although these results were less good in Brazil, mainly due to the instability of the Recife observatory, there is no doubt about MARGGRAFE's ability as a skilled observer.

In this regard, it is important to note that MARGGRAFE may be credited with being the first astronomer to ever record an observation of Mercury's phases. MARGGRAFE's case is further one of the earliest examples of the transfer of instruments, knowledge and practices between Europe and the Americas. In Recife, MARGGRAFE not only carefully copied the Leiden quadrant and sextant, but he also introduced new observational equipment. This included a telescope with a Dutch or Galilean configuration. For this instrument, he even designed a pedestal tray with a type of altazimuthal mounting. This usage is remarkable, because in the first decades of the seventeenth century telescopes still had no well-defined role as an astronomical instrument. Before the 1630s, their usage was rare, 'even among the learned'.[4] Moreover, MARGGRAFE's working papers demonstrate his skilled usage of this optical device in observing various celestial phenomena. His telescopic observations of sunspots, by means of the projection of a solar image on an opaque screen, demonstrate that he followed state-of-the-art procedures, introduced in the Netherlands shortly before (in the early 1630s) by ISAAC BEECKMAN and MARTINUS HORTENSIUS. Summarizing, we may conclude that MARGGRAFE's astronomical work in Dutch Brazil represents a remarkable milestone in the history of astronomy in the New World.

Unfortunately, MARGGRAFE's importance for contemporary astronomy was strongly reduced by the neglect of the Leiden professor

GOLIUS, who failed to keep the promise to publish the Brazilian observations. We can only speculate why GOLIUS did not pay more attention to sharing these astronomical data with the learned community of the Republic of Letters. Of course, GOLIUS's primary interest was the classic scholarship of the oriental world, but we cannot rule out the possibility that the fierce quarrel with MARGGRAFE's younger brother CHRISTIAN prevented him from publishing the astronomical achievements of his deceased brother. Furthermore, it was sheer bad luck, as we have shown, that after GOLIUS's death, the manuscript of MARGGRAFE's observations continued to be hidden behind a capricious succession of curtains, leaving him in limbo for a long time. And even when AUGUST-GUY PINGRÉ discovered the Paris manuscript in the mid-eighteenth century, he remained unaware of MARGGRAFE's Leiden notebook, with the result that these early Leiden observations are lacking in PINGRÉ's *Annales célestes*, the well-known eighteenth-century compendium of seventeenth-century observations, published in 1901. So, the first time that the Leiden observations were used for the benefit of astronomy was in 2011, when MARGGRAFE's Leiden sunspot observations gave rise to a new scenario for the onset of the well-known Maunder minimum.

Finally, largely because of the fall of Dutch Brazil in 1654, the MARGGRAFE episode remained an isolated and rather short event, with no follow-up in Brazilian scholarship. Only much later, in the early nineteenth century, did astronomy by European standards emerge again, first in the Portuguese colony and after Brazilian independence, as a national enterprise.[5]

THE PRECONDITIONS FOR MARGGRAFE'S SCHOLARLY WORK IN COLONIAL DUTCH BRAZIL

What may we conclude about the preconditions that facilitated MARGGRAFE's achievements in Brazil? And even as interesting, what contributed to his innovative 'thin' description of the things he observed and witnessed?[6] Are, for instance, the preconditions for knowledge gathering in an objective way, as described by HAROLD COOK in his *Matters of Exchange*, recognisable and applicable in MARGGRAFE's case?[7] Our conclusion is that this is *not* the case.

The first factor that undoubtedly must be coined as crucial for MARGGRAFE's scholarly achievements is the princely support of JOHAN MAURITS VAN NASSAU-SIEGEN. It was the count who agreed to erect an observatory on top of his first residence, and even paid for its construction, its instruments and further equipment. The founding of such an astronomical observatory fitted perfectly the self-fashioning of JOHAN MAURITS as a European-style semi-royal ruler. It was also JOHAN MAURITS who as governor-general ordered the expeditions in which MARGGRAFE participated, first to collect medicinal herbs for the colony's pharmacy, then to obtain plants and animals for his garden and menagerie, and later for charting the territory conquered for the Dutch West India Company. Without the European 'court culture' that JOHAN MAURITS imported into Brazil, the conditions for MARGGRAFE to achieve success in one of the areas for which he is known today were virtually non-existent.

The dark reverse of this support is that JOHAN MAURITS's fortune that made all this possible was based on the exploitation of enslaved Africans, a circumstance which is remarkably completely absent in the accounts MARGGRAFE has left us. In none of his documents does MARGGRAFE ever mention these enslaved Africans, who were present in large numbers in the Dutch colony. In the 'Historiae Rerum Naturalium Brasiliae', he wrote only once about '300 men' who were busy transplanting large mature coconut trees in the garden of *Vrijburg*. Not a syllable indicated the fact that these laborers were enslaved Africans, deprived of all liberty.[8] Equally remarkable is the observation – noted by NEIL SAFIER – that in none of MARGGRAFE's texts is there any sign of the circumstance that the Dutch colony was actually in a permanent state of war.[9]

That brings us to a second precondition, the very existence of the Dutch West Indian Company and its colony of Dutch Brazil. Aside from MARGGRAFE's astronomical ambitions, his other efforts to gain natural knowledge were certainly not free of values. His surveying practices and information gathering about the landscape, economy and wildlife of Pernambuco served the Dutch colonial war and trade efforts. That context should not be forgotten, when we consider MARGGRAFE's impressive scholarly achievements. Moreover, his achievements are by no means the work of one individual. During his expeditions, MARGGRAFE was facilitated by dozens of soldiers, porters, skippers, translators, indigenous Brazilians and Portuguese settlers. His observatory and most of the instruments used there also required the attention and craftsmanship of many artisans. This all was only possible thanks to the framework provided by the WIC.

A third important, and probably also crucial precondition for MARGGRAFE's achievements was his academic training, first in Germany, but especially at Leiden University. Our survey revealed that MARGGRAFE's apprenticeship in Leiden was crucial for the way he organized his astronomical observations in Brazil. In Recife, MARGGRAFE not only tried to re-create the Leiden location where he had practiced, but he also continued observing in a way he had become familiar with. He used instruments, methods and ideas that he learned at the Leiden Observatory. On the one hand, he continued to build on measuring methods introduced by BRAHE and elaborated in Leiden by WILLEBRORD SNELLIUS, but on the other hand, he introduced new observational equipment.

A fourth precondition, MARGGRAFE's religious Lutheran background as an underlying motive for his search of the heavens, should not be underestimated. In his Wittenberg thesis, as well in both his Leiden and Brazilian manuscripts, this personal motto Σύν θεω, meaning 'with God' or 'with God's help', is frequently present. GEORG's brother CHRISTIAN recognized this Greek motto even in one of the herbariums he saw during his first visit to the house of professor GOLIUS.

Finally, the role of MARGGRAFE's personal talent and passion is also an important factor that contributed to his success. His inquisitive mind and perseverance, his talent for careful observing, his drawing skills, and his command of several languages (German, Latin, French, Portuguese, and most likely also some Dutch and indigenous languages) are qualities that certainly contributed to his success. MARGGRAFE's 'way of seeing and knowing' – to use HAROLD COOK's terminology – produced a new kind of 'objective' way of describing of nature and cartography, hardly seen before. However, COOK's attribution of this style of knowledge-gathering to the Dutch merchant community, represented by trading companies such as the West India Company, seems totally misplaced in the case of MARGGRAFE. The West India Company, as the highest authority in the Brazilian colony, provided of course the arena in which everything took place, but this trading company does not seem to have emanated any organizing principle that influenced MARGGRAFE's scientific output. The way in which MARGGRAFE recorded his botanical and zoological observations, and how he designed his maps, can in our view be attributed entirely to his own critical mind, whereby his academic training and his experience as an astronomical observer may be regarded as decisive.

Intercultural cooperation across several language barriers: the missing Go-betweens

Finally, some last remarks on how MARGGRAFE gathered this vast load of information about the local flora and fauna as well as the places he charted. In his manuscripts, letters or the 'Historiae Rerum Naturalium Brasiliae', MARGGRAFE never mentions any assistant, informant or 'Go-between', this despite the fact that – as noted above – he must have received assistance from various sides. The *Vitae* only mentions that on his journeys he was 'much conversant, chiefly with the Jesuit Fathers'.[10] Most indigenous villages in the interior of the country (*aldeia*'s) were commanded or supervised by these religious officials. So, these Luso-Portuguese priests knew many details about the local topography and living conditions. Much information may have come from them.

In any case, it is striking that MARGGRAFE's part of the *Historia Naturalis Brasiliae* contains not only a description of the various Brazilian plants and animals, but also a lot of information about the medicinal and nutritional properties attributed to these species.[11] Such details can only have come from local residents, either Luso-Portuguese or indigenous. Nevertheless, it is obvious that MARGGRAFE must have encountered numerous communication barriers during his expeditions. In Dutch Brazil, with its citizens and soldiers coming from the Low Countries, Portugal, Spain, Germany, France, England, Scandinavia and Poland, added to the large group of Jews, enslaved Africans and two native Brazilian indigenous tribes, MARGGRAFE experienced a cacophony of languages around him. As remarked by WIESEBRON, the Dutch colony constituted in effect a linguistic 'Tower of Babel'.[12] As a result, part of the communication went through sign language.[13] But besides that, it can hardly be otherwise than that a kind of 'pidgin' language (used to negotiate certain local exigencies) arose in the 'Trading Zone' that represented the Dutch colony with its many nationalities and tongues. This comparable to what PAMELA LONG has described in other cases of early modern knowledge acquisition.[14] The case of the Norwegian mercenary JACOB JENSSØN NORDMAND (1614-1695), who between 1633 and 1640 served as a common soldier in Dutch-Brazil, demonstrates the language difficulties experienced in the field. Almost all names of places and persons NORDMAND remembered from his time in Dutch Brazil were badly mangled.[15]

In terms of language proficiency, MARGGRAFE was well equipped. In addition to his German mother tongue, his letters show that he also mastered Latin, French and Portuguese. The latter language he will have acquired during his two months crossing of the Atlantic. It is significant for his linguistic interest that he owned a booklet with phrases in Latin, French, Italian and German.[16] For his notes on Brazilian nature MARGGRAFE was also able to design 'a secret alphabet, with signs he invented', which forced the editor of the *Historia Naturalis Brasiliae,* JOHANNES DE LAET, to first to interpret and transcribe these notes, before any processing could start.[17] In addition, MARGGRAFE seems to have delved into the indigenous Brazilian languages, quite remarkable, as interpreters were rather scarce.[18] It is telling that DE LAET found among MARGGRAFE's papers a scheme that provided a chapter on these indigenous languages.[19] It is true that DE LAET found no details, reason why about the native languages he inserted a grammar from a printed version compiled in 1595 by the Jesuit priest JOSEPH DE ANCHIETA, and a linguistic glossary that he had received from a former Luso-Portuguese priest, a certain MANUEL DE MORAES.[20] Anyway, from the above it should be clear that in addition to language skills, MARGGRAFE must also have possessed the necessary social skills to gather all the diverse information that we find in his work.

THE TRANSMISSION OF GATHERED FACTS INTO CANONIZED KNOWLEDGE

This case study shows how much the process of converting collected observations and facts into canonized printed knowledge depends on chance and benevolent transformers, such as JOHANNES DE LAET as editor of the *Historia Naturalis Brasilia* and JOHANNES BLAEU as the transformer of MARGGRAFE's manuscript maps into popular cartographic commodities. On the other hand, the failure of GOLIUS to do the work that was assigned to him greatly marginalized MARGGRAFE's significance for contemporary astronomy.

It is all too well known to historians of science and culture that this transformation process, in which third parties are involved, also causes changes in the original content. In this case, for example, BLAEU added figurative elements to MARGGRAFE's sober maps, using illustrations by the painters in JOHAN MAURITS's retinue; DE LAET, too, expanded MARGGRAFE's text with his own annotations, which however he kept apart from the original text, something that WILLEM PISO deliberately forgot to do in his *Indiae Utriusque re Naturali et Medica* (1658), with all later consequences for his (posthumous) reputation.

EPILOGUE

At the end of our survey into the life and work of GEORG MARGGRAFE, it is good to realize how special it is to be able to reveal so many particulars about someone of such a low social status. For example, about MARGGRAFE's nephew JURRIAN, a son of his brother CHRISTIAN MARGGRAFE, who went to South Africa and Asia for six years (between 1709 and 1715) as an employee of the Dutch East India Company (VOC), there is nothing to be found other than the mere fact that he was there.[21] It is good to realize that historiography is very dependent on the contingency of what documentation has been preserved. As for GEORG MARGGRAFE's astronomical activities, we were close to the circumstance of knowing as little about them as we do about the adventures of his nephew . But fortunately, thanks to the vagaries of history, MARGGRAFE's astronomical manuscripts have survived in Leiden, Paris and Lisbon. These precious documents, together with many non-astronomical sources, have enabled us to sketch the astonishing story of MARGGRAFE's life and work. The level of detail we were able to report, thanks to these documents and sources, is truly astonishing. We really could witness 'science in the making', in a time long before the era of the scholarly societies or learned journals. Being able to chronicle these early modern developments also made us grateful. Grateful also to all those who have helped us in recent years with our investigations: too many to mention.

Finally, it is good to know that already decades ago a heavenly monument was erected for MARGGRAFE and his patron JOHAN MAURITS. Thanks to the efforts of the Palomar–Leiden Survey (PLS), a collaboration between the U.S Palomar Observatory and the Dutch Leiden Observatory, two minor planets discovered in September 1960 by C.J. (KEES) VAN HOUTEN and INGRID VAN HOUTEN–GROENEVELD on Palomar Schmidt plates taken by TOM GEHRELS were named after these two astronomical pioneers, namely asteroid no. 11238 (*Johanmaurits*) and no. 11239 (*Marcgraf*).[22] The challenge now is to ensure that this latest asteroid (*fig. 118*) is eventually renamed after the astronomer's actual proper name 'Marggrafe'.

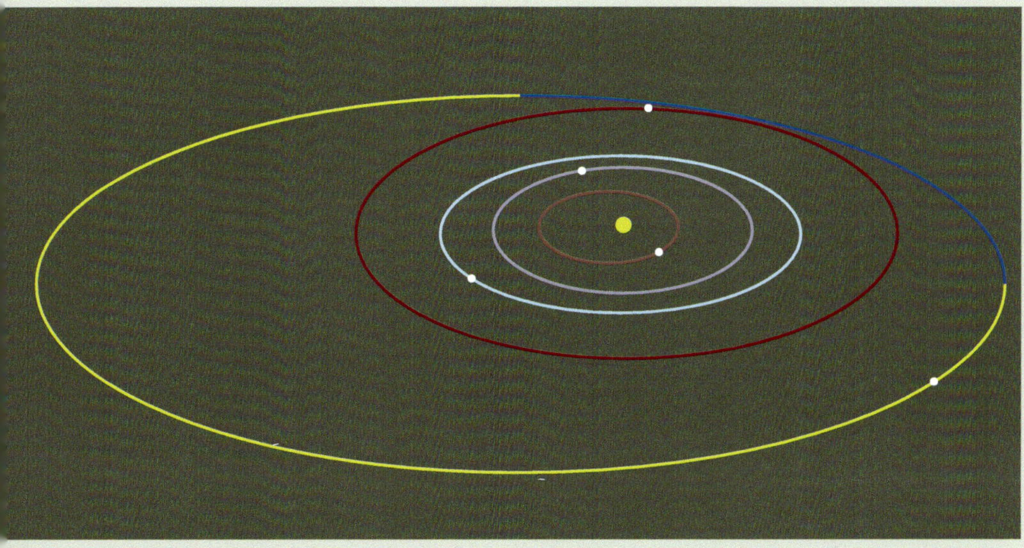

FIG. 118. Sketch of the approximate orbit for the minor planet (formerly asteroid) named **'Marcgraf'**, which orbits the Sun in 4.04 years in an elliptical orbit beyond the orbit of Mars. The colours yellow and blue indicate where the object's orbital plane is above or below the plane of the ecliptic, respectively. The solid coloured arcs show the orbits of the planets Mercury to Mars. (MPC Orbit Sketch)

APPENDICES

APPENDIX 1

CHRONOLOGY

(key dates according to the Gregorian calender)[1]

1610	30 September	GEORG MARGGRAFE born in Liebstadt.
1626		Brother CHRISTIAN MARGGRAFE born in Liebstadt.
1627	April	GEORG leaves his parental home and begins a ten-year period of travel and study throughout northern Europe.
—	May	Registration at Wittenberg University.
—	November	Registration at Strassbourg University.
1628	April	Registration at Basel University.
1629-1633		Unregistered student in Ingolstadt, Altorf (near Nürnberg), Erfurt, and Leipzig.
1633		Return to Liebstadt for family visit.
1634	March	In Wittenberg: obtained the degree of 'Medical Candidate' with an alchemical thesis written under direction of DANIEL SENNERT.
—	Fall	In Rostock, following lectures from the botanist SIMON PAULI.
1635	February/ September	In Stettin: observations of Jupiter with LORENZ EICHSTÄDT.
1636	September	Registration at Leiden University.
1637	January–November	Regular observations at Leiden Observatory.
—	November	MARGGRAFE leaves Leiden for Amsterdam and Texel.
1638	January	Departure from Texel to Brazil.
—	March	Arrival in Recife after a two months' crossing of the Atlantic.
—	8 April	[***Trip 1***] 'Armada' with JOHAN MAURITS and his troops (including MARGGRAFE) leaves Recife for São Salvador in Bahia.
—	15 May	MARGGRAFE writes a letter in Portuguese to JOHAN MAURITS during siege of São Salvador in Bahia.
—	17-18 May	Failed attack on São Salvador. Many casualties. MARGGRAFE almost killed by a canon ball.

—	5 June	Returned in Recife.
—	25 June	Failed observation of a lunar eclipse. The next day first letters to JACOB GOLIUS and FROM.
—	July	Baptism of ELISABETH SILNAER (1636-1639), daughter of HANS SILNAER, for whom MARGGRAFE constructs a horoscope.
—	August	MARGGRAFE receives a live boa constrictor. (*Hist. Nat.* 238).
—	September	First observational series in Recife without the observatory.
—	October	[***Trip 2***] Letter from the island of Itamaracá to JACOB GOLIUS in Leiden.
		Receives a large dead boa constrictor, which he buries, disinterring the skeleton three months later. (*Hist. Nat.* 238).
—	20 December	Recife: succesful observation of a lunar eclipse with a 1-foot quadrant.
1639	February	Description of a fish (*Hist. Nat.* 159).
—	18 May	Letter in German to SAMUEL KECHEL in Leiden: '*Ich habe alfür in Astronomie noch wenig observiren können, ...*' Letter in Latin to FROM, from '*nova civitade Mauricia est in Isla Anton Vaz*'. Letter in French to JEAN and ANTOINE VAN HUYCK.
—	18–21 May	Second observational series without observatory.
—	21/28 June–6 Aug.	[***EXP-1***] MARGGRAFE participates in an expedition to Ceará.
—	20 July	Acquires during the expedition a female tortoise who survived twenty-one months without food. (*Historia*, p. 241–242).
—	August	Returned home from the expedition to Ceará, MARGGRAFE planted in his garden on Antonio Vaz seeds from the *Ambaiba Brasiliensibus*, a plum tree that has almost no branches and the *Serpentaria Maior*, or dragon lily, with a toxic-smelling flower. (*Hist. Nat.* 27 & 92).
—	September	Description of a fish (*Hist. Nat.* 159).
—	15 September	Start of regular astronomical observations on the newly built observatory in Antonio Vaz, lasting until 18 March 1640.
—	22 October	[***Trip 3***] 'I went away to the field'.
—	4 November	and following days: [***Trip 4***] 'I was away'.

—	28–30 December	[***Trip 5***] 'the sky was fickle and I also went away to the countryside'.
1640	January	Recife: start of daily meteorological observations, noting wind direction, precipitation, and the incidence of thunderstorms, continued to the end of 1642, finally leading to conclusions on the seasons in north-eastern Brazil. During his absences from Recife, these were carried on by an assistant. (*Hist. Nat.* 265)
—	10 January	[***Trip 6***] 'several times I went away'.
—	6 February	'Mauricia' (Recife): letter in Latin and Portuguese to JOHANNES DE LAET in Leiden. MARGGRAFE has described and drawn about 350 plants, and a number of animals.
—	March	[***Trip 7***] at the Aquapetida (Aguapetiba) cove, 330 km south of Recife. (*Hist. Nat.* 119).
—	18 March	'Mauricia': The house shared by PISO, MARGGRAFE, and five others collapses. No one dies, but MARGGRAFE sustains a dislocated shoulder, which forces him to rest in Recife until 19 May.
—	April	Flowering of a stone apple in garden on Antonio Vaz. (*Hist. Nat.* 114).
—	28 May	MARGGRAFE kills and dissects an iguana. (*Hist. Nat.* 236).
—	11 June – 17 Oct.	Daily observations resumed at the observatory.
—	July	Observation of insects (*Hist. Nat.* 255).
—	27 August – 4 Sept.	Legislative Assembly in Recife
—	11 September	'In the evening I was prevented by friends (from doing the observations)'.
—	20 October- 10 Nov.	[***EXP–2***] Departure 'on horseback' for a three week expedition 'for the sake of chorography and topography'. During this journey 400 birds nests were observed in one tree.
—	13 November	Recife: partial observation of a solar eclipse; observations resumed up to 1 December
—	21 November	Description of a plant, which was planted in January. (*Hist. Nat.* 29).
—	December	Many moths observed in the garden on Antonio Vaz. (*Hist. Nat.* 250).

—	8-19 December	[***EXP–3***] *'I went out again in the following weeks'*: MARGGRAFE departs for a short natural history expedition into the interior. *'Finish'd this also on Horseback, being guarded with a company of Soldiers'*. No meteorological observations recorded between 13-19 December.
—	20 December	Recife: Observations resumed in a series that continued up to 7 February 1641.
1641	9 February	[***EXP–4***] MARGGRAFE departs for nine months for the purpose of gathering geographical information along São Francisco river.
—	2 July	On the high mountain Itapuamuru.
—	16 July	[***EXP–4***] Receives assignment to map the region around the Rio São Francisco (See also *Hist. Nat.* 204. no date).
—	21 September	MARGGRAFE named in message to the High Council about mapping the area around the Rio São Francisco.
—	18 oktober	Partial solar eclipse. Not observed during expedition.
—	November	Returned in Recife. MARGGRAFE included in list of persons receiving 'a free meal' at court.
—	27 November	[***Trip 8***] Assignment to map the area of Machioppe ('busy doing field work until 9 December').
—	10 December	[***EXP–5***] 'I departed to the southern part of Brazil for the sake of Chorography and Natural History'.
—	December	Dissects an iguana, probably during the expedition, removing a stone the size and shape of a chicken egg from its stomach. (*Hist. Nat.* 237).
1642	January	[***EXP–5***] Irregular registration of meteorological observations.
—	14/15 April	Fortress Ceulen: partly observation of a Lunar eclipse, because of clouds.
—	16 June	Returned to Recife: 'In the year of Christ 1642, on 16 June, I came back from that journey started in December of the past year'.
—	7 October	Recife: partly observation of a lunar eclipse with the 5–foot quadrant, hindered by clouds.
—	2–22 November	Third period of astronomical observations, among others on the planet Mercury
—	11 December	Arrival of CASPER SCHMALKALDEN in Recife.
—	20 December	Isolated observation of the meridian height of the Sun in Recife.

1643	12 January	Departure of SCHMALKALDEN with the WIC Expedition to Chili, returning only in December 1643.
—	13 Februari	Purchase of an instrument for map making, used by MARGGRAFE for his map of Brazil
—	12 March	Birth of FREDERICK FALLOO, born Mauritia, for whom MARGGRAFE constructs a horoscope.
—	10 March	[***Trip 9***] '... the following days I went out for matters concerning the *Historia Naturalis*'.
—	April	MARGGRAFE included on the list of 'domestiquen'.
—	4 April	Recife: partly observation of a lunar eclipse with the 5-foot quadrant, hindered by clouds.
—	3 May	Birth of HENRY BONJOUR, born Mauritia, for whom MARGGRAFE constructs a horoscope.
—	22 June	Last recorded astronomical observation in Brazil.
—	August	Last recorded botanical observation in Brazil. (*Hist. Nat.* 127)
—	14 August	[***EXP–6***] MARGGRAFE charged with mapping St Thomé and Angola.
		Leaves that week with the yacht *De Brack*
—	Fall	MARGGRAFE dies in Angola within five weeks after arrival in Africa.

==

1644	28 February	Bridge between Recife and Mauritsstad opened
—	11 May	JOHAN MAURITS departs from Recife. On 22 May he leaves Paraíba for Holland on board the *Amsterdam*. The ship has loaded 'a casket with astronomical instruments'.
1646	1 July	DE LAET informs WORM, about the arrival of MARGGRAFE's herbarium and other papers.
—	November	Printing of the *Historia Naturalis Brasiliae* started.
1647	10 July	Five Chapters of *HNB* are printed: three about botany, one about fish and another about birds.
1648	February	Publication of the *Historia Naturalis Brasiliae*.
1651		Brother CHRISTIAN MARGGRAFE contributes to a bundle of funerary poems of a Lutheran vicar from Altenburg, murdered in Copenhagen in December 1649.
1652	3 June	CHRISTIAN MARGGRAFE registers as a student at Leiden University.

—	20 July	Christian Marggrafe writes a letter to Johannes Hevelius in Danzig
1653	July	Villum Worm purchases Georg Marggrafe's Brazilian herbarium and sends it to Copenhagen.
1655	February	Christian Marggrafe requests an audience with Johan Maurits in The Hague.
—	March–May	Legal procedures about Georg's inheritance.
—	June–Febr. '56	Lawsuit before Academic Court ('academische vierschaar') about Georg's inheritance.
1658		Publication Willem Piso's *Indiae utriusque re naturali et medica.*
1659	20 January	Christian Marggrafe registers as 'Medical Candidate' at Franeker University, and becomes a Medical Doctor ten days later.
1667		Death of Jacob Golius.
1668		Death of Samuel Kechel.
1674	August	Prosecution of Christian Marggrafe for blasphemy and defamation of the States of Holland.
—	December	Christian Marggrafe marries Adriana Willemsdr van Ga(m)meren from Babeloniënbroeck.
1677	July	Birth of the first child, Georgius Marggrafe.
1685		Christian Marggrafe publishes the *Vita Georgii Marggravii.*
1687	January	Death of Christian Marggrafe, being buried in Valkenburg.
—	May	Auction in Leiden of Christian Marggrafe's library.

APPENDIX 2

MARGGRAFE'S HOROSCOPES

MARGGRAFE's horoscopes among the Leiden manuscripts are listed here more-or-less in chronological order of production. For the coordinates of the European birth places mentioned in the horoscopes, MARGGRAFE used the *Catalogus Civitatum Germaniae* as published in EICHSTÄDT's *Ephemeridum* (1634), pages 65–75. This catalogue was mostly copied after a similar list published earlier by TYCHO BRAHE in which he stated the longitudes with respect to a prime meridian belonging to the so-called *Insulis Fortunatis*, in English, the 'Fortunate Isles' or 'Isles of the blessed'. These were legendary islands in the Atlantic Ocean (positioned west of the Canary Islands), described in the Greek mythology as an earthly paradise. CLAUDIUS PTOLEMY introduced these imaginary islands as a reference point for the measurement of geographical longitude. This practise was continued through the Middle Ages until the seventeenth century. Compared to the current Greenwich meridian, a correction of 24° 03' West is needed. For the horoscopes of the four children born in Recife, MARGGRAFE used a longitude he had calculated from results he obtained during his observation of the total lunar eclipse of 21 December 1638. The result was a value of 343° 15' west of Uranienborg, or in present day coordinates 40° 48' west of Greenwich, which compared to the current longitude of 34° 52' 37" is in fact almost six degrees too far.[1]

IDENTIFIED HOROSCOPES

1. North doc. no. 86vs [list of planetary positions], 88, 89, 90, horoscope 10.
   Probably constructed in Wittenberg, ca. 1634.
   Birth date 20/30 September 1610 [o.s./n.s.] for [no name provided].
   Given location: 51° 06' N, 38° 05' E (= corrected 14° 02' E)
   Identified as **GEORG MARGGRAFE (1610-1643)** [the author], born in **Liebstadt**, south of Dresden.
   Current location of birth place: 50° 51' N, 13° 51' E (difference 0° 17' Lat., 0° 11' Long.)

2. North doc. no. 86, horoscope 7. Probably constructed in Wittenberg, ca. 1634.
   Birth date 21 September 1607 [o.s.] for 'M. Christof. Notnagel', born 'Hilperhusii' / 'Coburgii'.
   Given location: 50° 18' N, 35° 10' E (= corrected 11° 07' E)

Identified as **Christoph Nothnagel (1607-1666)**, born in **Hildburghausen**, north of Coburg, professor of astronomy at Wittenberg University in 1634, having taught there already since 1632.
Current location of birth place: 50° 25' N, 10° 45' E (difference 0° 07' Lat., 0° 22' Long.)

3. North doc. no. 83, horoscope 5. Probably constructed in Stettin, ca. 1635.
Birth date 26 May 1594 [o.s.] for 'J.F.', written also 'J Fabr. Husii, Hol'.
Given location: [not filled in] | Not in Eichstädt (1634)
Identified as **Johann Fabricius (1594-1652)**, born in **Husum** (Holstein), *Obersekretär* of the city of Stettin.
Current location of birth place: 54° 28' N, 09° 03' E

4. North doc. no. 87, fol. 1r–3r, 7vs; horoscope 8. Constructed in Leiden, ca. 1637.
Birth date 11 February 1609 [n.s.] for 'G.H. à B.', born **Altenburg** [south of Leipzig].
Given location: 51° 00' N | Eichstädt (1634): 51° 00' N
Identified as **Georgius Henricus à Beerensteyn (1609-1670)**, student of law in Leiden since August 1636. Before Leiden he studied in Jena. His German name was Georg Heinrich von Bernstein. He became dean of the cathedral of Magdeburg.
Current location of birth place: 50° 59' N, 12° 26' E (difference 0° 01' Lat.)

5. North doc. no. 87, fol. 3vs–5vs, 7vs; horoscope 12. Constructed in Leiden, ca. 1637.
Birth date 20 January 1613 [o.s.] for 'G.K.', born in **Stettin**.
Given location: 53° 27' N | Eichstädt (1634): 53° 30' N
Identified as **Georgius Kirstenius (1613-1660)**, or in German Georg Kirsten, from Stettin, student of medicine in Leiden since October 1634. He became professor of medicine in his birth town Stettin.
Current location of birth place: 53° 25' N, 14° 33' E (difference 0° 02' Lat.).

6. North doc. no. 87, fol. 6r–8r; horoscope 9. Constructed in Leiden, ca. 1637.
Birth date 22 February 1610 [n.s.] for 'A.F. von H.', born 'Gustrovius', modern **Güstrow**, south of Rostock.
Given location: 53° 42' N | Eichstädt (1634): 53° 42' N
Identified as **Adolph Friedrich von Hagen (1610–after 1667)**, student of law in Leiden since May 1636. He matriculated in Rostock in June 1626, where he stayed for ten years. After his study he became an influential lawyer in Rostock. In Leiden he lived in the same house as Georg Heinrich von Bernstein. (see no. 4).
Current location of birth place: 53° 47' N, 12° 10' E (difference 0° 05' Lat.).

7. North doc. no. 87, fol. 8vs; horoscope not listed in North. Constructed in Leiden, ca. 1637.
Birth date 18 October 1547 for [name not listed].
Given location: 50° 48' N | Not in Eichstädt (1634).
Identified as **Justus Lipsius (1547-1606)**, born at that date in Overijsse, professor of history at Leiden University between 1579 and 1590 and after that in Louvain until his death.
In 1634 Marggrafe used a citation from Lipsius in the *Album Amicorum* of Andreas Sennert.
Current location of birth place: 50° 47' N (difference 0° 01' Lat.).

8. North doc. no. 92, horoscope 13. Constructed in Leiden, ca. 1637.
Birth date 27 May 1614 [o.s.] for 'Th. Fuiren, Hafnia'.
Given location: [not filled in] | Eichstädt (1634): 53° 30' N
Identified as **Henrik Fuiren (1614-1659)**, born in Copenhagen, indeed on 27 May 1614.
(At first sight one would think that his brother Thomas Fuiren was meant, who in the years 1636-1638 also lived in Leiden, but he was born on 1 June 1616).
Current location of birth place: 55°40' N, 12°34' E.

9. North doc. no. 91, horoscope 11. Perhaps constructed in Leiden, ca. October 1637.
Birth date 15 June 1611 [o.s.] for [name not filed in], born in **Rewalia**, current Tallinn.
Given location: 59° 00' N | Not in Eichstädt (1634)
Possible identification: **Marcus Longorius** (or in German Marcus Lange) from Tallinn, student of theology in Rostock in 1634, who enrolled in Leiden in September 1637, telling he was 26 years of age.
Current location of birth place: 59° 26' N, 24° 46' E (difference 0° 26' Lat.).

10. North doc. no. 93, 98, horoscope 18. Constructed in Brazil, before 1639.
Birth date 1 July 1636 [n.s.] for 'Filiola (= 'little daughter') Steler?', born on Antonio Vaz.
Given location: 8°00' S, 343° 15' E (= corrected 40° 42' W)
Identified as **Elisabeth Silnaer (1636-1639)**, daughter of a certain Hans Silnaer.
Current location of birth place: 8°04' S, 34°52' E (difference 0° 04' Lat., 5° 50' Long.)

11. North doc. no. 84–85, horoscopes 6 & 22. Constructed in Brazil, before August 1639.
Also North doc. no. 85, 100, 'Revolutio thematis'.
Birth date 28 June 1607 [n.s.] for 'J.B.B', born **Loosdrecht**.
Given location: 52° 15' N, 27° 00' E (= corrected 2° 57' E) | Not in Eichstädt
Identified as **Johannes Bodecher Benning (1607-1642)**, member of the *Politieke Raad* in Recife and professor of ethics and physics at Leiden University.

Current location of birth place: 52° 12' N, 5° 8' E (difference 0° 03' Lat., 2° 11' Long.)

12. North doc. no. 89, horoscope19. Constructed in Brazil, 1639.
Birth date 22 May 1639 [n.s.] for [name not filled in].
Given location: [not filled in]
Identified as **Maria Falloo**, daughter of the army captain and military engineer Andries Falloo, commander of fortress De Bruijn, and his wife Anna Gariges.
Current location of birth place: 8° 04' S, 34° 52' E
13. North doc. no. 99, horoscope 20. Constructed in Brazil, 1643.
Birth date 12 March 1643 [n.s.] for 'Frederic Valloy int Fort De Bruijn'.
Given location: [not filled in]
Identified as **Frederick Falloo**, son of the army captain and military engineer Andries Falloo, commander of fortress De Bruijn, and his wife Anna Gariges.
Current location of birth place: 8° 04' S, 34° 52' E

14. North doc. no. 88, horoscope 21. Constructed in Brazil, 1643.
Birth date 3 May 1643 [n.s.] for 'Henry Bonjour, natus in Mauricia in Brasilia'.
Given location: [not filled in]
Identified as **Henry Bonjour**, son of Pierre Bonjour, the French *Maitre d'[h]ostel* of Count Johan Maurits, and his wife Maria Gariges.
Current location of birth place: 8° 04' S, 34° 52' E

NOT IDENTIFIED HOROSCOPES

15. North doc. no. 7 and no. 87, horoscope 15. Constructed in Leiden? (because written on North doc. no. 87)
Birth date 10/20 July 1617 [o.s./n.s.] for '**D. à K.**', born half a mile from 'Citzâ' (= Ceitza, Czeitz, currently the city of Zeitz, south of Leipzig). (North wrongly read the city of Gera).
Given location: 51° 06' N, 36° 20' E (= corrected 12° 17' E) | Eichstadt (1634): 51° 00' N, 36° 20' E
Current location of birth place: 51° 03' N, 12° 08' E (difference 0° 03' Lat.; 0° 09' Long.)
Perhaps this horoscope is related to an acquaintance of Henricus Kralitz, who as Marggrafe's colleague was hired in 1636 as Piso's assistant, but who drowned during the journey to Brazil (or shortly after his arrival). Kralitz was born in Potewitz, a small village near the town of Zeitz.

16. North doc. no. 95 and no. 96, horoscope 2. Perhaps constructed in Stettin, ca. 1635.
Birth date 12 September 1632 for '**D.K.**', no birth place given.
Given location: 53° 30' N. | Eichstädt (1634): 53° 30' N = Stettin
Current location of birth place: 53° 25' N, 14° 33' E.

17. North doc. no. 91, horoscope 14. Perhaps constructed in Leiden, ca. 1637.
    Birth date 7 February 1615 for '**S.K.**', born Prague.
    Given location: 50° 06' N. | Eichstädt (1634): 50° 06' N
    Current location of birth place: 50° 05' N, 14° 25' E. (difference 0° 01' Lat.).

18. North doc. no. 94, horoscope 16. (Booklet with several horoscopes belonging to one family.
    No names provided. Only the indication 'filia' (daughter) or 'filius' (son).

I.	Birth date 16 September	1622,	daughter,
II.	Birth date 6 November	1623,	son,
III	Birth date 10 December	1624,	daughter,
IV	Birth date 30 September	1626,	son,
V	Birth date 7 December	1627,	daughter,
VI	Birth date 18 January	1629,	son,
VII	Birth date 15 June	1630,	son,
VIII	Birth date 12 March	1635,	daughter.

    Given location: [not filled in]
    Current location of birth place: –

19. North doc. no. 50 and no. 88, horoscope [unrecorded by North].
    Birth date 13/23 & 14/24 April 1590 [o.s./n.s.] for [no name provided].
    Given location: [not filled in]
    Current location of birth place: –

20. North doc. no. 50, calculations for a horoscope [unrecorded by North].
    Birth date 11 March 1600 [n.s.] for [no name provided], from 'Goesa', probably Goes in Zeeland.
    Given location: [not filled in]
    Current location of birth place: 51°30'N, 3°53'E

APPENDIX 3

THE FALSE ATTRIBUTION OF AN 'ASTRONOMICAL' MANUSCRIPT

FIG. 119. Parchment carving representing *Vrijburg* palace in Dutch Brazil, inspired by FRANS POST's depiction of *Vrijburg* as published in 1647. However, the terrain around the palace is worked out more in detail. Figures of soldiers and other servants are added, as well as a lot of birds and other animals. JOHAN MAURITS's coat of arms was also added, being held in the air by an angel. (Rijksmuseum Amsterdam, object no. NG 1053).

GEORG MARGGRAFE's fame has led to several incorrect assumptions. For example, the question has been raised whether he was the maker of a piece of parchment carving representing *Vrijburg* palace (*fig. 119*).[1] Although this assumption could immediately be discarded, because FRANS POST's depiction of *Vrijburg*, published long after MARGGRAFE's death in 1647, was taken as starting point, this piece of clever craftsmanship still needs our attention. This because MARGGRAFE's authorship was motivated by the fact that the carving in question was mounted on a paper quire consisting of five sheets, full of astronomical notes written in a seventeenth-century hand (*fig. 120*).

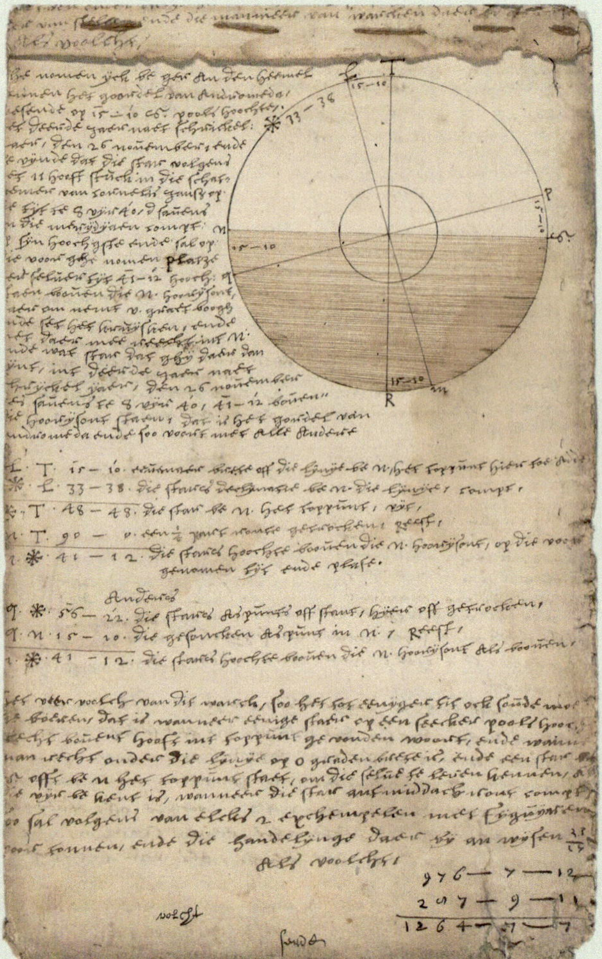

FIG. 120. Fragment of the nautical manuscript in a seventeenth-century handwriting that acted as support to the parchment carving of *Vrijburg* palace. (Rijksmuseum Amsterdam, object no. NG 1053).

An article discussing this carving of *Vrijburg* stated that this manuscript contained 'astronomical observations', so our strong interest in this piece of material was raised. However, a close examination by one of us (HJZ) revealed that the manuscript did not contain any observations, but personal notes about the determination of the polar height.[2] The notes refer several times to *Hooftstuck 11* ('Chapter 11') from a book named the *Schatkamer* by a certain CORNELIS JANSZ. By this reference undoubtedly is meant *De schat-kamer des grooten seevaerts-kunst* (The treasury of the great maritime art), published in Amsterdam for the first time in 1621.[3] Its author, CORNELIS JANSZ LASTMAN, was a former skipper from the Dutch island of Vlieland. After returning from his sea voyages, he started a school on navigation in Amsterdam that would become rather well known. Today, only very few copies of of LASTMAN's *Schatkamer* have survived, but an online available copy of the book shows that *no* Chapter 11 refers to the subject discussed in the manuscript. How is this possible? The solution to this riddle is to realize that at the time LASTMAN's book was so widely used as a manual for Dutch skippers that the name *Schatkamer* became generic for all Dutch navigational manuals.[4] In later years LASTMAN reworked his book and renamed it into the *Beschrijvinghe van de kunst der stuer-luyden* (Description of the art of the helmsmen).[5] It is actually this last book that is meant here. It has indeed a chapter discussing the determination of the polar height, or more generally the problem 'how to find the latitude?'. Then, this chapter is NOT numbered eleven, but it is Chapter 2! Now it becomes clear that the seemingly Arabic number 11 in the manuscript must in fact be read as a Roman number II. This is definitely the chapter in question.[6] The first known edition of LASTMAN's revised book is published in 1642, but it seems that an earlier – now lost – edition was issued in 1634.[7] So, this defines the *terminus post quem* of the manuscript to 1634.

The author of the above named article, EVELINE SINT NICOLAAS, consulted the late JOHN NORTH about the manuscript in 2007. His comment was that the notes concerned some northern stars (among others Capella (= alpha Aurigae) and Regulus (= alpha Leonis) seen from various geographic latitudes, that may be associated with the region of Angola.[8] However, a closer look of the actual manuscript showed this to be wrong. A polar height mentioned in the manuscripts as 8° 36' (Luanda in Angola has a polar height of minus 8° 50') was used only as an arbitrary example, without mentioning any location. Nor is there an indication that the manuscript notes did have

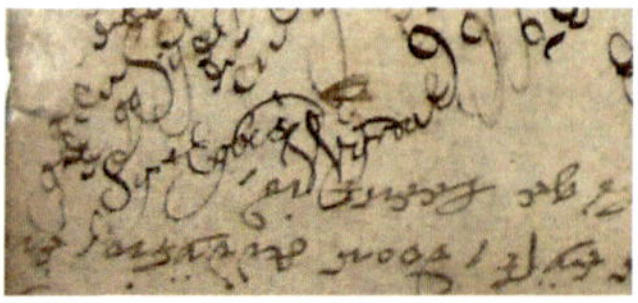

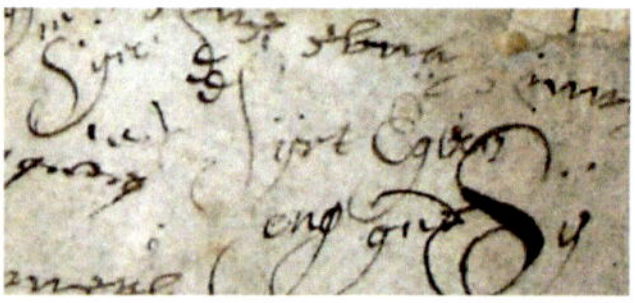

FIG. 121. Fragment of the above named manuscript, revealing the name SYRT EGBERS and the surname WY[N]DAL. (Photo by the authors).

any relation to Dutch Brazil. The notes could easily be written in the Netherlands. The text only contains exercises to elaborate the subject discussed in LASTMAN's book.

Finally, SINT NICOLAAS overlooked that one part of the manuscript was used several times to try the performance of a goose pen, by writing the name 'SIJRT' , 'SIJRT EGBERS' or 'SYRT EGBERS, and on one occasion also the name (?) 'WY(N)DAL' (*fig. 121*). Whether this is the author of the manuscript is not clear. It could also be a former owner. Today, the name 'Siert' is a name still used in the north of the Netherlands.[9] The only SIERT EGBERS found in the Dutch digital archives lived around 1700 in the province of Groningen (Oldenzijl and Uithuizermeden). The surname 'Wijndal', 'Wyndal' or 'Wijndaal' is known to have existed in the Netherlands in the early eighteenth century.[10] The family name only survived in the Caribbean parts of the former Dutch colonies.[11] A small village called 'Wyndal' existed also in the former Dutch colony of South-Africa. Later this place in the western Cape Province was renamed 'Franschhoek'. A similar spot called 'Wyndal' is located in in Fayette County, West Virginia (USA), in an area where many people of Dutch descent live.[12]

NOTES

Chapter 1

1 For details about Marggrafe's lost and surviving manuscripts, see Chapter 6 'legacy'.
2 De Laet, 'Ad Benevolos Lectores' (1648), 2.
3 Christian Marggrafe to Johannes Hevelius, 20 July 1652: 'Fratris mei Georgij observationes Astronomicas iam demu[m] vidi. Sunt Theoriae novae Planetarum, praecipue Mercurij, qui eo in loco, quo vixit, melius quam apud nos conspici potuit. Edentur brevi a clariss. Golio una cum alijs observationibus Astronomicus ex Arabia transmissis'. Observatoire de Paris, Hevelius correspondence.
4 Golius, *Muhammedis Ferganensis [...] Elementa astronomica* (1669).
5 North, 'Georg Markgraf, an astronomer in the New World' (1979). Reissued with an addition in 1989.
6 Gudger, 'George Marcgrave, the first student of American Natural History' (1912), 255.
7 North, 'Georg Markgraf' (1979), 394.
8 *Ibidem*, 399.
9 De Laet, 'Ad Benevolos Lectores' (1648).
10 Polman, *Markgraf e o Recife de Nassau* (1984).
11 Matsuura and Zuidervaart, 'America's earliest (European–style) astronomical observatory, founded and used by Georg Marggrafe in Dutch colonial Brazil' (2014).
12 Matsuura, *O observatório no telhado* (2011). See also his 'Um observatório de ponta no Novo Mundo' (2014).
13 Whitehead, 'The Biography of Georg Marcgraf' (1979), 311, note 1; Hantzsch, 'Georg Marggraf' (1896), 201.
14 ELO, bibliotheek sign. LB 18677.
15 Universitäts- und Landesbibliothek Bonn, sign. O 4' 236/1.
16 Album Amicorum of Andreas Sennert (1606-1689). Universitäts- und Landesbibliothek of Sachsen–Anhalt, sign. Misc. 55 oct.
17 Having established how Georg Marggrafe spelled his own name, it is interesting to note how Marggrafe's youngest brother Christian spelled his own name during his Leiden years. In 1652 he signed his surname 'Marggraviij'; his disputation (1654) was signed 'Marggravius', identical to the signature used in his correspondence with Leiden University in the years 1655 and in poems he wrote for students at Leiden University (1661, 1669). Christian's last will (1667)

and other notarial deeds (1667-1674) are signed again 'Marggraviij' or 'Marggravius'. But the books he published between 1674 and 1689 bear the name 'Marggravio' or 'Marggravii'.

Chapter 2

1 Heijer, *De geschiedenis van de WIC* (2002).
2 Brendecke and Friedrich, 'Introduction' to *Transformations of Knowledge in Dutch Expansion* (2017), 2.
3 Huigen e.a., *The Dutch Trading Companies as Knowledge Networks* (2010).
4 Cook, *Matters of Exchange* (2007), 40: 'The intellectual activities we call science emerged from the ways of knowing valued most highly by the merchant-rulers of urban Europe'.
5 Mary Lindemann, review of *Matters of Exchange*, in *Medical History* 52:2 (2008), 277–278.
6 Steven Paul Matthews, review of *Matters of Exchange*, in *Journal of World History*, 21:1 (2010), 336–338.
7 This theme is discussed extensively in Jorink, *Reading the Book of Nature in the Dutch Golden Age* (2010).
8 Rienk Vermij, review of *Matters of Exchange* in *European History Quarterly* 41:2 (2011), 302–304.
9 Van Berkel, 'Rediscovering Clusius' (2008), 232.
10 Den Heijer, 'Bewindhebbers, gouverneurs en raden van bestuur' (2005), 35–36; Odegard, *Colonial Careers* (2018), 62.
11 The current stadtholder of Friesland and Groningen, Count Hendrik Casimir van Nassau–Dietz (1612-1640), was Johan Maurits's full cousin. Tthe stadtholder of Holland and Zeeland, Frederik Hendrik van Nassau, Prince of Orange (1584-1647), was a full cousin of his father.
12 Roscam Abbing, *Brazilië* (2021), 16-17, 161.
13 Odegard, *Colonial Careers* (2018), 67–72. Van de Tol, *Lobbying in Company* (2018), 64. In the 1640s, the name of the Political Council would change to Council of Justice.
14 'Conditien voorgedragen ende gepresenteert aen sijn genade Graeff Maurits van Nassau bij de vergaderinge der XIX wegens de West Indische Compagnie', §10 and §11. OWIC, 2 (5 August 1636).
15 Schmidt, *Innocence abroad* (2001), 251.
16 Boxer, *The Dutch in Brazil* (1957), 114.
17 Wätjen, *Das holländische Kolonialreich in Brasilien* (1921), 331.
18 The literature on the Dutch painters present in Brazil in the 1640s is vast. See for instance for Post, Larsen, *Interprète du Brésil* (1962); De Sousa-Leao, *Frans Post* (1973); Krempel, *Maler des verlorenen Paradieses.* (2006) and Corrêa do Lago, *Catalogue Raissoné* (2007). For Eckhout, see Brienen, *Visions of savage paradise* (2006) and Buvelot, *Albert Eckhout* (2004). For Willaerts, see Van Groesen 'Marine Painter in Dutch Brazil' (2019).

19 The silversmith JEREMIAS SPOREN is mentioned in the WIC accounts of December 1641 (OWIC 69). His colleague ANDRIES KETELAER, in November 1642. The 'glaesmacker' and 'glaesschrijver' PIETER VAN CONINXLOO in December 1641 for the delivery of glasses since August 1640. He had requested his passage to Brazil on 13 November 1636 (OWIC 14, fol. 209vs). See also Starcke, 'Die Coninxloo's' (1898).
20 Mout, 'The youth of Johan Maurits' (1979).
21 Graf, 'The Collegium Mauritianum in Hesse-Kassel' (1997).
22 Whitehead and Boeseman, *A portrait of Dutch 17th century Brazil* (1989), 20.
23 Barlaeus, *Rerum per octennium in Brasilia. Et alibi nuper gestarum, sub Praefectura Illustrissimi Comitis J. Mauritii Nassoviae* (1647). [= History of the recent activities in Brazil and elsewhere over a period of eight years under the governorship of Count Johan Maurits]. See the preface to the English translation by BLANCH VAN BERCKEL–EBELING KONING, *Barlaeus, the History of Brazil* (2011).
24 Van der Veldt, 'An Autograph Letter of John Maurice of Nassau' (1947). Harmsen, 'Barlaeus' descriptions' (1994).
25 Plante, *Mauritiados Libri XII.* (1647). See Eekhout, 'The Mauritias. A neo-Latin epic by Francis Plante' (1979). See for the publication history of PLANTE's poem also also Roscam Abbing, *Brazilië* (2021), 177, footnotes 345–349.
26 See below, Part II, Chapter 3.
27 See for instance De Quelen, *Kort verhael van den staet van Fernanduc* (1640), [19]: '... en indien het sijn Excellentie niet dickwijls weder te recht en braght, ick gheloove dat het alles soude verlooren gaen'. AUGUSTUS DE QUELEN was a broker in Brazil and Amsterdam. Also positive in his assesment of the count's performance as governor was the Portuguese friar MANOEL CALADO, in his *O Valeroso Lucideno* (1648). See the translated passages in Boxer, *The Dutch in Brazil* (1957), 115-117.
28 Boxer, *The Dutch in Brazil* (1957), 114.
29 See for instance Monteiro and Odegard, 'Slavery at the Court of the "Humanist Prince"' (2020) and Van de Vinde, *Shifting image. In search of Johan Maurits* (2019).
30 Wätjen, *Das holländische Kolonialreich in Brasilien* (1921), 311. He estimated the total yield of the sale of these enslaved Africans at almost seven million guilders.
31 Caland, 'Hofhouding' (1898) 558.
32 OWIC 57, no. 39 (undated c. 1642).
33 Van Berckel-Ebeling Koning, *Barlaeus, the History of Brazil* (2011), 142.
34 See about *Vrijburg*'s garden: Diedenhofen, 'Johan Maurits and his garden' (1979); Da Silva and Alcides, 'Collecting and framing the wilderness. The garden of Johan Maurits' (2002) and Almeida, Borstelmann and Meunier, 'Animais e plantas do horto zoo-botânico do palácio de Friburgo' (2011).
35 Diedenhofen, 'Johan Maurits and his garden' (1979), 197.
36 CALADO, *O Valeroso Lucideno* (1648), 52–53. See the translated passages in Boxer, *The Dutch in Brazil* (1957), 115-117.

37 Marggrafe, 'Historiae Rerum Naturalium Brasiliae' (1648), 140: 'Vidi tamen grandes viginti quatuor aut plurium annorum transferri feliciter anno 1640. Maritiae, adhibita multitudine hominum circiter trecentorum'.

38 Moreau, *Klare en waarachtige beschrijving* (1652), 32–33: 'Zy hieuwen af de schone en heerlijke bomen van Brasilienhout palmenhout, ebbenhout, cederhout, hout zo wit als sneeu, violettehout, marmerhout en ander ruikend hout, die de ruime en lange wandelryen, verder dan men uitzien kon, vercierden, en het heerlijk en prachtig Lusthuis, 't welk Graaf Maurits daar had doen timmeren, omringden'.

39 See also this testimony from his later period in Cleve: 'He has been a great lover of plants, who [...] taught the citizens of Cleve how to plant avenues and forests'. *Kleefsche Waterlust* (Amsterdam: Buyn, 1752), 29: 'Hy is een groot Liefhebber van Planten geweest, welke niet alleen deeze, en eenige andere Alléen geplant, maer ook de Kleefsche Burgers, het planten van Alléen en Bosschen geleerd, en zich daerdoor by Hen een eeuwige Nagedachtenis gesticht heeft'.

40 Johannes Grevingh, steward of the count's finances, to the Chamber Zeeland, 1 September 1643 (OWIC 58, no. 105): 'Het is een lust om te sien is wat sijn Excellentie in desen deele voor een licht aen de heele werelt verleent heeft met boomen, cruijden ende alderleij gewas hier op te queecken te saeijen en te planten, Hij practiseert int planten dagelijcks wat nieuws ende mainteneert boven al toe cultivatie vant landt, de opqueeckinge van alderley raere beesten en gevogelte [en] de onderhoudinghe van visscherijen'.

41 Van Berckel-Ebeling Koning, *Barlaeus, the History of Brazil* (2011), 143.

42 *Ibidem*, 144.

43 Calado, *O Valeroso Lucideno* (1648), 53. English translation by Boxer, *The Dutch in Brazil* (1957), 115-117.

44 Wiesenbron, *Brazilië in de Nederlandse archieven* (2005), 12.

45 See about the concept of the 'Trading zone', applied to Early Modern Science, Long, *Artisan/Practitioners and the Rise of the New Sciences* (2011).

Chapter 3

1 Hantzsch, 'Georg Marggraf' (1896), 201: 'Teufflinge Anno 1610': 'Georgius, Georgii Marggravij Söhnlein, tum temporis Schulmeister allhier, ward auf diese Welt geboren 20. septemb. d. vesperi intra sextam et septimam horam, getauft am 23. ejusdem'. Date in old style, then with a difference of 10 days with the Gregorian calendar.

2 *Ibidem*: 'Trauregister Liebstadt, 27 nov 1609 (O.S.): 'Foedus connubiale contrahebat cum honesta et pudica virgine Elisabellia, reverendi et ductissimi viri Dei .M. Johannis Simonis, pastoris hujus ecclesiae fidelis, Georgius Marggravius, scholae ejusdem tum temporis moderator, 27. die Nov. ao. 1609'.

3 Küchenmeister, *Etwas von Liebstadt* (1743), 15.

4 Kreuszig, *Album* (1898), 112.

5 See Pfarrersuche Sachsen, https://pfarrerbuch.de/sachsen (consulted June 2020).

6 *Ibidem.*
7 There seems to be no relationship of the Liebstadt family with the imperial clockmaker CHRISTOPH MARGGRAF, working in Prague at the court of RUDOLPH II. See Von Bertele & Neumann, 'Der kaiserliche Kammeruhrmacher Christoph Marggraf und die Erfindung der Kugellallfuhr' (1963).
8 Küchenmeister, *Etwas von Liebstadt* (1743), 12. See the online *Corpus Inscriptorum Vitebergense* (CIV) and the *Album Academiae Vitebergensis (1587-1610)*, fol. 134r, registration 15 July 1598: 'Georgius Marggraffe, Libstadensis Misnicus', together with 'Johannes Simon, Witebergensis'. (See: http://digital.bibliothek.uni-halle.de/hd/content/ pageview/2530432. Consulted 28 July 2020). MARGGRAFE's grandfather JOHANN SIMON reveals his place of origin ('Iuchsensem Francum' [= Jüchsen]) on the title page of his eulogy of the lord of the Liebstadt castle, RUDOLPH VON BÜNAW. See Simon, *Christliche Leichpredigt* (1603).
9 Küchenmeister, *Etwas von Liebstadt* (1743), 12.
10 Christian Marggrave, 'Vita Georgii Marggravii' (1685): '... ac pictor non contemnendus evaserit'. An English translation by the apothecary JAMES PETIVER (1663-1718) is published by Whitehead, 'The Biography of Georg Marcgraf' (1979).
11 Weissenborn, *Album Academiae Vitebergensis* (1934), 308. See also the online *Corpus Inscriptorum Vitebergense* (CIV), entry on 10 May 1627 (o.s.): 'Nicolaus Rebhun, Liebstadio Misn.' and 'Georgius Marggraffius, Liebstadio Misn. Gratis receptus'. The classmate was NICOLAUS REBHUN, most likely the son of the former school assistant and organ player in the Liebstadt church, the late NICOLAUS REBHUN Sr. († 1610). See: Küchenmeister, *Etwas von Liebstadt* (1743), 17.
12 We changed the order of these cities a bit, to fit the chronology better with the verifiable data, discussed below. See Marggrafe, *Vitae* (1685). The list of cities is shortened in Whitehead, 'Biography'(1979).
13 Cf. Whitehead, 'The Biography of Georg Marcgraf (1979), 308 and 311–note 4 and Meijer, 'De omstreden nalatenschap' (1972), 75.
14 Knot, *Die alten Matrikeln der Universität Strassburg* (1897), 288 (22 November 1627). 'Georgius Marggrafius, Liebstadio-Misnicus'. JOHN NORTH, in his account of MARGGRAFE's astronomical activities (1979), suggested that in Strasbourg, MARGGRAFE had contact with the astronomer JAKOB BARTSCH, the assistant and son-in-law of JOHANNES KEPLER. However, this supposition is false. BARTSCH had left Strasbourg in 1624, living in Italy until 1627, when settled in Lauban in German Silesia (nowadays Luban in Poland). He only returned to Strasbourg in 1630, to marry KEPLER's daughther SUZANNA.
15 ELO, North no. 27vs. (list of books owned by Marggrafe). This list mentions Habrecht's *Planiglobium coeleste ac terrestre* (Argentorati [Strasbourg], 1628).
16 Wackernagel, *Die Matrikel der Universität Basel* (1962), 306 – April 1628: 'Georgius Margraffius, Liebstadio-Misnicus — nihil propter paupertatem'.

17 SIMON PAULLI (1603-1680) studied medicine in Rostock (1617), Leiden (matriculation, 16 May 1623), Paris (1629-1630) and Wittenberg (matriculation, 8 May 1630). Two month later, on 10 July 1630 he received his doctorate there, with a thesis entitled *De Arthritide Disputationem* (1630). His promoter was the professor of medicine, DANIEL SENNERT, whereas his future father in law, the Rostock professor of medicine JACOB FABRICIUS (1576-1652), acted as co-promotor. So, in theory MARGGRAFE could already have met PAULLI in Wittenberg. In April 1634 PAULLI was appointed professor of medicine and botany at Rostock University. A year later he married ELISABETH FABRICIUS, daughter of the aforementioned JACOB FABRICIUS. In 1639 he moved to Denmark, being appointed Professor of anatomy, surgery and botany in Copenhagen.

18 Paulli, *Quadripartitum botanicum* (1667), 551: '... discipulus olim meus in Rostochiensi Academia p.m. Georgius Marggravius ...' ('... a student of mine at the University of Rostock, GEORGE MARGRAVE ...'). As SIMON PAULLI was appointed professor in Rostock in April 1634, coming from Lübeck where he had worked as a physician after 1631, MARGGRAFE must have been among his first students.

19 MARGGRAFE, *De Causa continente morborum* (1634). Halle, Universitäts- und Landesbibliothek Sachsen-Anhalt Call number: 05 A 1570. See: http://dighist.fas.harvard.edu/projects/sennertcollab/items/show/2436. (Consulted 28 July 2020).

20 *Ibidem*, proposition 5: '... cum Excellentiss. Dn. Sennerto, Patrono nostro, & Praeside eximio ... '.

21 *Ibidem*, p. 1 'Prologus'.

22 Σύν θεω is the Old Attic or Homeric form of the expression 'with God'or 'with Gods help', also written as σύν θεω or ξύν θεω. JOHN NORTH erroneously interpret this phrase as 'study carefully'. North, 'Markgraf'(1979), 218.

23 Marggrafe, *De Causa continente morborum* (1634), 17: '*Omnia ad Dei Opt. max. gloriam proximi salutem*'.

24 *Ibidem*, proposition 58: 'Scio qujdem judiciorum varietatem. In eodem prato bos quaerit gramen, canis leporem, ciconia ranam, sus pro suaveolente amaraco lutum amat ...'.

25 *Ibidem*: 'Tuum tamen erit, benivole Lector, haec omnia in meliorem partem interpretari Si quid vero novisti rectius, candidus imperti: si non, HAEC RODERE NOLI'.

26 This motto is a variation on a well known saying from the polyhistor PHILIPP CAMERARIUS 'Absentem & fato defunctum rodere noli', meaning 'About the dead, one says nothing but good'. See Camerarius, *Historischer Lustgarten*, vol. 3 (Leipzig: Wachsmann, 1630), 356.

27 FRIDERICUS GOEBELIUS matriculated at Wittenberg University on 16 May 1631, as coming from 'Katzwango Noricus', or Katzwang, a part of the city of Nürnberg. In the spring of 1634 he registered at the University of Frankfurt (Oder) as coming from 'Galroth, Noricus', being a 'pauper'. He would obtain his medical degree there in 1635, with a thesis entitled *Disputatio Medica Inauguralis De Apoplexia*

(Francofurticis Viadrum: Kochius, 1635). See: Friedlaender, Ältere Universitätsmatrikeln (1887), p. 728. In later years GOEBEL settled as a Rector in Melldorf in Ditmarschen (Schleswick).

28 'MARGGRAVIO Phoebus genium formavit; at Hermes | Externis Doctorum addit amicitiis | Pro se quitque trahit, duo certant numina in uno | Discere seu maleat sive docere velit. | FREDERICUS GOEBELIUS Galeota, | Noricus, Med. Stud.'.

29 Justus Lipsius, 'De ratione cum fructu peregrinandi et praesertim in Italia'. This letter was addressed to the nobleman PHILIPP DE LANOY. See for instance: Lipsius, *Epistolarum selectarum tres centuriae* (Antwerp, 1691), no. XXII, pp. 23–29. In relation to this quotation it is very curious that between the horoscopes in the Leiden manuscripts also some astrological calculations are preserved relating to a person born on 18 October 1547 at a location with a latitude of 50° 48' NB. The Louvain humanist JUSTUS LIPSIUS (†1606) was born on this very date in Overijse (Belgium), which has a latitude of 50° 47' NB. Apparently MARGGRAFE was so impressed with LIPSIUS that he still wanted to draw up his horoscope. See ELO, North no. 87.

30 Francis Schiller, 'An eighteenth century view of the Placebo effect', *Clio Medica. Acta Academiae Internationalis Historiae Medicinae* 19 (1984), 81–86.

31 Krohn, *Christoph Nothnagels Lehr- und Forschungstätigkeit* (2014). NOTHNAGEl was officially appointed on 28 March 1634 (O.S.).

32 North, 'Markgraf' (1979), appendix, horoscope no. 10.

33 JACOB FABRICIUS (1576-1652) worked on Hven from 1592-1596, before going on a study trip through Holland (matriculation Leiden, 21 May 1598), England and Germany (Jena 1602). In 1603 he settled in Rostock as a physician, where he was appointed professor of medicine and higher mathematics at Rostock University in 1612. From 1637 until his death in 1652 he became the personal physician of the Danish king in Copenhagen. His daughter ELISABETH married his former student SIMON PAULLI in 1635. See: Christianson, *On Tycho's Island. Tycho Brahe and his assistents* (2000), 276–277.

34 Lauremberg a.o., ΕΠΑΙΝΕΤΙΚΑ. *Mitrae Doctoralis* (1635). MARGGRAFE's fellow student SALOMON FISCHER was a son of THOMAS FISCHER, surgeon to the Elector of Saxony. FISCHER studied medicine in Leipzig [1630-1632] with JOHANNIS ZEIDLER (1596-1645) and then in Rostock [1632-1635] with JACOB FABRICIUS and SIMON PAULLI. His adventures after graduating in Rostock are unknown.

35 Eichstadt, *Prognosticon* (1622).

36 LORENZ EICHSTÄDT (1596-1660) started his studies at the gymnasium in Stettin; in 1612-1613 he visited the University of Greifswald, went in 1614 to SENNERT in Wittenberg, and continued his study during the years 1617-1619 at different places in Germany and the Netherlands. He obtained his medical degree with a thesis titled *De Pleuritide Disputatio* (Wittenberg, 1621), after which he became city physician of Stettin in 1624. He lived in Stettin until 1645, when he was appointed professor of mathematics and astronomy in Dantzig.

37 Eichstädt, *Ephemerides Astro-Meteorologicae [...] Oder, Teutsche Allmanach* (1631).
38 Eichstädt, *Pars Prima Ephemeridum Novarum et Motuum Coelestium* (September 1634), unnumbered page before page 1. See also the lemma 'philochymiater', in: Ramminger, *Neulateinische Wortliste* (used on 26.7.2020).
39 Eichstädt, *Pars Altera Ephemeridum Novarum et Motuum Coelestium* (1636; 1639^2), 109.
40 ELO, doc. no. 65. Next to the date, the note also bears a small scribble 'Stetin'.
41 North, 'Markgraf' (1979), appendix, no. 16: a number of horoscopes, probably made for a family 'K' from Stettin, with eight children, born between 1622 and 1635. See also the next note.
42 *Obersekretär* JOHANN FABRICIUS was born in Husum (Holstein) on 26 May 1594, as the son of LAURENTIUS FABRICIUS, citizen of Husum and ANNA NICOLAI. He was probably related to the poet OTTO FABRICIUS, who was Husum's city secretary from 1582-1584. See North, 'Markgraf' (1979), appendix, horoscope no. 5. See: http://encyklopedia.szczecin.pl/wiki/Johann_Fabricius. (Consulted 28 July 2021).
43 FABRICIUS's *Album Amicorum* (with some 120 entries from Amsterdam, Königsberg, Leiden, Paris, Stettin, Uppsala and other places) is preserved in the Kungliga Biblioteket in Stockholm. See the digital *Repertorium Alborum Amicorum*. See also: Davidson, 'Deutsche Stammbücher in swedischen Sammlungen' (1981), 90.
44 See the digital *Repertorium Alborum Amicorum*: Stammbuch Christoph Loth, Stadtbibliothek Königsberg, S 64 / 12° (Verlust), fol. 87r (Eichstadius) and 159r (Marggrafe). LOTH collected also entries of the Wittenberg professor of astronomy AMBROSIUS RHODIUS (fol. 148r) and the Danzig mathematician ad astronomer PETRUS CRÜGER (1580-1639). See also: Bogun, 'Die Stammbuchsammlung in der Stadtbibliothek zu Königsberg, III' (1909), 32. CHRISTOPH LOTH registered as a student in Königsberg in the summer of 1629. There, he was a class mate of CHRISTOPH NOTHNAGEL, who in 1634 became MARGGRAFE's Wittenberg professor of astronomy. See: Erler, *Die Matrikel und die Promotionsverzeichnisse der Albertus Universität zu Königsberg* (1910), 313–314. In 1635 CHRISTOPH LOTH became one of the students at the Stettin Gymnasium. On 20 May 1639 'Christophorus Lothus Regiomontanus' also registered as a student of medicine at Leiden University.
45 Marggrafe, in: *Epicedia ad clarissimum, [...] Laurentium Eichstadium* (1636).

CHAPTER 4

1 Noordegraaf & Valk, *De gave Gods* (1996), 59, 99, 105, 111. See also *De Navorscher* 76 (1927), 76.
2 Du Rieu, *Album Studiosorum* (1875), 280. As the Leiden 'recensie register' for the year 1637 is missing, his repeated registration in the following years could not be checked.

3 DIRCK EDUWAER was a tailor, who in 1631 had married LYSBETH ARENTSDR, from Haarlem. At that time the couple already lived in the 'Choersteech'. In MARGGRAFE's time the couple had two young children: EDUAERT (*1632) and ARENT (*1635). A third child – HILLEGOND – was born in 1638, after he had left the country.
4 Whitehead, 'Biography of Marcgraf' (1979), 309.
5 In two letters by WILLEM PISO to JACOB GOLIUS, dated 12 and 16 May 1655, he says that MARGGRAFE lived in the boarding house of 'Mrs GRIETTIE BAILLY, the former housekeeper of Professor BRONKHORST'. The latter was EVERARD BRONCKHORST (1554-1627), professor of law at Leiden University, who during his life had been a close friend of WILLEBRORD SNELLIUS. GRIETTIE BAILLY lived on the Rapenburg, directly opposite the academy building (today no. 54). See: See: Lunsingh Scheurleer, *Het Rapenburg* 5 (Leiden, 1990), 721 and Meijer, 'De omstreden nalatenschap'(1972), 68–69.
6 SAMUEL KECHEL AB HOLLESTEYN (1611-1668) matriculated in Leiden on 16 October 1632 as a student of philosophy. See about him further: Zuidervaart & Beurze, 'Kechel ab Hollenstein' (2014).
7 Howse, 'The Greenwich List of Observatories' (1986). Cf. De Wreede, *Snellius* (2007), 101, 147. Only a few years earlier MARGGRAFE could have preferred Franeker University in the Northern part of the Netherlands. Here, until his death, ADRIAEN METIUS (1571-1634) lectured in mathematics and astronomy. As a former apprentice of TYCHO BRAHE, METIUS also performed astronomical observations in the open air with instruments after his own design. In 1629 RENÉ DESCARTES was one of his students. Cf. Christianson, *On Tycho's Island* (2000), 322.
8 For a crisis in this policy around 1620, see Grafton, *Athenae Batavae* (2003).
9 Huisman, *The Finger of God* (2009), 128-133.
10 Lunsingh Scheurleer & Posthumus Meyjes, *Leiden University in the seventeenth century* (1975); Ruestow, *Physics at seventeenth- and eighteenth-century Leiden* (1973); Wiesenfeldt, *Leerer Raum in Minervas Haus* (2002).
11 Otterspeer, *Groepsportret met Dame*, 1 (2000), 326.
12 *Ibidem.*
13 Bouwman (et al), *Stad van Boeken* (2008). See also: Verbeek & Bos, *Descartes en Leiden* (2003).
14 De Wreede, *Snellius* (2007).
15 Christianson, *On Tycho's Island* (2000), 358–361; De Wreede, *Snellius* (2007), 47. In Prague SNELLIUS also had made his acquaintance with JOHANNES KEPLER.
16 Christianson, *On Tycho's Island* (2000), 254–256.
17 For more details on SNELLIUS's triangulations, see: Haasbroek, *Frisius, Brahe and Snellius and their Triangulations* (1968).
18 In 1625 SNELLIUS ordered two similar brass quadrants for his French correspondent PIERRE GASSENDI, because 'the Paris artisans were not able to make an instrument with the desired degree of precision'. In July 1625 this quadrant was at SNELLIUS's house. Cf. De Wreede, *Snellius* (2007), 93–95.

19 Du Rieu, *Album Studiosorum* (1875), 97 (13 March 1610). For his second entry as a student of law, see page 135 (26 March 1618).
20 Wiesenfeldt, 'The *Duytsche Mathematique*' (2019), 60.
21 Juynboll, 'Jacob Golius' (1931), 128.
22 *Ibidem,* 145-146 .
23 Molhuysen, *Bronnen,* 2 (1916), 143. See for GOLIUS's work as a mathematician, Dijksterhuis, 'Moving around the ellipse' (2011).
24 The remaining part of SNELLIUS'S instruments (making up a list of some fifteen mathematical instruments, including another 'quadrans ferreus amplissimus') WAS PUT UP FOR AUCTION WITH HIS LIBRARY IN MARCH 1629. CF. *Catalogus [....] Willebrordi Snellii* (1629), 68–69.
25 Leiden University Library, Archive of the Curators (AC) I: *Resolutien 1628-1637,* 6 August 1632. See also AC-I, inv. no. 22, fol. 78: purchase quadrant for 125 guilders. Cf. Molhuysen, *Bronnen,* 2 (1916), 177. SNELLIUS's first large quadrant remained in GOLIUS's possession. On 4 June 1668 it was sold at auction with GOLIUS's library. Cf. *Catalogus Golii* (1668), 'Instrumenta mathematica', no. 7: 'Cl. Snellii quadrans, quem adhibuit in dimensuratione ambitus globi Terreni, cum requisitis'.
26 Heineman, *Het Academiedak* (1979), 4.
27 Resolution 9 August 1634. Molhuysen *Bronnen,* 2 (1916), 190. See the specifications for the building contractor of the turret, dated 14 June 1634. Cf. Univ. Library Leiden, AC I, inv. 22, fol. 151-153.
28 See for the early history of Leiden Observatory: Kaiser, 'Geschichte und Beschreibung der Sternwarte in Leiden' (1868).
29 Whitehead, 'The Biography of Georg Marcgraf' (1979), 308. ADOLPH VORSTIUS (1597-1663) was professor of medicine and botany since 1624. He lived in the official service residence of the professor of botany in the Nonnensteeg, the alley next to the academy building and adjacent to the *Hortus Botanicus.*
30 ELO, Marggrafe papers, North doc. no. 63.
31 Juynboll, 'Jacob Golius' (1931), 138.
32 *Catalogus Golii* (1668), 'Instrumenta mathematica', no. 18: 'Figuræ Mathematicæ ad Apollonium Pergæum de sectionibus conicis pertinentes, cura J.Golii in buxino sculptae'.
33 After 1635 FRANS VAN SCHOOTEN THE ELDER (1581-1646) was assisted in his lessons by his talented son FRANS VAN SCHOOTEN THE YOUNGER (1615-1660). See for these academic mathematics lectures in the vernacular: Van Winter, *Hoger beroepsonderwijs avant la lettre* (1988). See also: Muller & Zandvliet, *Admissies als landmeter in Nederland voor 1811* (1987).
34 Witkam, 'Jean Gillot, een Leids ingenieur' (1967) and (1969), esp. 39. GILLOT enrolled at Leiden University in 1630 as a student of mathematics.
35 For the 'Hollandse cirkel', see Pouls, *De landmeter Jan Pietersz. Dou* (2004) and Cocquyt, 'The Holland circle' (2014).
36 ELO, North no. 64 (calculations); no. 22 (draft observations 19 January) and no. 52 (observation register).

37 North, 'Markgraf' (1979), appendix, horoscope no. 8: GEORG HEINRICH VON BERNSTEIN (†1670 in Magdeburg) became dean of the cathedral in Magdeburg and 'Domherr' in Naumburg. He was also the owner of a manor in Saxony-Weissenfels. He married MARIA SOPHIA VON BENNIGSEN IN 1648. Their marriage Bible is preserved in Halle. See: http://www.inschriften.net/halle-an-der-saale/inschrift/nr/di085-0516.html#content (consulted 1 August 2020).

38 North, 'Markgraf' (1979), appendix, horoscope no. 9 for 'A.F. von H.', born on 22 October 1610 at Gustrovius [*today Güstrow, about 35 km south of Rostock*]. Entry in the Leiden *Album Studiosorum* on 20 May 1636: ADOLPHUS FRIEDRICH VON HAGEN, Megapolitanus [= Mecklenburg], 24 years [* ~1612], 'in de Molensteegh int gecroonde schaep'. He became an influential lawyer in Rostock and married there to ELISABETH HAHN (1598-1667) [her third marriage] on 2 August 1642.

39 North, 'Markgraf' (1979), appendix, horoscope no. 12 for 'G.K.', born on 20 January 1613 [o.s.] in Stettin. Entry in the Leiden *Album Studiosorum* on 31 October 1634: 'GEORGIUS KIRSTENIUS, Stetinensis. 22 years [= * c. 1612], at 'GRIETJE MOMBERS' [= Mommers], together with PHILIPPUS OESELER, also from Stettin. This is the only 'G.K.' from Stettin in the Leiden *Album* in the period 1630-1638. According to his biography, he indeed was born on 20 January 1613 in Stettin. See: Wigartho, *Des mit allerley Merckwurdigkeiten Gelehrter Leute aus Alt- und Neue Zeiten*, 1 (1738), 37–38. GEORGIUS KIRSTENIUS studied earlier in Greifswald (entry 19 September 1627) and Strasburg (entry 14 October 1630). See: Friedländer, Ältere Universitätsmatrikeln. II. Universität Greifswald (1893), 488; Knot, *Die alten matriklen der Universitat Strassburg* (1897), 297. But according to his biography, cited above, he studied also in Jena, Tübingen, Leiden (1634), Franeker, Groningen and Utrecht, where he attended the inauguration of the University (26 March 1636). He graduated in Leiden on 16 July 1638 with a *Disputatio medico-philosophica inauguralis de lumbricis* (Leiden: Elzevier, 16 July 1638). Towards the end of his life KIRSTENIUS remembered the days he spent as a student in Leiden in a poem for another study friend, JOHANNES CUNRADUS DIETERICUS (entry Leiden, 14 January 1637). See: *Antiquitates Biblicæ* (Gissae Hassorum: Seyler, 1671), [unnumbered page in the beginning of the book]: 'Ita gratulur Amico suo Lugduni in Batavia, GEORGIUS KIRSTENIUS, Med. D.'. He was married to CHRISTINA ELISABETH RÖCHLIN (1634-1680).

40 Kirstenius, *Adversaria et Animadversiones* (1648). See also Thorndike, *A History of Magic and Experimental Science* (1958), 197.

41 North, 'Markgraf' (1979), appendix, horoscope no. 13 for 'Th. FUIREN' [North read 'Th Pursen'], born on 27 May 1614 in Hafnia (= Copenhagen). At first sight one would think that THOMAS FUIREN was meant, who in the years 1636-1638 also studied in Leiden, but the *Dansk Biografisk Leksikon* confirms that his brother HENRIK was born on 27 May 1614.

42 HENRIK FUIREN arrived in Leiden on 15 October 1633, where he registered at the university three days later: 'HENRICUS FUIRENUS Danus

Hafniensis', student in medicine, 'apud viduam Chakelij, conrectoris in de Lochorststrate'. PETRUS GAKELIUS was a conrector of the Leiden Latin School. At the registration FUIREN gave as 21 as his age, so he pretended to be two years older than he really was. See for his study of mathematics Bartholin, *Oratio in obitum Henrici Fuiren* (1659), 24.

43 During these years FUIREN lived in Leiden (1633-1635), Franeker (1635-1636), Leiden (January 1636–October 1637), Paris (1638-1639), Padua (1639-1645), Basel (1645) and again Leiden (1646). See for his *peregrinatio academica*: Grell, *Centres of Medical Excellence?* (2010), 184-189. HENRIK and THOMAS FUIREN's *Albora Amicorvm* over these years are preserved in the Kongelige Bibliotek in Copenhagen, Signatur NKS 372 &373 oktav. and Thot, 561–562. oktav. See Helk, *Stambogs-skikken* (2001), 315.

44 Henrik Fuiren, *Bibliotheca Medica quam patriae academiae hafniensi legavit* (Copenhagen, 1659) and Thomas Fuiren, *Rariora musaei Henrici Fuiren quae Academiae Regiae Hafniensi legavit* (Copenhagen, 1663).

45 Registrations in the Leiden *Album Studiosorum*: 10 February 1636: 'THOMAS FUYREN, Hafniensis, Danus. 20 [years], M[edicine]. 'Habitans apud HENRICUM BRANDIS op het Steenschuyr'; 11 July 1637: 'BARTHOLUS CASPARII BARTHOLINUS, 23 [years], studiosus philologia, & THOMAS CASPARII BARTHOLINUS, 21 [years], studiosus philologia & medicinae, ejusdem frater, apud ANNELIEN BAUDIUS'.

46 HENRIK FUIREN to ØLE WORM, 26 August 1636 and 10 September 1636. Scheperlern, *Breve* (1965), no. 605 and 606.

47 ØLE WORM to HENRIK FUIREN, 11 June 1636; HENRIK FUIREN to ØLE WORM, 31 August 1636. Scheperlern, *Breve* (1965), no. 596 and no. 606.

48 Citation from the Roman writer CATULLUS, *Carmen* 13: 'You will dine well with me, my Fabullus, in a few days, | if the gods favor you, if you bring with you a good and large dinner, | not without a dazzling girl and wine and wit and all your loud laughter. | If you bring these things, I say, our charming one, | you will eat well; for the wallet of your Catullus is full of cobwebs'.

49 OTTO HEURNIUS [OTTO VAN HEURN] (1577-1652), Professor of Medicine and Anatomy after 1601.

50 ADRIANUS FALCOBURGIUS [VAN VALCKENBURG] (1581-1650), Professor of Anatomy after 1625. In one of his works THOMAS BARTHOLIN recalls and illustrates FALCOBURGIUS's dissection in 1637 of a Brazilian pig, a *Tajacu Caaigoara*. This animal had been transported to the Netherlands alive and had been kept by JOHANNES DE LAET in his garden on the Rapenburg. See Bartholin, *Twee hondert getal seltsame ondervindingen* (1657), and DE LAET's 'annotatio' in the *Historia Naturalis Brasiliae* (1648), second part, 229.

51 THOMAS BARTHOLIN to ØLE WORM, 6 March 1638. Scheperlern, *Breve* (1965), no. 702.

52 KECHEL's first known observation is one of the solar (annular) eclipse of 1 June 1639, which was partial in Leiden. But according to a letter

of JØRGEN FROM to LONGOMONTANUS, dated 1638, KECHEL and FROM already made observations together since 1637. See Pingré, *Annales* (1901), 129.

53 Du Rieu, *Album Studiosorum* (1875), 367 (10 april 1646): 'THOMAS BARTOLMUS Danus. 29'; 'HENRICÜS FUIREN. Danus. 31', and 'ERASMUS BARTHOLMUS, Danus studiosi matheses, ann. xxi, habitat apud Samuel Kechel'. See also the 'recensieregister' for the year 1646, in which their accommodation with KECHEL is noted. (Univ. Libr. Leiden, ASF 10). After the death of KECHEL's wife in 1647, the BARTHOLIN brothers moved to other lodgings. See Zoeteman, *De studentenpopulatie van de Leidse universiteit* (2011), 299. During his study in Leiden, ERASMUS BARTHOLIN would become co-editor of Descartes' *Geometry* (1651), an edition that KECHEL also owned. Returned to Denmark he became a professor of mathematics in Copenhagen. As astronomer he observed the comet of 1665. ERASMUS BARTHOLIN was also engaged in the editing of the observations of TYCHO BRAHE. This work was never finished, but the notes were brought to Paris by JEAN PICARD in 1671, being copied there in the beginning of the eighteenth century at the request of DE L'ISLE by the same scribe who would copy MARGGRAFE's Brazilian observations.

54 Witnesses at the baptism on 3 October 1646 of ANNA ELISABETH KECHEL, daughter of SAMUEL KECHEL AB HOLLENSTEYN and ALIDA LUS were HENRICUS FUIREN; ERASMUS BARTHOLIN; ARENT VAN 'S GRAVESANDE [the Leiden city architect]; MARIA [VAN LOO], widow of the professor of philosophy GILBERT JACCHEUS (†1628), and ANNA LUS [born BAUDIUS] (the child's grandmother). (ELO).

55 Du Rieu, *Album Studiosorum* (1875), 287 (11 July 1637): 'BARTOLUS CASPARI BARTOLINUS, Hafniensis Danus. 23, P. apud ANNELIEN BAUDIUS. | THOMAS CASPARI BARTOLINUS, ejusdem frater. 21, P. et M., eadem'. Idem: 25 July 1639): CASPARUS BARTHOLINUS Hafniensis. 21, J. apud ANNA BAUDII. (Lodgings added from Univ. Libr. Leiden, ASF 9).

56 The two Danish student who JØRGEN FROM accompanied were CHRISTIAN BROCHMAN († London, 1638), son of the Copenhagen Professor of Theology JESPER RASMUSSEN BROCHMAND (1585-1652) and JOHANNES MATTHIAS [= HANS MECKELBORG (1619-1663)], son of the rich merchant and Mayor of Copenhagen MATHIAS HANSEN (1555-1628).

57 His name does not appear in the official registration records of these universities.

58 FROM to JESPER BROCHMAND, 3 September 1636, cited in: Kragh, *Jørgen From* (2014), 10-11. A draft letterbook by FROM from the years 1636-1638 is preserved in the Kongelige Bibliotek in Copenhagen: *Epistolae Latini et nonnullae Gallicæ ad viros in Dania claros ex Belgio annis 1636, 1637 et 1638 a Georgio Fraem perscriptae.* Sign. Rostg. 157 kvart. See also Breve [Jørgen (Georgius) From (Fraem), matematiker astronom professor Afskrift efter 157 eller 158 kvart i Rostg. Saml. Sign. NKS 2007 kvart.

59 Du Rieu, *Album Studiosorum* (1875), 281 (30 September 1636).
60 Today Rapenburg 69A, southern part. See: Lunsingh Scheurleer, *Het Rapenburg* 6b (1992), 607. FROM told in an undated letter to JESPER BROCHMAND that for his Leiden lodgings he paid 96 florins per month for all three, or eight florins per week per person (draft letter no. 17).
61 Entry 8 October 1636: 'GEORGIUS ANDREAE FRISIUS, Nobilus Danus. 26, Mathesis', and his butler 'ANDREAS LAURENTIUS, Danus, famulus Nobilis FRISII, 25'. His *praeceptor* was HANS JØRGENSEN MULE (1605-1669), who matriculated a few days later (on 15 October 1636) as 'JOHANNES MOLENIUS, Danus. 26, Litt..' Their boarding house was *In Salomons Tempel op de Lange Brugge.* See: Kragh, *Jørgen From* (2014), 13.
62 FROM to CHRISTEN FRIIS, undated [?] (draft letter no. 24).
63 FROM to BROCHMAND, undated [?] (draft letter no. 41).
64 FROM to BROCHMAND, 9 September 1636 (draft letter no. 4).
65 FROM to BROCHMAND, 7 May 1637. (letter number not mentioned by Kragh).
66 FROM to LONGOMONTANUS, date unknown. (letter number not mentioned by Kragh).
67 FROM to LONGOMONTANUS, 1 May 1637 (draft letter no. 54). With the latter statement FROM referred to three well known astronomical tables that corresponded to the three rival world systems of the time (excluding the Ptolemaic), namely, Keplerian, Tychonian and Copernican. See Kragh, 'Georgius Frommius' (2015), 51.
68 MARGGRAFE to FRAEM (= FROM), 26 June 1638. (ELO).
69 Du Rieu, *Album Studiosorum* (1875), 281 (30 September 1636).
70 FROM to LONGOMONTANUS, January 1638 (draft letter no. 46).
71 MARGGRAFE to FRAEM (= FROM), 26 June 1638. (ELO)
72 FROM's gift to KECHEL was the book by the Italian engineer GUIDOBALDO DEL MONTE, *Mecanicorum liber.* (Venice: Evangelisto Deuchino, 1615). See: Sothebys, *The library of the Earls of Macclesfield removed from Shirburn Castle,* Part 5: Science I-O (14 April 2005), no. 1440.
73 Van Berkel, 'De illusies van Martinus Hortensius' (1997/1998).
74 Vermij, *The Calvinist Copernicans* (2002) 104n. Some of HORTENSIUS's observations are published in the works of PHILIPPUS LANSBERGEN.
75 Pingré, *Annales Célestes* (1901), 78-182.
76 FROM to HORTENSIUS, undated [1636] (draft letter no. 6). ERIK OLUFSEN TORM (1607-1667) a Danish mathematician. He would marry SØSTER WORM (1623-1685) in 1640, a daughter of ØLE WORM.
77 FROM to BROCHMAND, undated [1638] (draft letter no. 13).
78 FROM to LONGOMONTANUS, undated [1638] (draft letter no. 51).
79 HORTENSIUS to BOULLIAU, 20 May 1637. BNF, Paris, Ms Fr. 13037, fol. 94. See the transcription and Dutch translation by AD DAVIDSE, online at https://adcs.home.xs4all.nl/hortensius/briefw.html#20-3-1637 (consulted April 2021): 'Maar een of andere Duitser in Leiden zegt dat hij met voldoende toewijding heeft waargenomen, en dat Mars nergens meer is verschenen ...'.
80 See the marginal note on the flipside of MARGGRAFE's draft letter to FRAEM (=FROM), 26 June 1638. (ELO)

81 Du Rieu, *Album Studiosorum* (1875), 280 (11 September 1636): 'OTTO UXKUL, Nobilis Livonus. 21 [=* c. 1615], Pol.' = [Otto von Uexküll, son of JOHANN VON UEXKÜLL, 'Freiheer zu Padenorm' (1586-1648), 'Oberstleutnant' in the Swedisch army, and ANNA VON MAYDELL (1590-1648). He died shortly afterwards in Rocroi on his *Grand Tour* in France. See: *Genealogisches handbuch der baltischen Ritterschaften, Band 2-1: Estland* (Gorlitz,1930), 501, note 26.

82 Du Rieu, *Album Studiosorum* (1875), 287–288 (13 July 1637): (1) CONRADUS UXKUL, Nobilis Livonus. 20, J. et Pol.; (2) RAINOLDUS LIEVE, Nobilis Livonus. 19, Philos. Stud.; (3) GULIELMUS TAUBE, Nobilis Livonus. 22, J. et Pol.; 'bij GRIETGE BAILLY.' These entries concerned (1) KONRAD VON UEXKÜLL (c. 1617-1663), brother of OTTO. Later in life he became an 'Oberst' in the Swedish army. He was included in the Swedish nobility in 1648 with the name Von Yxkull-Gyllenband; (2) his full cousin REINHOLD VON LIEVEN [or VON LIEWEN] (1621-1665), Freiherr zu Eksjö, son of BERNHARD VON LIEVEN (1595-1649) and GERTRUDE ELISABETH VON ÜXKÜLL (1590-1666). Later in life REINHOLD became the Swedish governor of Oesel (Saaremaa). He was included in the Swedish nobility in 1653; (3) their common relative WILHELM VON TAUBE. [* c. 1615], son of BERENDT VON TAUBE von Carlö und Maidel (1592-1666) and SOPHIA VON UEXKÜLL (c. 1592-1662). See: *Genealogisches handbuch der baltischen Ritterschaften, Band 2-1: Estland* (Gorlitz,1930), 495, 501.

83 BERENDT VON TAUBE served as 'Ritterschaftshauptman' from 1624-1629 and JOHANN VON UEXKÜLL from 1640-1643. See Martin Klöker, *Literarisches Leben in Reval in der ersten Hälfte des 17. Jahrhunderts (1600-1657)* (Berlin: De Gruyter, 2011), 637.

84 FROM to BROCHMAND, 17 December 1637 (draft letter no. 14): JØRGEN FROM taught his two pupils 'the first six books of Euclid, and the individual parts of geometry and their application in architecture'.

85 HANS JØRGENSEN MULE (1605-1669) matriculated first on 3 June 1631 as 'JOHANNES MOULENIUS, Danus. 26, Histor. et T.', at that time as tutor for the Danish nobleman ERIK ROZENCRANTZ. MULE (as MOLENIUS) registered again on 15 October 1636, then as tutor for GEORG ANDREAS (= JØRGEN) FRIIS, but he left Leiden shortly afterwards, fleeing the plague. MULE returned for a third time on 12 December 1636, now also as companion for the two Danish noblemen HENRIK and JACOB LINDENOV. For his assistance during JØRGEN FROM's illness, see: Kragh, *Jørgen From* (2014), 13 (15 April 1637).

86 'CHRISTOPHORUS URSINUS, Lipstadio-Misnicus, servus. 20' registered in the Leiden *Album Studiosorum* on 16 February 1635 as servant of four German students from Hamburg. He returned to Leiden and registered again on 19 February 1639, then as servant for three boys from Leipzig. There is no indication that he ever met his fellow villager GEORG MARGGRAFE. He probably fled the city of Leiden after the outbreak of the plague in the fall of 1635. CHRISTOPHOR probably was related to GABRIEL URSINUS (1591-1639), and his namesake

nephew (†1674), who both served as 'Pfarrer von Reinhardtsgrimma, 12 km from Liebstadt'.

87 Among the ten noblemen who followed this course were Konrad von Uexküll and his cousin Reinhold von Lieven. See Buck, *Lebensbeschreibing .... Christian Otters* (1764), 249. Otter's *Album Amicorum* from the former city library in Köningsberg survived WW-II and is now in the library of the Lithuanian academy of sciences. It shows that Otter had contact with several prominent scholars in the Netherlands, among whom were Jacob Golius, René Descartes, Martinus Hortensius and many others.

88 Rainoldus Lieve (= Reinhold von Lieven) for example, reported 19 years as his age, when it was actually 17.

89 In the Leiden *Album Studiosorum*, the only possible fit for this horoscope is for the student Marcus Longorius, Livonus, who registered at Leiden University on 9 sept 1637 as a student in theology, living in the boarding house of Heijlien Pieters. At his registration he told that he was 26 years of age, which life span exactly fits the required birth year of 1611. According to Böthführ this Marcus Longorius (in German Marcus Lange) is the same as Marcus Langius, Revaliensis, who entered Rostock University in May 1636. Nothing further is known about his life. See: Du Rieu, *Album Studiosorum* (1875), 288 and Böthführ, *Die Livländer auf auswärtigen Universitäten* (1884), 160.

90 UB Leiden, BPL 2014: *Gazophylacium Heroum ill. bonarumque mentium Filiis Augustanae Confessioni Invariatae in Batavorum Academia addictis ab eiusdem Confessionis ecclesiae Presbyteris dicatum consecratum.* A modern copy with a name index is made in the early 20th century, see BPL 1450.

91 In the later years 1653 and 1659, Georg's younger brother Christian Marggrafe donated twice to the Leiden congregation.

92 Marggrafe to De Laet, 6 February 1640 (ELO). See Rammelman Elsevier. *Inventaris*, vol. 2 (1865), 23.

93 Jorink, *Het 'Boeck der Natuere'* (2006), 307–311. Cook, *Matters of Exchange* (2007), 210–216, 410–411. See also: Bremmer & Hoftijzer (eds.), *Johannes de Laet* (1998) and Schmidt, 'Space, time, travel: Hugo de Groot, Johannes de Laet, and the advancement of Geographic Learning' (1998).

94 Whitehead, 'The Biography of Georg Marcgraf' (1979), 308 [with a modernization of the spelling].

95 Johannes de Laet lived in a large house at the location of the present day houses Rapenburg 29–31 (both built in 1664). See: Lunsingh Scheurleer, *Het Rapenburg* 3a (1988), 201–202; 215–217.

96 Boukema, *Het Leven van theoloog-bewindhebber Johannes De Laet* (2018), 24, referring to meetings of the board of the WIC, where it was discussed sending settlers to Brazil, voluntarily or otherwise.

97 OWIC 52, no 58 (letter sent 6 May 1637 and received in July 1637), fol. 11: 'Wij connen hier niet naerlaten U.Ed. te adviseren dat den Docktor Wilhelmus Milanen op den tocht is overleden, soo believen

U ed naer een ander bequaem persoon uijt te sien, die deese plaetse waerdichlijck moege suppleeren'. VAN MILANEN died on an expedition to Porto Calvo area on 1 April 1637, and was buried at the coast in Barra Grande in the Church of São Bento. OWIC 68 (6 and 9 April 1637). His death is also recorded in the Utrecht archive, DTB registers, 17 July 1637: 'Guiljame van Milanen [...], gestorven op de custen van Brasil'.

98 In 1637 the count fell ill for quite some time. See Wätjen, *Das holländische Kolonialreich in Brasilien* (1921), 89. In BODECHER BENNING's *Epigrammata Americana* (Leiden, 1639), devoted to the life of JOHAN MAURITS, paragraph XXI is devoted to this episode. See: Van den Besselaar, 'Breve introdução' (1984), 29.

99 OWIC 52, no 103 (26 August 1637), fol. 32–33, cited by Stokvis, *La médicine coloniale* (1884), 26: 'Wij hebben UEd. voor desen geschreven, aengaande het overlijden van den heer WILHELM VAN MILANEN, medicinae doctor, soo gelieven U Edele naer een ander bequaem ende ervaren persoon uyt tesien ende ons wel spoedelyck te sturen, want hier niemant is, die men de cure van eenige sware ofte periculeuse sieckte soude mogen vertrouwen. Soodat sijn Excellentie bij gebreck sich tenwoorlig dient van eenen Portugeschen doctor. Wij hebben wel een goet chirugijn ofte twee, maar dienen niet boven haren muyl te gaen, want sij onderwinden haer lichtelijk meer, als se konnen uitvoeren'.

100 See PISO's preface in his 'De medicina Brasiliensi', the first part of the *Historia Naturalis Brasiliae* (1648), i. According to the historian ALEXANDER BICK these two *bewindhebbers* BURGH and DE LAET were the most influential persons in the Amsterdam chamber of the WIC. Bick, *Governing the Free Sea* (2011), 83.

101 Registration 6 May 1607: 'HERMANNUS PISO, Clivensis. 27, M'. Du Rieu, *Album Studiosorum* (1875), 86.

102 PISO's parents were HARMEN PIES (1580-1645) and CORNELIA VAN LIESVELD (†1663). His father, a 'luytenist' from Cleve (Germany), came to Leiden to study medicine in 1607, but did not finish this study. Instead he became the regular organ player of the Leiden 'Hooglandse Kerk'. The couple lived in a house on the Rapenburg (present no. 44), but moved in 1634 to a house on the Nieuwsteeg (present nr. 12). Their son WILLEM PISO matriculated in 1623, at the age of 12, at Leiden University as living with JACOBUS LETTING, also on the Rapenburg (present no. 69). LETTING was extraordinary professor of law, but foremost rector of the Latin School. PISO's entry was only for reasons of tax freedom; in fact he went to the Latin School. See: Lunsingh Scheurleer, *Het Rapenburg*, 5b (1992), 429; 437–438.

103 ELO, North nos. 28, a 29 and 30.

104 OWIC 2 (5 August 1636): '§ 10. Verder soo sal de Compagnie sijn Genade besorgen een Godtsalig bedienaer des Goddelijcken Woorts, een doctor in de medicine ende een secretaris op haren costen. Ende sijn Genade vordere dienaren met militair gages onder de Compagnie van sijn Genade's guarde'.

105 OWIC 14 (29 August 1636). The same day, probably before the directors of the WIC, the Utrecht city council already knew that VAN MILANEN would leave his job as a city doctor, because 'he soon would leave for Brasil' as personal physician of Count JOHAN MAURITS VAN NASSAU. See De Vrijer, *Henricus Regius* (1917), 8–9.
106 Utrecht Archive, 34–4. U020a001 (notary H. Ruysch), deed 231 (16 March 1634) – testimony of JOOST VAN HAEFTEN, who bequeathes 100 guilders, in recognition, to his doctors GUILLAUME MILAENEN and ANTHONIS PELT). Idem, U024a001 (notary G. van Zyderveld), deed 4 (21 June 1634) – testimony by VAN MILANEN and the surgeon DIRCK MANSON about the treatment of the gonorrhoea of HENRICK WILLEMSS OUTHEYNEN.
107 Utrecht Archive, DTB registers, 21 July 1618: 'Mr. WILLEM DE MILANEN, doctor in de medicynen, ende ELISABETH POTS, j.d. beyde wonende tot Utrecht'.
108 Utrecht Archive, 34–4. U022a006 (notary G. Houtman), deed 33 (30 March 1633).
109 OWIC 14 (1 September 1636): 'De Heren [MATHIAS] VAN CEULEN ende DE VRIES refereren dat sij met de heer CONRADUS [BURCH] in conferentie geweest sijnde met den doctor MILANEN, deselve nae veel woorden over en weer, om alle de dienaers van de compagnie voor sijn tractement te dienen, geconsenteert heeft te bedienen alle officieren ende soldaeten die in diens van den compagnie gequest werden, ende maer daer buijten soude moeten betaelen. Ende gecomen was tot 150 gulden ter maent, met een dienaer tot Adelborst gagie, ende een distillateur dear boven. Ende is geresolveert dat men de conditien alvoren eens sijnde de Gagie sal connen toestaen, gelijck naederhant twee dienaers toegestaen sij, d'eene tot 10 gulden ende d'ander tot 8 gulden s'maents'.
One of the servants hired by VAN MILANEN was the surgeon ABRAHAM DUURCOOP, who entered the service of the WIC as 'barbier' in October 1636 and worked in Brazil as 'opperchirurgijn' and 'apotheecker' until June 1642. In Brazil, he is frequently named as working with with WILLEM PISO. In the summer of 1643 DUURCOOP probably returned to his wife MARRITGEN PIETERS [RUBBENS] in Amsterdam. When she was buried in December 1647, she was registered as the widow of 'Abram Duijrcop'. She left a bakery for the processing of Brasilian sugar that was sold in 1650. See OWIC 14 (10 October 1641) and the Amsterdam city archive, DTB, as well as several notary deeds and 'kwijtscheldingen'.
110 OWIC 14 (15 September 1636): 'Is gelesen een missive van sijn Genade GRAEFF MAURITZ VAN NASSAU in dato 7 september 1636. Waerop is geresolveert den Dr. WILLEM VAN MILANEN te accomoderen ende toe te staen, dat sijn Edele van de officieren ende anderen in dienste van de Compagnije van middelen sijnde, in natuerlijcke sieckten vervallen wesende, soor sijn visitatien een redelijck salaris sal mogen genieten. Vertrouwende zijn Edele's discretie den armen, ende die niet connen betaelen, gratis te sullen bedienen'.

111 STEPHAN CARL BEHAIM, 3/13 May 1636 'Military pay is eighteen gulden a month, to which is added my regular plunder'. Ozment, *Three Behaim Boys* (1990), 275.
112 HEINRICH KRALITZ (1607-1638) [or HENRICUS CRALITZ] matriculated in Leiden as a student of medicine on 27 September 1632, living in the home of a certain BARBARA CLAES. His origin was noted as '*Zizensis Misnicus*', showing he came from the city of Zeitz in the Meissen-region in Saxony.
113 OWIC 14 (15 September 1636): 'HENRICUS CRALI[T]S, Candidatus Medicinae, versouckende voor vrijman over te gaen naer Brasil. Is aengeseyt dat hy sijn kist in 't packhuys moet brengen, om gevisiteert te werden, ende sich dan naer Texel te vougen om op het schip vanden heer VAN CEULEN te embarqueren'.
114 Bittcher, *Pförtner Album* (1843), 132. See also KRALITZ's inscription (1627) in the 'Stambuch' of GEORG JENTSCH, one of his fellow students in Pforta. (Weimar : Herzogin Anna Amalia Bibliothek, Stb 279).
115 Erler, *Die jüngere Matrikel der Universität Leipzig*, 1 (1909), M 114 (1628).
116 Kralitz, *Memoria vitae novissima [...] Barbarae Vogeliae* (1631). BARBARA VOGEL (1568-1631) was the widow of JOHANN FRIEDRICH SCHRÖTER (1559-1625), professor of medicine in Jena. In PISO's preface of the *Historia Naturalis Brasiliae*, both MARGGRAFE and KRALITZ are denoted as 'Medical Candidates', indicating that KRALITZ, too, had obtained this lower academic degree. See Piso, 'De medicina Brasiliensi' (1648), i.
117 Kralitz, *Catalogus plantarum horti academici Lugduno-Batavi* (1636). Although the name of the professor of botany, ADOLPH VORSTIUS, is on the title page, HEINRICH KRALITZ is credited as the editor of this small booklet by the Amsterdam physician, and later Franeker professor of botany, JOHANNES ANTONIDES VAN DER LINDEN (1609-1664), in his *De scriptis medicis libri duo* (1637), p. 197. The *Catalogus* itself has a Greek note at the beginning, signed with the initials 'H.K.'. On 25 May 1638 (n.s.) THOMAS BARTHOLIN shipped a copy of this catalogue from Leiden to ØLE WORM in Copenhagen, who replied on 19 June 1638 (o.s): 'I have extracted some things from the catalog and if you can get some plants, I would love to receive them next spring, but be careful don't let the gardener deceive you!' Scheperlern, *Breve* (1965), no. 716 and no. 722. This garderner was HENDRICK CHRISTOFFELS [VAN CATHAGEN], who worked in the garden from 1626 until 1653. In 1637 his salary as 'dijenaer van des universiteijts cruythof' was 200 guilders a year. See Sluijter, *Bestuur, instellingen, personeel en financiën van de Leidse universiteit* (2004), 290 and UB Leiden, AC–I, 315, fol. 260.
118 See the introduction of 'De medicina Brasiliensi' (1648), where PISO states about KRALITZ: 'immatura morte suffocates' ('premature death by drowning').
119 UB Leiden, BPL 1450, fol. 16r.
120 In Leiden GOLIUS observed lunar eclipses in 1635, 1638 and 1642. Pingré, *Annales Célestes* (1901), 104, 106, 123.

121 Bernard Varenius to Joachim Jungius, 20 March 1646. See Avé-Lallemant, *Jungius Briefwechsel* (1863), 318. See also Vermij, *The Calvinist Copernicans* (2002), 25.
122 The following persons have been appointed successively as *opsieners* of the observatory: (1) Pieter Vincentsz van Heemskerck (1634-1652); (2) Eusebius Meisnerus (1652-1653) and (3) Claes Simonsz. Schouten (1653-1669). See: AC I, nr. 22, fol. 149 & nr. 315, fol. 264. See also Sluijter, *Bestuur, instellingen, personeel en financiën van de Leidse universiteit* (2004), 175 and Otterspeer, *Groepsportret* 1 (2000), 205.
123 Varenius to Jungius, 20 March 1646. Avé-Lallemant, *Jungius Briefwechsel* (1863), 318.
124 Unfortunately, Kechel's observation register has been lost. This manuscript with his astronomical observations was auctioned together with Kechel's library in 1668. Fortunately, the astronomer Ismael Boulliau, who visited Kechel in 1662, has copied some observations from it. See Zuidervaart & Beurze, 'Kechel' (2014).
125 Nikitinski, 'Petrus Cunaeus über Astrologische Wahrsagerei' (2014), 252. Adrianus Hovius published in 1652 in Utrecht a now lost tract on how to construct a goniometric table. He died in 1655 as conrector of the Latin School of Utrecht.
126 Bengt Hedraeus (1608-1659) enrolled at Leiden University on 2 October 1641 as a student in practical mathematics. In 1648 he went back to Sweden, where he later became professor of mathematics at Uppsala University.
127 In 1642, Utrecht University founded an astronomical observatory on top of the medieval 'Smeetoren'. It also received a rectangular platform with an octagonal turret on it. But instead of a quadrant, the Utrecht instrument maker Johannes Sneewins made a large sextant for this site ('het groot instrument seckxtant'). He also delivered a pair of globes and a telescope. Cf. Deiman, 'Het instrumentarium van de Utrechtse Sterrenwacht' (1987). For the observatory of Bengt Hedraeus in Uppsala, with a Tychonic quadrant on top of his house, see Kallinen, 'Lectures and Practices' (2006).

Chapter 5

1 Marggrave, *Vitae* (1685), 3; Whitehead, 'Biography' (1979), 308: 'as he himself has daily noted in his Itinerary'.
2 ELO, North nos. 28, 29, 38 and 40: Notes and calculations starting at the end of December 1637, or in January 1638. The astronomical calculations in North nos. 30 and 31 are not dated, but these could also belong to the notes made during the crossing of the Atlantic.
3 After his return to the Netherlands, Abraham de Vries, one of Nassau's Brazilian officials, who served as 'Raad and Thesaurier' in Recife, made a statement before notary Martin Beeckman in The Hague, on 1 June 1655. He then said that he had known Georg Marggrafe very well, and that he knew that letters from the WIC director De Laet had strongly recommended that he would be

employed in the art of mathematics. (Municipal archive The Hague, Notary archive, inv. no. 267, deed 219. For a copy see University Library Leiden, ASF 290).

4 In a letter, dated February 1655, WILLEM PISO asserts that in Brazil, in the first two years (=1638-1639), MARGGRAFE served as his 'domestique'. Cf. Meijer, 'De omstreden nalatenschap' (1972), 68

5 MARGGRAFE to JOHAN MAURITS VAN NASSAU, 15 May 1638. (ELO).

6 In his letter to the *Heeren XIX* of 29 June 1638, JOHAN MAURITS states he had 3,300 soldiers to his disposal and 1,000 Brasilians. (OWIC 52, no. 92 & 93). Odegard, *Colonial Careers* (2018), 102 gives the number of 4,600 troops. ZACHARIAS WAGENER, who was an eye witness of the event, speaks of 'a considerable fleet of 47 ships', with forces consisting of 8,000 whites and 3,000 Brazilians'. Wagener, *Thierbuch and Autobiography* (1997), 222. CUTHBERT PUDSEY, a soldier who also contributed in the attack, mentions 5,000 men on 20 ships. Pudsey, *Journal of a Residence in Brazil* (2000), 125.

7 Maggrave, *Vitae* (1685), 3; Whitehead, 'Biography' (1979), 308.

8 This in contrast to Barlaeus, who claims that the island had the name of a former owner. (Barlaeus, [English translation] (2011) 140). The Portuguese name 'Vaz' or 'Vaaz' was rather common in Recife. In March 1648 even a slave called ANTONIO VAEZ was baptised in the local church. See: Wasch, 'doopregister' (1889), 2.

9 Nederveen, *Recife* (1989), 264–265; Van Oers, *Dutch Town Planning Overseas* (2000), 139-151 .

10 CHRISTOFFEL MARCKGRAEFF received as *wachtmeester* a salary of 24 guilders, supplemented by a *costgelt* of 16 guilders for rent and food. (OWIC 68, 24 September 1637). His place of origin is given as 'Huppin', which seems to refer to the Brandenburg town of Neuruppin, the origin of a large family named 'Marggraf' which has produced many well-known pharmacists. The oldest verifiable representative of this family is CHRISTOPH MARGGRAF, who is mentioned in 1623 in Neuruppin as a 'Musquetirer' (musketeer). In Recife, CHRISTOFFEL MARCKGRAEFF stood witness at the baptism of MARIA BICKER, daughter of GERRIT BICKER. This father was probably the 'soldaat te water', who in August 1636 in Amsterdam had married AELTIE LAMBERTS from Utrecht. He was *not* related to the influential GERRIT BICKER, one of the founders of the VOC. See Wasch, 'Doopregister' (1889), 169 and the weblog 'Die andere Gerrit Bicker' on the website of alleamsterdamseakten.nl (consulted November 2020). In a later year (25 July 1649) the baptism register of Recife also names a certain ASCANIUS MARCKGRAEFF. In 1650 he is named in Rotterdam as a servant of admiral WITTE DE WITH. This admiral was appointed commander of an expeditionary force to preserve Dutch Brazil in 1647. (Rotterdam archive, notary Jacob Duyfhuysen jr, 15 September 1650).

11 Nederveen, *Recife* (1989), 95.

12 De Laet, *Historie ofte iaerlijck verhael* (1644), 194–202.

13 Nederveen, *Recife* (1989), 95.

14 Van Oers, *Dutch Town Planning Overseas* (2000), 149.

15 According to ZANDVLIET, *Mapping for Money* (1998), 246 PEETERS made this view of the bay of Recife guided by a map published earlier that year (see *fig. 8*).
16 De Laet, *Historie ofte iaerlijck verhael* (1644), 400, describing the situation in 1634: 'Het volck daghelijcx uyt Hollandt meer ende meer aen-kommende wert aen't klooster op Antonio Vaaz een Loogies gestelt daer men omtrent twee hondert man onder konde laten'.
17 SIMON STEVIN (1548-1620) was an influential Dutch mathematician and military engineer born in Flanders. He made various contributions in mixed mathematics, both theoretical and practical. For his ideas on city planning, see Van Oers, *Dutch Town Planning Oversees* (2000) and Van de Heuvel, *'De Huysbou'* (2005).
18 Nederveen, *Recife* (1989), 104-106. Van Oers, *Dutch Town Planning Overseas* (2000), 148-149. ELO, MARGGRAFE to FROM, 18 May 1639.
19 In 1654, after the surrender of the Dutch colony to the Portuguese, the house and various other properties in its surroundings were claimed by the captain BELCHIOR ALVEZ CAMELLO, as being the former property of his father BELCHIOR ALVEZ Senior. According to JOSÉ LUIZ MOTA MENEZES they were the heirs of the former owner PEDRO ALVARES, who had seen his house confiscated by the Dutch in 1630. See Nederveen, *Recife* (1989), 164-165 and Menezes, 'A casa de sua excelencia' (2005).
20 ZACHARIAS WAGENER (1614-1668) from Dresden came to Brazil as an ordinary soldier in 1634. He was assigned to the company of Major CORNELIS BAIJERT (also called Bajert), based in Fort Ernestus on the island of Antonio Vaz. When his military contract ended in 1637, WAGENER entered the service of Count JOHAN MAURITS, in a role he himself called *Küchen Schreiber*. The Dutch phrase was indeed *Keuckenschrijver* (See OWIC 69, 30 November 1641, about WAGENER's successor HESSEL HEDDENSE). WAGENER left Brazil in April 1641. After a fascinating career in the Far East, he became governer of the Dutch VOC colony in South Africa from 1662-1666. Eventualy he died in Amsterdam in 1668. See: Wagener, *Thierbuch and Autobiography* (1997), 222–230.
21 OWIC 69, DN, 20 January 1643: The city council of Mauritsstad was granted the usage of the front part count's old mansion, 'up to the stone staircase, where one climbs up at the rear'. ('het oude hoff oft wooninge van sijn Excellentie toe te staen het gansche voorste gedeelte tot aende steene trap daer men van achteren op gaet om hetselve tot haeren dienst te gebruijcken').
22 Pereira da Costa, *Inventario das armas* (1940), no. 363 (main house) and the surrounding buildings no. 349, 353–362 and 368–371.
23 Other watercolors in WAGENER's *Thierbuch* are related to works by the painter ALBERT ECKHOUT (C. 1610-1665).
24 Corrêa Do Lago, *Frans Post* (2007), no. 22.
25 *Ibidem*, no. D. 55.
26 ROSCAM ABBING has recently pointed to the fact that FRANS POST returned to the Netherlands *before* JOHAN MAURITS, who left Brazil

in May 1644. That month Post already worked in The Hague. See: Roscam Abbing, *Brazilië* (2021), 162, footnote 35. See also Krempel, 'Biographische und stilistische Notizen' (2006).

27 *Ibidem*, 146.

28 'Doctor Piso & Mr. Abraham [Duurcoop] verthonen bij lijste diversche drogos inde apothecque te wesen, die daer langen tijt gelegen hebben ende niet versien werden ende pericul loopen van te bederven. Soo sijn deselven geauthouriseerd om die ter meesten profijte van de compagnie te vercoopen ende goede reckeninge daer van te houden' (OWIC 68 – 29 July 1638). Also mentioned by Stokvis, *La médicine coloniale* (1884), 26 with the wrong date and lacking a transcription. The mentioned list is not preserved. Another list of pharmaceutical medicines, dated 1639 (OWC 54, no. 39), seems not to be in Marggrafe's handwriting.

29 Marggrave, *Vitae* (1685), 3; Whitehead, 'Biography' (1979), 308, 310:

30 OWIC 68 (16 september 1638): 'Verkocht aan de heer Willem Piso, medicinen doctoor, een vervallen erve staende op Anthoni Vaes, voor eenige jaeren gedient hebbende tot een gasthuijs, met de vervallen muiren ende soodaniche hout als bij 't selve nach te vinden is, daer voor hij sal betalen de somma van duijsent drie hondert gulden, te betalen in drie Jaeren'.

31 Piso refers also to Marggrafe as *meo domestico* in his dedication of 'De medicina Brasiliensi' (1648). See also Piso to Jacob Golius, 16 May 1655: 'Want Markgraef de drie laeste jaren niet meer mijn domestique was, maer bij sijn Vorst[elijcke] Gh[enade]'.

32 Johan Maurits van Nassau to the trustees of Leiden University, 6 February 1655: 'mijnen, in Angola overledenen Mathematicus, Marggravius genoemt'. Letter printed in Molhuysen, *Bronnen 2* (1918), 107. See also Meijer, 'De omstreden nalatenschap' (1972), 66–67

33 Statement by Abraham de Vries, 'Raad and Thesaurier' of Johan Maurits, before notary Martin Beeckman in The Hague, 1 June 1655 (University Library Leiden, ASF 290). Marggrafe's predecessor, Cornelis Goliath, received as *caertemacker* the same salary (OWIC 68, DN, 18 November 1639). In 1641 this salary was 42 guilders a month, that is 504 guilders a year (OWIC 69, accounts December 1641).

34 Whitehead, 'Biography' (1979), 308.

35 Marggrafe (from 'San Anthonio' (Bahia) to Golius (in Leiden), 26 June 1638.

36 Marggrafe (on the island of 'Tamaraca' [=Itamaracá]) to Golius (in Leiden), 4 October 1638. This letter was a response to the one written by Golius on 22 June 1638, which Marggrafe received in Brazil more than three months later, on 3 October 1638.

37 *Ibidem*, 'ex beneficio Conte'.

38 Whitehead, 'The Marcgraf Map of Brazil' (1987), 17. Goliath worked in Brazil as a cartographer between 1634 and May 1641, when Johan Maurits sent him to Lisbon as a representative to attend the coronation of the new Portuguese King João IV. However, Goliath's ship was captured by pirates who brought him as a captive to Algiers. Being

freed in 1643, he returned to Holland, where he married. After the uprising in Brazil in 1645, he returned there but stayed only a year. GOLIATH's best known map of Dutch Brazil was made after his return in 1647. See for his biography Mostert, 'Cornelis Goliath' (2018).

39 Barlaeus, *Rerum* (1647), 196-198; idem, *Brasilianische Geschichte* (1659), 551–552; idem, *History of Brazil* (2011), 190-192.

40 See for instance the small note about the solar eclipse of 13 November 1640 by 'schijper' (pilot) LEENDERT REIJNSEN VERHOCHEN preserved among MARGGRAFE's Leiden manuscripts. With regard to the later lunar eclipse of 15 April 1642 skipper JACOB ABRAHAMSEN of the barque *Scheveling* confirms that JOHAN MAURITS personally had ordered to make notes about the eclipse. See his letter to JOHAN MAURITS: '... alsoo U Ed mijn order gegeven hadde om op het eclipsen der mane acht te geven ...' ('... because Your Noble had given me orders to heed the eclipses of the moon ...'). (ELO, North no. 1 and no. 2)

41 Odegard, *Colonial Careers* (2018), 98.

42 This controversial, although magnificently executed book discussing the determination of the value of π is denoted by BIERENS DE HAAN as *een waar prachtstuk* ('a truly superb piece'), but the Leiden humanist professor JOSEPHUS JUSTUS SCALIGER made a huge blunder in it by assigning to π the value of 3.1622777. At that time far better values were already abundantly available. see: Bierens de Haan, 'Josephus Scaliger als cirkel-quadrator' (1878).

43 Odegard, *Colonial Careers* (2018), 98.

44 See the letter by JOHANNES GREVINGH, steward of the count's finances, to the Chamber Zeeland, 1 September 1643 (OWIC 58, no. 105).

45 See for a number of disputes about the cost of the count's princely court, OWIC 56, no. 298 (Nov–Dec. 1641); OWIC 68 (DN, 8 March 1642); OWIC 69 (DN 6 March 1643); OWIC 58, no. 205 (Dec 1643).

46 OWIC 56, no. 298-C. A similar list, made in April 1643 (OWIC 58, no. 206) was published by Caland, 'Hofhouding' (1898).

47 North, 'Markgraf' (1979), appendix, horoscope no. 21. No baptism is recorded in the Recife baptismal register between 28 April and 20 May 1643. See: Wasch, 'doopregister (1888/1889), 226.

48 North, 'Markgraf' (1979), appendix, horoscope no. 18 (with the wrong decryption as 'Fitiola Heide'). Our reading of MARGGRAFE's scratchy handwriting is 'FILIOLA STELER?' (Note that the question mark is MARGGRAFE's). So, the name SILNAER fits reasonably with MARGGRAFE's scribble. ELISABETH SILNAER is the only girl baptised in Recife between 8 June and 16 July 1636). One of the witnesses, ELSIEN THIELEMANSS, also stood witness at baptisms on 18 September 1637 and 21 November 1638 for persons with similar unidentifiable professions. See: Wasch, 'doopregister' (1888/1889), 142.

49 As a military engineer, ANDRIES FALLOO was for instance responsible for the construction of the fortress Van der Dussen at the Cabo Santo

Agostinho, south of Recife (OWIC 68, 30 May 1639). He also made three batteries with a large hornwork ('drie batterijen met groot hoornwerck') on the Island of Antonio Vaz. (A 'hornwork' is part of a bastion). (OWIC 68, 2 July 1639). When the new town of Mauritsstad was designed, where – as in every other Dutch city – canals were projected, FALLOO became one of the contractors for the digging of these new canals (OWIC 69, 2 December 1641). He died in August 1645, after being shot during the revolt of the Portuguese settlers (moradores). See Van den Broeck, *Journael* (1651), 5.

50 North, 'Markgraf' (1979), appendix, horoscope no. 20; Wasch, 'doopregister (1888/1889), 225–226. The other witness was the *Politieke Raad* HENRICK HAMEL (appointed December 1639).

51 North, 'Markgraf' (1979), appendix, horoscope no. 19 (with the wrong date November instead of May 1639); Wasch, 'doopregister' (1888/1889), 169.

52 Wasch, 'doopregister (1888/1889), 198.

53 MULHEISER registered at Leiden University as a student of theology on 23 August 1629. See: Van Groesen, 'Van vaandrig in Brazilië tot dichter in Dordrecht' (2008).

54 JOHANNES BODECHER BENNING (1609-1642) was a son of NICOLAAS BODECHER BENNING, a minister from Loosdrecht, who was impeached for his Remonstrant sympathies in 1618. The family then settled in Leiden, where NICOLAAS rented a house on the Rapenburg (no. 50) between 1629 and 1633. See: Lunsingh Scheurleer, *Het Rapenburg* 5 (Leiden, 1990), 615. For his BODECHER's time in Brazil, see: Gonsalves de Mello, 'Bodecher Benning' (1984). He was the first to identify him as the 'I.B.B.' in MARGGRAFE's horoscope.

55 Bodecher, *Poemata* (1637)

56 Molhuysen, *Bronnen*, 2 (1916), 222–223.

57 MARGGRAFE to FROM, 26 June 1638. Draft letter (ELO). BODECHERUS BENNING had brought with him FROM's letter to MARGGRAFE, dated 2 April 1638.

58 The High Council in Recife to the *Heeren XIX*, 5 March 1639: 'De Heer JOHANNES BODECHERUS BENNING heeft een tijt lang een wonderlijcke melancholie gehadt met vreemde imagination, doch wij hopen dat het metter tijt sal beteren'. (OWIC, inv. no. 54, no. 2).

59 The High Council in Recife to the *Heeren XIX*, 10 april 1639: 'Den E. JOANNES BODECHERUS BANNING blijft noch seer naer ouder gewoonte, sijn melancholie wel wat minder, maer evenwel seer bediens in de vergaderinge van de Politieque Raden, soo hare besognes door hem swacklijck bevorderd worden; hij soude misschien in de academie meerder diens hebben kunnen doen; sijn confraters hebben geen volcomen genoegen in sijn assistentie en vallen altemet wat clachtig over eenige inpertinente bejegeningen. Somma, hij waar beter bij sijn vrunden gebleven'. (OWIC, inv. no. 54, no. 79).

60 VAN DE VOORDE, 9 July 1639: 'continuelijck heeft ons confrater BODECHIUS sich in sijn gansche leven, door drincken, hoereren, en andersints soo aengestelt dat ons collegie daer door onder alle de

inwoonderen wierd gevilipondeert, en de sententien geblasmeert; niet tegenstaende dat hij hier over van t avontmael des heeren is afgehouden; van Godt in sijn verstand en sinnen somwijls geslaegen; en ondertusschen seer dikwils met harde vermaningen van ons is onderricht geweest; sulcx dat wij eijndelich genootsaect sijn geworden uijt respect voor ons collegie, dit sijn onbehoorlijk leven aen sijn Excellentie en de heeren Hoogen Raade, [...] voor te draegen, en remedie te versoecken; waerop hij buijten twijfel met de naeste schepen naer huijs sal vertrecken'. (OWIC, inv. no. 54, no. 153). See also VAN DE VOORDE, 28 July 1639: 'wij sijn versekert dat hij in Hollant sijn oudt leeven haest sal openbaeren' (OWIC 54, no. 175).

61 Bodecher Benning, *Satyricon* (1631). Discussed in De Smet, 'Town and Gown in the Dutch Golden Age' (2000).

62 The High Council in Recife to the *Heeren XIX*, 9 July 1639. (OWIC, inv. no. 54, no. 164, fol. 18).

63 The High Council in Recife to the *Heeren XIX*, 5 August 1639 (OWIC, inv. no. 54, no. 193, fol. 16). The return fleet consisted of four ships, the *Blauen Engel* from Amsterdam, the *St Michiel* from Zeeland, the *Mercurius* from the Noorderkwartier and the *St. Jan* from the chamber Groningen. The journey was not easy, as one of the ships was captured by an enemy ship (possibly a Dunkirk hijacker) and two others strayed from the convoy. See Gonsalves de Mello, 'Bodecher Benning' (1984), 174.

64 ELO, North nos. 84. 85, 100 and 101.

65 ELO, North no. 101, fol. 1.

66 *Ibidem.* According to these documents, JOHANNES BODECHER BENNING was born in Loosdrecht, on 18 June 1607 on the Julian calendar = 28 June on the Gregorian calendar.

67 Molhuysen, *Bronnen*, 2 (1916), 249, 270. See also Bodecher, *Epigrammata Americana* (1639), discussed in Van den Besselaar, 'Breve introdução' (1984). BODECHER's library was auctioned in Leiden in 1642. See: *Catalogus ... Bodecheri Benningii* (1642).

68 See Volume 2, notes of 25and 26 June 1638.

69 MARGGRAFE from 'San Anthonio' (another name for 'Antonio Vaz') to GOLIUS in Leiden, 26 June 1638. Draft letter in Latin (ELO).

70 MARGGRAFE to FROM, 26 June 1638. Draft letter in Latin (ELO). In this letter MARGGRAFE refers to two letters dispatched by FROM, one dated 13 March 1638, sent to Brazil with the ship *Amersfoort*, and another dated 2 April 1638, brought to him in June 1638 by BODECHER BENNING. The draft of this last letter is preserved in a bundle of FROM's correspondence in Copenhagen.

71 See also the marginal note, dated 18 July 1638, on the flipside of MARGGRAFE's draft letter to FROM of 26 June 1638. (ELO).

72 At the end of the draft of MARGGRAFE's Latin letter to FROM, dated 18 May 1639, there is another draft of a small letter – written in French – to the 'Messieurs JEAN et ANTOINE VAN HUYCK', asking their cooperation to deliver the letter to 'FRAEM', because MARGGRAFE has no idea where he lives. About the brothers VAN HUYCK no specifics could be

found. They probably belonged to a Dutch family who later settled in 'Nieuw Amsterdam' (currently New York). We have not been able to consult the book by Sylvia E. Wilson, *Huyck, Howk, Houck in America, 1600's-1982* (Medina, Washington, 1982).

73 MARGGRAFE to KECHEL, 18 May 1639. Draft letter in German (ELO).

74 Location mentioned in MARGGRAFE to GOLIUS, 4 October 1638. Draft letter in Latin (ELO).

75 'Brasilians' is a term referring to those native Americans who spoke the Tupí language. Tapuyas belonged to a different tribe of native Americans. 'Tapuya' means 'from a different tongue' or 'enemy' in the Tupi language. The 'wild Tapuyas' belonged to still another tribe of native Americans. The latter group supported the Portuguese side, whereas the Tupís and the 'civilized' Tapuyas supported the Dutch. Cf. Dams, *Descriptions* (2015), 85–86.

76 Marggrave, *Vitae* (1685), 4; Whitehead, 'Biography' (1979), 309.

77 Brienen, 'Georg Marcgraf'(2001), 96.

78 'Historiae Rerum Naturalium Brasiliae' (1648), 241–242.

79 On that occasion the proponent JAN MICHIELS was assigned to this task. (A proponent was a minister not yet confirmed in office). Wasch,'Classicale Acta' (1873), 51. On MICHIELS, see Schalkwijk, *The Reformed Church* (1998), 130. MICHIELS returned to Serooskerke in Zeeland in 1640 without being ordained as minister.

80 Unfortunately, the curious, but also somewhat cluttered report of one of the soldiers who served in VAN MANSVELT's company does not apply to the journeys in which MARGGRAFE took part. This soldier, the Norwegian JACOB JENSEN NORDMAND, who came to Brazil in 1634, left MANSVELT's company just after the battle of Bahia in 1638, being transfered to the Fortress Ceulen near Natal. NORDMAND left Brazil in the summer of 1640. Returned to Denmark, he achieved some fame as an ivory turner. See: [Nordmand], 'En ubenaevnt Dansk kunstners Levnets-historie' (1784). See also Bering Liisberg, 'Jacob Jensen Nordmand' (1893).

81 See for instance OWIC 68 (11 December 1638), mentioning as coming from Ceará '140 Brasilians and 100 women with children who came to serve with us in the war' ('140 Brasiliaenen, 100 mans vrouwen als kinderen, die comen om ons inden oorloge dienst te doen ....'). A week after the end of the expedition, some Brasilians were paid for their service. A Brasilian captain received 30 guilders a day and an ordinary soldier four guilders a day. (See OWIC 69, 17 August 1639).

82 Brienen, *Information* (2002), 10.

83 After his return, Major VAN MANSVELT immediately received a new order, to go with his company to the Alagoas, 'where the enemy had invaded'. OWIC, DN, 15 August 1639. See for details about this trip, Van Liebergen, *Apologiae, ofte waerachtighe verantwoordinghe* (1643).

84 Returned home in August 1639, MARGGRAFE planted seeds from the *Ambaiba Brasiliensibus*, a plum tree that can become extremely tall, and has generally almost no branches. Another plant that he took with him from Ceará was the *Serpentaria Maior*, or dragon lily, with a toxic smelling flower. *Hist. Nat.* 92 and 27.

85 See Volume 2, note made after the entry of 9 January 1640.
86 ELO, LB 18677: 'Ao Senhor Joam de LAET, o quĕ mora em Leida | Senhor, | O grande favor de V[osso] S[enhor] paracon mi muytas, vezes amostrado nas cartas enviadas ao ill[ustrissi]mo Conde Mauricio a Nassovio e tambem ao senhor Piso. Amoesta-me mandar este papel ao V[osso] S[enhor] para testemunhar meu acatamento, e fazer à saber que eu estou senyme de V[osso] S[enhor] muy humilde servador. Por tanto eu pido [= peço], V[osso] S[enhor] de tomar em boa parte o gual fala brevemente este meu messageiro: o muito falar naŏ quero, porque eu sey bem como V[osso] S[enhor] cada dia de muitos negocios està carregado. Nossa saude (graças à deos) està ainda bem, et nosso estudo naŏ passa seu previsto, todo estando posto no rastrejar as ralezas [=belezas] da naturaleza, aqual aqui assi[m] no ceo como na terra està muyto rica. Pelo presente temos trezentas mais cincoenta [= cinquenta] et pouco mais, prantas [= plantas], com as letras eo pincil diligentimente debuxadas; Temos tambem muitas animales, pássaros, peixes, eo outros animaes sem sangue, osquaes insecta chamamos. Cada dia accreintase o numero, para a obra de senh[or] Piso et diligência de mi mesmo. Daremos pella graça de deos em algum tempo toda esto para usar ao mundo sabio et agradecido, quando nos mesmos dara à nossa patria nosso Senhor. Ja enviamos algũs animaes, peixer, et tambem diversas sementes, et fruitos, rayzes,gomas, et outras cousas naturais. Em algum espaço de tempo mandaremos mais. As causas presentes estaĕ para amostrar nosso estudo, et comprar o bem querer de nossos favoreidores et amigos. Eu me vou acabando as palavras, e regando mayormente V[osso] S[enhor] de ser meu favonavel amigo, porque eu nunca desfalecerey pagar – possivelmente todo – com obra bem feito, | Sendo et continuadamente de V[osso] S[enhor] | O muytofiel servidor | Marggrafe | Nacidade Mauricía no Brasil | 6 de Fevereiro, de 1640'.
87 Thomas Bartholin to Øle Worm, 1 June 1639. Scheperlern, *Breve* (1965), 126 (letter no. 779).
88 This garderner was Hendrick Christoffels [van Cathagen]. See Chapter 4, note 117.
89 'Historiae Rerum Naturalium Brasiliae' (1648), 119: Marggrafe named the tree 'Ibacuru-pari' (currently named *Rheedia macrophylla*). The Aquapetida (=Aguapetiba) cove on the coast of Alagoas was located in the vicinity of Lagoa do Pau, some 330 km south of Recife. It is one of the places named on the *Wegwijser*, a travel guide along the coast of Dutch Brazil, discussed elsewhere in this volume. See: Pickel, *Flora do nordeste do Brasil* (2008), 150; Almeida, *Historiae* (2016), 34.
90 'Historiae Rerum Naturalium Brasiliae' (1648), 119 (March 1640), 36, 114 (April 1640), 236–237 (28 May 1640), 255 (July 1640).
91 See Volume 2, note after the entry of 9 January 1640. See also the draft version of the observation register, ELO, North no. 53. The daily minutes of the Polical Council for 1640 are missing between 3 February and the beginning of September, so more details of this accident are not known.

92 In the *Rekeninge* over the year 1643, submitted by the count's *rentmeester* JOHANNES GREVINGH, it is stated that the count was forced to build his own house *Vrijburg*, because of this collapse of the front part of his earlier residence: 'te vooren t'gedeelte van selver woninge daer sijn Excellentie in logeerde ingevallen ende vervallen was, dat eenige sijnere domistijken woonden, en wel den selfs den fiscael MONCHEROEN [...] drije hooft wonden creeg'. (OWIC 58, no. 205).
93 PISO stood witness at the baptism of the young COSMO DE MOUCHERON on 29 September 1641 and of his brother HENRICK on 20 June 1643. See: Wasch, 'doopregister' (1889).
94 Eekhout, 'The Mauritias' (1979), 377.
95 Marggrafe, *Vitae* (1685), 5. Whitehead, 'Biography'(1979), 309.
96 Amburger, Erik, *Die Anwerbung ausländischer Fachkräfte für die Wirtschaft Rußlands vom 15. bis ins 19. Jahrhundert* (Wiesbaden, 1968), 28.
97 Du Rieu, *Album Studiosorum* (1875).
98 ELO, archive notary CORNELIS DIRCX VAN GROTELANDE, 23 and 25 July 1637. Bodecher Benning, *Poemata* (1637), 36–37, 140.
99 Wiesebron, *Brazilië in de Nederlandse Archieven*, 4 (2011), 156. In 1642 DE MOUCHERON became also a member of the *Politieke Raad.*
100 J.E. Almquist, lemma 'Hendrick de Moucheron' in *Svenskt biografiskt lexikon* (online – consulted August 2020).
101 Archief Oude WIC, inv. nr. 69, scan 1036 (25 August 1642) 'no. 16. de vervallen huijsinge naest het oude logement van sijn excellentie in Mauritsstadt, sijn[de een] vervallen huis die lange hebben gelegen, ende voor desen geordonneert waeren op te bouwen voor de amonytie van [den] oorlog'.
102 According to Marggrave, *Vitae* (1685), 4 (Whitehead, 'Biography' (1979), 309), the first expedition was from 20 October until 10 November 1640; the other from 8 until 19 December 1640. The astronomical observations are interrupted, first between 18 October and 13 November 1640, and then between 1 and 20 December 1640.
103 Marggrave, *Vitae* (1685), 4. Without any context, the tree 'Uti' is described as *Uty Brasiliensis* in the 'Historiae Rerum Naturalium Brasiliae' (1648), 120. See Whitehead, 'Biography' (1979), 312.
104 OWIC 68 (20 November 1640): '12 Brasiliaenen mettens een Sargeant uijt geweest met de Majoor MANSFELT de tyt van vier weken met een partie soldaten, sijn toegestaenen gagie bedragende 8 Lb betaelt te werden in lijnwaet tot 12 stuijvers [per] deel'. ('Twelve Brasilians with a sargeant have been out with Major MANSVELT during four weeks with a batch of soldiers. Their allowed salary, by the amount of 8 pounds is payable in linnen of 12 stuvers [per] part').
105 Marggrave, *Vitae* (1685), 5; Whitehead, 'Biography' (1979), 309.
106 Zandvliet, 'Johan Maurits and the Cartography of Dutch Brazil' (1979), 502. See also idem, *Mapping for Money* (1998), 205–206.
107 'It would have been appropriate if these lands, and also other Captaincies, were to be accompanied by a single map: but we shall restrict ourselves to the many drawings of this coast and the many maps that have been despatched to Your Honours until a suitable

opportunity presents itself to design a general map of this entire sea coast, its rivers, harbours, bays, its depths and shallows, and also the course, situation, and distances of the rivers, towns, and settlements in the interior'. Cited after Zandvliet, 'Johan Maurits and the Cartography of Dutch Brazil' (1979), 502.

108 Zandvliet, *Mapping for Money* (1998), 294.

109 *Ibidem*, 205.

110 Mostert, 'Goliath' (2018).

111 Chorography: phrase refering to a description of a country.

112 See Volume 2, note of 7 February 1641.

113 OWIC 68 (22 December 1640): 'Is ordonnantie gepasseert aen de Majoor MANSVELT bij ordre van zijn Excellentie gaende op een expeditie te landinwaerts de somme van 800 gulden daer van hij hem op de tocht gedient hebbende tsijner wedercomste zal reeckening geeven'. On 22 January 1641, MANSVELT received an additional amount of 400 guilders (OWIC 69, account 31 Jan.).

114 OWIC 68 (24 February 1641): letters from Porto Calvo mention the approach of Major VAN MANSVELT.

115 OWIC 69 (11 April 1640). That same day he received another 1200 Guilders: 'Aen Majoor HENDRICK FREDERICK MANSVELT om op den tocht te gebruijcken in twee ordonantien 1200' (OWIC 69, Account 30 April 1641).

116 This appointment was already announced on 21 January 1641 (OWIC 69, DN).

117 VAN MANSVELT arrived in Amsterdam in August 1641. There he called himself 'HENDRICK FREDERICK VAN USINGEN, GESEIJT MANSVELT'. (Usingen is a German town in the county of Hessen). He bought a house in Amsterdam at the Nieuwezijds Voorburgwal for 3,000 guilders on 16 February 1642 and another one on the same canal 'bij de St Claesbrugh' for 4,200 guilders on 24 September 1642. MANSVELT's wife, ELISABETH DE CLERCK, is named in the Amsterdam notary archives on 14 December 1641. In Recife she stood witness at a baptism on 19 December 1639. VAN MANSVELT had been active in Brazil since 1633. See De Laet, *Historie* (1644), 365 and *passim*.

118 GIJSBERT DE WITH (*1611) from Dordrecht, studied in Leiden and Orleans, where he obtained a degree in law in 1635. He then joined the WIC and became *Politieke Raad* in September 1640. After the fall of the colony in 1654 he settled first in his birthplace Dordrecht, but moved soon to The Hague where he became a *schepen* (alderman) in the local government. In 1645, in Recife, he married the Luso-Portuguese ANNA GONSALVES PAES DE AZEVEDO (1612-1674), owner of a large sugar plantation. See: Hulsman, 'Gisberth de With en Anna Paes' (2005) and Wasch, 'Braziliaansche pretentiën' (1887), 75.

119 OWIC 69 (16 juli 1641): 'Ontdeckinge van de revier van St Francisco belast te doen. Is de heer GIJSBERT DE WITTE, directeur in Rio St Francisco, aengeschreven, dat hij met dese gelegentheijt dat aldaer niet gevaceert wiert, mochte het hostiels bij de Portugesen geattenteert wiert, de commandeur van het vaertuygh, ende eenen Mathematicum

MARCKGRAEF, soude de rivier opwaerts aen soude voorsien wesende met genoegsame matrosen ende soldaten om voor de Tapuyes bevreijt te sijn. Ende soo danige gereetschappen. Ende nodicheden als sij oordeelen dienstich te sijn om over het val komende hooger op te mogen varen, ende in dier voegen de Westlijcke contrijen van Brasil te ondersoecken, op dat wij eenmael partinent bericht mogen hebben, wat gelegentheijt van mineralen ende andersins sich aldaer bevinden'. Also in De Sousa-Leao, *Frans Post* (1973), 161. The desire for an expedition to the São Francisco river with the aim to investigate the presence of *salpeter* (potassium nitrate) was already expressed by the *Heeren XIX* in their meeting at Middelburg on 24 June 1639. (OWIC 8). See Wiesebron, *Brazilië in de Nederlandse Archieven*, 4 (2011), 150.

120 OWIC 69 (Account 7 August 1641): 'Aen CLEMENT LE CAMP voor passagie van diverse personen met sijn barck nae Rio St. Francisco f 30. 'Dito voor vracht van sijn pasagiers met haer bijhebbende bagagie nae Rio St. Francisco f 40'. Also in De Sousa-Leao, *Frans Post* (1973), 161.

121 'Historiae Rerum Naturalium Brasiliae' (1648), 203: 'I kept a specimen alive for a long time, in Fort Maurício, next to the S. Francisco river'. See: Almeida, *Historiae* (2016), 34.

122 OWIC 69 (21 September 1641): 'Uijt Rio St Francisco over de ontdeckinge vande westlijcke contreijen. Noch uijt Rio St Francisco van de heer GIJSBERT DE WITTE over de gestaltenisse der saecken aldaer, ende particulierlick belangende de ontdeckinge der westlijcke landen door die rivier hem voor desen in last gegeven, waartoe hij vertoonde noodich te sijn 300 koppen, die hij te deser tijdt uijt dat garnisoen niewen koste uijt maecken, ende meer andere nootsaeckelickheden, die bij hem ende de Mathematicum MARCKGRAEF in een lijste gespecificeert wierden, op welck stuck naerder gelet sal worden'. Also in De Sousa-Leao, *Frans Post* (1973), 161.

123 OWIC 69 (30 September 1641): 'Dhr GIJSBERT DE WITTE tot Rio St. Francisco gesonden tot betalinghe van kosten en leningen, 1000,-'. He returned with the same skipper: 'Met dit bescheijt is na Rio St. Francisco geordeneert te gaen de Barcque van CLEMENT LE CAMP, wesende vol vivres geladen, ende hebbende duijsent gulden in contanten'.

124 OWIC 69 (2 January 1642): 'Heeft de heer GIJSBERT DE WIT, gewesene directeur in Rio St Francisco, ter vergaderinge mondelingh rapport gedaen belangende de gelegentheijt van het selve district ende is sijn edele aengeseght dat hij het zelven in schrift soude vervatten ende overleveren opdat men des te beter van alles geinformeert mochte sijn'.

125 Marggrafe, 'Historiae Rerum Naturalium Brasiliae' (1648), 266; idem, 'Tractatus' (1658), 11.

126 Den Heijer and Teensma, *Nederlands Brazilië in kaart* (2011), 55.

127 Teensma, 'Onherkend' (2011), 70.

128 Storms, 'De kaart' (2011), 39.

129 CASPAR HONINCKHUIJSEN, 'Commandeur der Brasilianen', was also a trader of slaves (see OWIC 54, 106, (May 1639: 'Verkooprekening van

negros aangevoerd door het schip St. Michiel van Sonho, in Congo'). He was killed in Tamarica on 29 April 1646 during an uprising of Portuguese settlers in the Dutch colony. See: *Journael ofte kort discours nopende de rebellye* (1647).

130 OWIC 69 (27 November 1641): 'De schout Honinckhuijsen van Iguaracú belast de Mathematico Marckgraef alle behulp te geven datter een perfecte kaerte van de landerijen van Machioppe ende Caruba de Compagnie toebehorende op het papier mach gebracht worden op dat de scheydinge ende limiten met de groote ende gelegentheijt derselve landerijen te beter mogen aengewesen worden ende inde vercoop, die men daer mede voorheeft om alle questie te beschouwen, near het ordinaries gebruijck hout [= hoort?] gespecificeert [te] warden'. Marggrafe must have finished this job in February, as the sale of these lands in Machioppe and Caruba was announced in the minutes of the Political Council of 18 February 1642 (OWIC 69), being auctioned in a public sale on the following 7 March.

131 *Historia Naturalis Brasiliae* (1648), 237 (December 1641).

132 *Ibidem*, 252 (January 1642), 115 (February 1642).

133 See Volume 2, note of 14 April 1642. See also two letters by Dutch navigating officers, probably sent to Johan Maurits van Nassau, now preserved among Marggrafe's legacy in the Leiden archive (ELO). The largest account of the eclipse is an anonymous letter from Luanda (Angola), dated 29 April 1642. On this letter Marggrafe has put the remark: *Haec observaõ nullius pretii est*, or, in English, 'This observation is worthless'. The other letter, dated 15 April 1642, was written by Jacob Abrahamsen, pilot of the barque *Schevelingen*, in the mouth of the river he called 'Upanema'. As today a city called 'Upanema' is located at the Rio do Carmo, a tributary of the Rio Apodi, probably this river near the border of Ceará is meant.

134 Marggrafe probably returned from Fortress Ceulen with the barque of Jan Willemsz Barcquer, who was paid for his trip to and from the Rio Grande del Norte the day after Marggrafe returned in Recife. OWIC 69, account 17 June 1642.

135 '13 februari 1643: Joos Coppijn tot inckoop van een instrument ende eenige verve tot het maken van eenige caerten voor Edele Heeren Hoge Raden 100,- ' (OWIC 69, DN, 1 March 1643). Johannes, also known as Jan or Joos Coppin is named as an ensign-bearer in 1644 and 1645. He baptised a son at Recife in October 1648. Wiesenbron, *Brazilië in de Nederlandse archieven* 4 (2011), 537; 5 (2013), 345. Wasch, 'Doopregister' (1889).

136 Marggrafe's four maps have the following titles: 1. Praefectura De Cirîiî vel Seregippe del Rey cum Itâpuáma; 2. Praefectura Paranambucae pars Meridionalis; 3. Praefectura Paranambucae pars borealis una cum Praefectura de Itâmaracâ, and 4. Praefectura de Paraiba, et Rio Grande. See for a bibliographic description of all editions of these cartographic highlights, Whitehead and Boeseman, *A portrait of Dutch 17th century Brazil* (1989), 151-161.

137 Storms, 'De kaart van Nederlands Brazilië door Georg Marcgraf' (2011). See also Den Heijer and Teensma, *Nederlands Brazilië in kaart* (2011), 54–57.
138 Zandvliet, 'The Cartography of Dutch Brazil' (1979), 504.
139 OWIC 69 (DN, 15 March 1641): 'Missive van GUIDEON MORRIS, commandeur in het Fort Siara', about the 'Gelegentheijt van de Soutpannen', which were 'bij hem op nieu ontdeckt aen de riviere Upanemma'.
We did not come across the report by JOHANNES VAN WALBEECK and ELBERT GERRITSZ SMIENT, dated 5 October 1641, mentioned by ZANDVLIET. However, the minutes of the Political Council of 2 October 1641 do mention a report by ELBERT SMIENT about salt pans near the river Salinas. Preserved is also a testimony of THIJS TEUNISSEN VAN VEENHUISEN, the cook of commander ELBERT SMIENT, about these same salt pans, west of the river 'little Salinas' (OWIC 58, no. 280). This river, however, is far south in Brazil, even below Rio de Janeiro.
140 OWIC 56, no. 299: letter of the Political Council to the *Heren XIX*, 20 December 1641.
141 OWIC 69, minutes of the Political Council, 22 May 1641: Message of commander PHILIPS ANDRIES of the Dutch fortress in Seregippo del Reij, stating that HERCKMANS would bring with him a plan of the recently completed fortress and its surroundings ('bracht een plan mede van het fort, ende de omleggende situatie').
142 OWIC 69, minutes of the Political Council, 6 November 1641: 'Op huijden is de heer HERCKMANS van Paraiba wederom gekomen rapporterende dat hij bij westen het geberchte van Capaovar niet als weste landen ende groote wildernissen ontdeckt hadde, dan dan soude eerst een pertinente kaerte van sijn gegane partasie maecken, ende die hebbende aen de vergaderinge van sijn gants wedervaeren volkomen relaes doen, omdat hij op wech wesende noch tijdt noch gelegentheijt gehadt hadde om sulcks te mogen volbrengen'.
143 Herckmans, 'Generale beschrijvinge vande Capitanie Paraiba' (1639, printed in 1879). See also Dams, 'Herckmans' (2011) and idem, *Descriptions* (2015), 217–227.
144 Barlaeus, *Rerum per octennium in Brasilia* (1647), 219; idem, *Brasilianische Geschichte* (1659), 593–620.
145 *Ibidem* (1659), 609: 'Ich sol die Lande und die Wüsten Copaoba fleisig durchsuchen, und nachforschen was dasselbe land vor eine Art habe und welcher Gestalt man desselben geniessen konnte'. This is in contrast to Dams, who asserts that 'His ambition was to find the silver mines reportedly lying in Copaoba'. (*Descriptions* (2015), 225).
146 OWIC 69, Minutes of the Political Council, 2 October 1641: 'Noch dat de Heer ELIAS HERCKMANS den 25 September op de ontdeckinge van de westelijcke landen gegaan was, geaccompagneert sijnde met de predikant VAN DER POEL omtrent 50 soldaten ende 70 of 80 Brasilianen'. CORNELIUS VAN DE POEL was the regular preacher in Paraiba after March 1637, where he lived in the town of Frederica (today the city of João Pessoa). See Wasch, 'Classicale Acta van Brazilië' (1873).

147 Barlaeus, *Rerum per octennium in Brasilia* (1647), 372; idem, *Brasilianische Geschichte* (1659), 616.
148 Barlaeus, *Rerum* (1647), 218; *Brasilianische Geschichte* (1659), 606–607. On his journey HERCKMANS also visited the ruins of this deserted town. On MARGGRAFE's map, further inland, also a *vervallen Tapuyers leger*, i.e a deserted village of the also dispelled Tapuya's can be seen. See for HERCKMANS account of this place, Barlaeus, *Brasilianische Geschichte* (1659), 613.
149 De Laet, *Historie* (1644), 456.
150 Wiesebron, *Brazilië in de Nederlandse Archieven*, 3 (2008), 52. See also Levy Pereira, lemma 'Duarte Gomes', in: *BiblioAtlas – Biblioteca de Referências do Atlas Digital da América Lusa.* online on: http://lhs.unb.br/atlas/Duarte_Gomes. (consulted 31 August 2020).
151 Whitehead, 'Biography' (1979), 309:
152 Barlaeus, *Brasilianische Geschichte* (1659), 601, 612. One monument was placed at the mountain *Iurupari-bacaí*, a native name that meant 'At this place the devel has looked back'. The precise location of the other monument is not given.
153 For instance the Dutch 'ingenho's' of Amstel and Middelburg.
154 Ormeling, 'Gebruik en effectiviteit van symbolen op oude kaarten' (2012).
155 Barlaeus, *Brasilianische Geschichte* (1659), 620: 'Sie haben mit den Waffen, mit bevorderung der kaufmanschafft mit durchsuching und durchgrubelung wert abgelegener fremder Lande dem Gewinst an Geld und Guth vor die West Indische Societat ernstig nachgestrebet. keine Walde, keine Wildenissen, keiden felsen, keine klippe, keine strome, keine Seen und Meere haben sie dich davon abhalten lassen'.
156 JOHANNES BLAEU placed an advertisement for the large wall map in the first Dutch newspaper *Courante uyt Italien, Duytslandt, &c.* of 28 September 1647. ('t 'Amsterdam by Jan Blaeu wort uytgegeven een nieuwe curieuse Caerte van Nederlandts Brasilien / gaen meten en maken door sijn Excell. Grave Maurits van Nassau / ten tijde van sijn Gouvernement in Brasil; met een Beschrijvinge daer onder aen').
157 This manuscript map is published in full in: Brommer and Den Heijer, *Grote Atlas van de West-Indische Compagnie* (2011), 178-179.
158 Zandvliet, 'Johan Maurits and the Cartography of Dutch Brazil' (1979), 504.
159 For a detailed explanation of the vignettes added by BLAEU to MARGGRAFE's map, see Boogert, 'A Well-Governed Colony' (2011).
160 Ferrâo & Soares, *The Voyage of Caspar Schmalkalden*, 2 (1998), 154.
161 Van den Boogaart, 'A Well-Governed Colony' (2011), 266. See also De Souza Leão Vieira, *Topografias Imaginárias* (2010).
162 Ibidem.
163 OWIC 58, no. 206, April 1643, published by Caland, 'Hofhouding' (1898)
164 Painting by Cornelis de Man (1621-1706), oil on canvas, 74.2 by 60.6 cm. Sold by Sotheby's in 2007.
165 Whitehead, 'The Marcgraf Map of Brazil' (1987), 20.

166 Zandvliet, *Mapping for Money* (1998), 206.
167 Storms, 'De kaart van Nederlands Brazilië door Georg Marcgraf' (2011), 46.
168 *Ibidem*. See also Storms, 'Brasilia qua parte paret belgis' (2013).
169 Teensma, 'Onherkend want anoniem en zonder jaar. Georg Marcgraf's Braziliaanse manuscriptkaarten uit 1643' (2011), a claim falsified by Storms, 'De kaart van Nederlands Brazilië' (2011), 38. The atlas 4.VEL Y mentions the cease-fire agreement between the Dutch Republic and Portugal made in 1641, so it was made after that date.
170 Pimentel and Pereira, 'A astronomia e o mapa Brasilia qua parte paret Belgis' (2014).
171 Whitehead, 'Markgraf and Brazilian zoology' (1979), 426. See also Pereira da Costa, 'Primeira observações meteorologicas no Brazil' (1900) and Morize, 'As mais antigas observações meteorológicas na America do Sul' (1900).
172 Wasch, 'Request' (1889).
173 *Vitae*, and Whitehead, 'Biography' (1979), 310, referring to a letter KECHEL received in 1643.
174 'Sommier Discours' (18 January 1638); printed in *BMHG* (1879), 292–293.
175 *Ibidem*: 292: 'die van Angola werden hier van de beste gehouden, deels om datse beter willen wercken, deels oock om datse, nieu overgecomen sijnde, beters vande oude negers leren, alsoo sij malcanderen van tale verstaen'.
176 Ratelband, *Nederlanders in West-Afrika* (2000), 102-114; 143-149.
177 *Ibidem*, 175-179.
178 'Wij recommanderen uwe's oock om de persoone van GEORGE MARKGRAEF alle behulp ende faveur te doen alsoo de selve derwarts gaet om een pertinente beschrijvinge ende kaert van dit quartieren te maecken'. Letter from Governor and Supreme Council in Brazil to the WIC directors on the South Coast of Africa (OWIC 58, no. 262, 14 August 1643). See also OWIC 70, the minutes of the *Hoge en Secrete Raad* of Dutch Brazil, 10 August 1643: 'Is gelesen het concept van de missive die met het jacht *De Brack* na de kust van Angola soude gaen aen de heeren directeurs NIEULANDT en MOLS'. CORNELIS NIEULANDT and HANS MOLS were the WIC commanders 'over het zuiderdistrict' of the coast of Africa [= Angola]. NIEULANDT had served earlier as *Politieke Raad* in Brazil.
179 OWIC 58, no. 205, 20 August 1643: 'per oncosten gemaect aen GEORGE MARCKGRAEFF gaende naer Angola om de Caerte vant lant aldaer te beschrijven .... fl 200'; and earlier in the same account: 'GEORGIE MARGRAFF op de reijse naar Angola ende St Thome fl. 200,-'.
180 'Historiae Rerum Naturalium Brasiliae' (1648), 122.
181 OWIC 70, 14 August 1643 'De brieven na Angola ende St Thome met het Jacht *De Brack* afgesonden ende met het selve jacht na St Thomé gesonden dese volgende vivres ende amoytie van oorloogh: 30 vaten meel, 15 vaten veurs, 3 vaten speck, 1 vat brandewijn, 30 alcquen

sout, 2 amen olis, 4 vaten Sp[aansche] wijn, 1 kist suijcker; ende ende amoutie [=Munitie] 1000 lb kruijt, 1000 lb loot, 500 lb mosquet loot [= musketkogels], 150 devers [= diversen]'. See also OWIC 70, 31 August 1643: (the account of that month): '18 Aug. 1643: ADRIAEN JACOBSZ voor de verkofte netten ten behouve vant Jacht *De Brack*, vaerende na St Thomé ende van daer na Angola, 62 / 16 / -'.

182 Ratelband, *Nederlanders in West-Afrika* (2000), 147: 'de ongesonde locht waer van dit eylant befaemt is'.

183 Nieuhof, *Gedenkweerdig Brasiliaense Zee- en Lant-Reize* (1689), 7–8: 'De meesten der onzen, die aldaer komen, en des nachts aen lant blijven, sterven er, door de ongewoonheit van de lucht'. [...] 'Onder hondert vreemdelingen, die daer komen, blijven naulix tien over en in 't leven'. [...] 'Onder [de] driehondert Brasilianen, die hy meenam, bleven geen zestigh in 't leven'.

184 Ratelband, *Nederlanders in West-Afrika* (2000), 147.

185 *Ibidem*, 149.

186 PISO to GOLIUS, 12 May 1655 (Leiden University Library, ASF 290): 'In Africa heeft hij vier of 6 weken geleeft'.

187 The WIC merchant JOHAN NIEUHOF also travelled from Recife to São Thomé in the same month as MARGGRAFE did. NIEUHOF was on the yacht *De Bruijnvisch* that left Recife on 26 August 1643, first to the coast of Paraiba (to a bay near the Ponta de Lucena, called by the Dutch *Het Roode Land*), to take up fresh water and firewood as dead weight. (OWIC, 69). The ship left Brazil on 31 August 1643 and arrived at São Thomé's roadstead on 9 September 1643. NIEUHOF only went ashore in the middle of the day when the fog that always surrounded the island had cleared. He therefore survived his visit to the island. His yacht began the return trip to Recife on 24 September 1643, loaded with a cargo of black sugar. See Nieuhof, *Gedenkweerdig Brasiliaense Zee- en Lant-Reize* (1689), 6–7.

188 Barlaeus, *Rerum per octennium in Brasilia* (1647), plates between pages 204–205 and 208–209. The maps are engraved by JAN VAN BROSTERHUYZEN (as all the plates in BARLAEUS's book).

189 VINGBOONS's manuscript map of the city of São Thomé is in the Atlas Blaeu-Van der Hem (Austrian National Library, Vienna) and his manuscript map of Luanda is in the Nationaal archief, The Hague, VELH0629.63. The cartographer JOHANNES VINGBOONS (ca. 1616-1670) worked frequently for JOHANNES BLAEU, the publisher of Barlaeus's *Rerum per octennium in Brasilia.*

190 Brommer and Den Heijer suggest that MARGGRAFE may have made a surviving rough sketch map of São Thomé. However, this sketch was clearly *not* made by someone with a surveying background. Moreover, the map is in Dutch, a language MARGGRAFE never used. See *Grote Atlas van de West-Indische Compagnie: I. De Oude WIC* (2011), 395.

Chapter 6

1 'Een ciste met astronomische wercktuyghen'. Mentioned in the account OWIC 59, no. 67. Transcribed in Sousa-Leao, *Frans Post* (1973), 161.
2 This instruction is mentioned in Johan Maurits's letter to the trustees of Leiden University, 19 March 1655: 'De schriften ende teijckeningen, betreffende de natuijrlicke historie van Brasil, ende de beschrijvinge ende afmetinge der voorsegde landen, sijn door ons gegeven aen den heere De Laet, Ende de schriften, aengaende de astronomische observatien, aen den heere Professor Gool, ten eijnde deselve saecken op onse costen ondersoecht, ende bijeen gebracht, mochten in 't licht gegeven worden, soo als ten deele is geschied'.
3 ELO, document without number given by North.
4 In the *Vitae*, Christian Marggrafe also asserts that his brother 'made this title [...], which he clearly set down in his itinerary'. Whitehead, 'Biography' (1979), 308.
5 De Laet, 'Ad Benevolos Lectores' (1648). Translation from Dams, *Descriptions of Dutch Brazil* (2015), 240.
6 Whitehead, 'Biography' (1979). 310.
7 Dams, *Descriptions of Dutch Brazil* (2015), 241.
8 Whitehead, 'Markgraf and Brazilian Zoology' (1979); Brienen, 'The zoological drawings of Eckhout and Marcgraf' (2007); Scharf, *Libri Principis* (2019).
9 De Laet, 'Ad Benevolos Lectores' (1648).
10 Two of the 15 woodcuts mentioned came from dried specimens outside the Copenhagen herbarium. Some Marggrafe specimens went at some moment to the herbarium of Jacob Breyne (1637-1697) in Danzig, and from there to the herbarium of William Sherard (1659-1728) in Oxford. See at length Andrade-Lima e.o., 'Marcgrave's Brazilian Herbarium' (1977), 130. See also MacBryde, 'Rediscovery of G. Marcgrave's Brazilian Collections' (1970) and Wagner, 'Das Markgraf-Herbarium' (2008).
11 British Library London, MS Sloane 1554: *Marggravii, Brasiliae Historia Naturali, Johannis de Laet, qui opus istud primum edidit manu descripta*, c 1648. This manuscript contains fourteen small drawings of flowers and fruits, executed with pen and pencil. See Joppien, 'The Dutch vision' (1979), 318–319; Andrade-Lima e.o., 'Marcgrave's Brazilian Herbarium' (1977), 128 and Wagner, 'Das Markgraf-Herbarium' (2008).
12 Scharf, *Libri Principis* (2019).
13 Brienen, 'Markgraf' (2001), 101-103; idem, 'The zoological drawings' (2007), 278; Scharf, *Libri Principis* (2019), 166. In his recent work Roscam Abbing suggests – without further substantiation – that the *Libri Picturati* were made by another (unknown) draughtsman than Marggrafe. Unfortunately he has left unnoticed the detailed study by Claudia Scharf, who has distinguished the hands of six different artists, but confirms the attribution of most of the pictures to Marggrafe. See Cf. Roscam Abbing, *Brazilië* (2021), 74.

14 After the publication of the *Historia* in 1648, the *Handbooks* were evidently returned to JOHAN MAURITS. Four years later, the count donated them to the Elector of Brandenburg, together with a large number of other pictures relating to the natural history of Brazil (mostly made by ALBERT ECKHOUT). In Germany, the loose material was bound into four large volumes, called *Theatrum rerum naturalium Brasiliae.* Somewhat later this set was supplemented by one other volume, called *Miscellanea Cleyeri.* These volumes with Brazilian valuables were added to a larger series already in the possession of the Elector, called *Libri Picturati.* All these illustrated manuscripts were kept in the Royal library in Berlin. At the end of WW II all *Libri Picturati* were transported to Poland, where they are now preserved in the Special Collections Department of the Jagiellon University Library, Cracow. See: Ferrao & Soares (eds), *Theatrum Rerum Naturalium Brasiliae* (1993). See also in detail, Whitehead and Boeseman, *A portrait of Dutch 17th century Brazil* (1989). The *Handbooks* can be consulted online on the website of the Jagiellon University Library. A recently published article on the botanical identifications of seventeenth century plant illustrations in the *Libri Picturati* was too late to be discussed in this volume. See: Alcântara-Rodríguez et al, 'Looking into the flora of Dutch Brazil'(2021) and their forthcoming article 'Nature portrayed in images in Dutch Brazil'.
See for Handbook I [= Libri Principis A 36]: https://jbc.bj.uj.edu.pl/dlibra/publication/193891 and Handbook II [= Libri Principis A 37]: https://jbc.bj.uj.edu.pl/dlibra/publication/193892. (both consulted September 2021).

15 Brienen, 'The zoological drawings' (2007), 279. See also MARGGRAFE's dedication to Count JOHAN MAURITS, in the 'Historiae Rerum Naturalium Brasiliae' (1648), where he states that he made all the illustrations 'ad vivum'.

16 *LP* A36, fol. 46. See Scharf, *Libri Principis* (2019), 207, 228.

17 The two *Handbooks* differ a bit in size, reason why BRIENEN supposed that they were composed for different purposes. According to her the first *Handbook* functioned as a record of the animals in JOHAN MAURITS's menagerie and as documentation of animals given to him while he was governor of Brazil. *Handbook II,* with images made by probably five to six artists (including MARGGRAFE), has no systematic sequence. Originally it was bound together with a German treatise on horse medicine, reason why BRIENEN guessed the book was aimed at an aristocratic audience. Brienen, 'The zoological drawings' (2007).

18 Safier, 'Beyond Brazilian Nature'(2014), 176.

19 Joppien, 'The Dutch vision' (1979), 318; Whitehead and Boeseman, *A portrait of Dutch 17th century Brazil* (1989), 28 They also conclude that as the size often differs, the woodcuts in the *Historia Naturalis Brasiliae* were not made by tracings from the *Handbooks. Ibidem,* 41.

20 Whitehead and Boeseman, *A portrait of Dutch 17th century Brazil* (1989), 28.

21 Wagener, *Thierbuch* (1997). This image of the Tamandua is also copied in the Brazilian part of CASPAR SCHMALKALDEN's travel notes which are

composed after his return to Gotha in 1652. As we will argue below, it is certain that he used the illustrations from the *Historia Naturalis Brasiliae* (1648) for that purpose. Heidhues's suggestion that Schmalkalden met the artists in the retinue of Johan Maurits during his service, and that 'they may have encouraged him to make his own drawings' therefore is wrong. See: Ferrão & Soares (eds.), *The Voyage of Caspar Schmalkalden* (1998), vol. 2, 54–55 and Heidhues, 'An Early Traveler's Compendium: Caspar Schmalkalden's Images of Asia' (2005), 154.

22 Scharf, *Libri Principis* (2019), 223.

23 See more in detail, the forthcoming study of Alcàntara Rodríguez, *et al*, 'Nature portrayed in images in Dutch Brazil. Tracing the sources of the plant woodcuts in the *Historia Naturalis Brasiliae* (1648)'. This preliminary text came into our hands only after the submission of our text to Amsterdam University Press.

24 De Silva and Alcides, 'Collecting and framing the wilderness' (2002), 172. The quote is from a letter by Johan Maurits to the Marquis de Pomponne, 21 December 1678, cited by Joppien, 'Dutch vision of Brazil' (1979), 326.

25 In his recent study of the pictorial legacy of Dutch Brazil, Michiel Roscam Abbing attributes the design of the title page of the *Historia Naturalis Brasiliae* to Jacob van Campen, the architect of the *Mauritshuis*. However, this in contrast to Borba de Moraes, who mentions that Theodore Matham's original drawing for the title page is preserved in the Albertina Library in Vienna. Unfortunately this drawing cannot be found today. Requests into this matter by Peter Whitehead did not yield any result. See: Borba de Moraes, *Bibliographia Brasiliana* (1983[2]), 677; Whitehead, 'Markgraf'(1979), 438; Whitehead and Boeseman, *A portrait of Dutch 17th century Brazil* (1989), 29 and Roscam Abbing, *Brazilië* (2021), 74.

26 Scharf, *Libri Principis* (2019), 135, 150.

27 *Catalogus bibliothecae ... De Laet* (1650), [libri medici folio]; Moreira and Mendes, 'Sloths of the Atlantic Forest' (2016), 7. This wrong picture of the *luiaard* as depicted by Carolus Clusius was also copied in a drawing by Caspar Schmalkalden from Gotha, who – among others – served as a soldier in Dutch Brazil between 1642 and 1645, and left an account of his travels. This proves that this account was made after the publication of the *Historia Naturalis Brasiliae* 1648, when Schmalkalden had returned to Gotha in 1652. See: Ferrão & Soares (eds.), *The Voyage of Caspar Schmalkalden* (1998), vol. 2, 64–65.

28 Dams, *Descriptions of Dutch Brazil* (2015), 290–296.

29 De Laet to Worm, 1 July 1646. Scheperlern, *Breve* (1965), no. 1430.

30 Worm to De Laet, 23 October 1646. Scheperlern, *Breve* (1965), no. 1431.

31 John Morris to De Laet, 29 October 1646 (old style). Bekkers, *Correspondence* (1970), no. 73. One of De Laet's annotations is also dated 1646. See 'Historiae Rerum Naturalium Brasiliae' (1648), 10.

32 Hoftijzer, 'The end of the officina Hackiana' (1996), 258.

33 *NNBW*, lemma 'Lodewijk Elsevier (3)'.

34 John Morris to De Laet, 17 May 1647 (old style). Bekkers, *Correspondence* (1970), no. 80. This concerned De Laet's book *De Gemmis et Lapidibus* (Leiden: J. Le Maire, 1647). The letter speaks about 'wrapped in a leaf of Marggrafe's herbarium', which we have interpreted as a proof-sheet of a text about plants.

35 De Laet to Worm, 10 July 1647 (old style). Scheperlern, *Breve* (1965), no. 1512.

36 Worm to Steffen Stepkanius. 29 February 1648 (old style). Scheperlern, *Breve* (1965), no. 1565.

37 Worm to Bartholin, 20 July 1647; Bartholin to Worm, 10 April 1648. Scheperlern, *Breve* (1965), nos. 1513 and no. 1572.

38 Worm to De Laet, 1May 1648 (= 21 April 1648 old style). Scheperlern, *Breve* (1965), no. 1577.

39 To Marggrafe's part De Laet added a short appendix, in which he discussed indigenous people, their language and also the expedition to Chile. This part is illustrated with woodcuts after paintings by Albert Eckhout. In his 'Ad benevolos lectores' De Laet explains this as follows: 'I composed the eighth book, of which the author only left some notes and for which he had outlined the titles, with at my disposal the different means he left and other resources that were handed over to me by the illustrious Count through various persons. Because of the similarity of subject, I added a short History of Chile, as it was recently reported by our men, when they perambulated the province and encountered a navigation passage to the Pacific Ocean'. Translation from Dams, *Descriptions of Dutch Brazil* (2015), 290.

40 Dams, *Descriptions of Dutch Brazil* (2015), 239.

41 *Ibidem.*

42 Alcàntara Rodrígueza *et al*, 'Marcgrave and Piso's plants for sale' (2020).

43 Dams, 'Transcriptions and Descriptions of Exotic Fauna'(2012), 343.

44 See also Scharf, *Libri Principis* (2019), 145.

45 Van Berkel, 'Cradle of the telescope' (2010), 70: 'It is hard to imagine a painter like Bosschaert studying a fly or a caterpillar without the help of a magnifying glass. In the sixteenth and seventeenth century such an instrument was not uncommon in the textile industry'.

46 Johannes Goedaert, *Metamorphosis Naturalis* (Middelburg: Fierens, n.d. [= 1660]), 101. Worldwide, the first published account of the use of a magnifying glass for the study of nature is Thomas Mouffet's posthumously printed book *Insectorvm sive minimorum animalivm theatrvm* (London, 1634). De Laet's copy of this book was bought in 1650 by the book collector Johannes Thysius (1621-1653), and is stil preserved in the *Bibliotheca Thysiana* in Leiden, built in 1655 two houses away from De Laet's home. See Hoftijzer, 'The library of Johannes de Laet' (1998), 215.

47 *HNB* (1648), 28: '... delineati ut per Megascopium ...'. Marggrafe gave as its Portuguese name *erva do vina*, a corruption of *erva divina* or divine herb. The illustration placed by De Laet concerns the *Plumbago scandens.* This confusion is copied by Piso in the edition of

1658. Because of this confusion, PICKEL says that 'we must not attach importance to what he [PISO] wrote on the subject'. See: Pickel, *Flora do nordeste do Brasil* (2008), 82 and Piso, *Indiae utriusque* (1658), 201, who calls the plant *Herva d'amor.*

48 *HNB* (1648), 257: 'Icon per megascopium fuit delineata'.

49 *HNB* (1648), 259.

50 Ashworth, 'Natural history and the emblematic worldview' (1990), 318.

51 ØLE WORM to VILLUM WORM, several letters, September–December 1653. See Schepelern, *Breve* (1965), nos. 1746-1748, 1759-1760. VILLUM WORM negotiated the printing of the *Museum Wormianum* with LODEWIJK ELZEVIER in Amsterdam. In the end the book was published by his relative JOHANNES ELZEVIER in Leiden. Only very few of the original woodcuts of the *Historia naturalis Brasiliae* were used in the *Museum Wormianum.* For example, the woodcuts of a spider (*HSN*, 248 = *MW*, 244) and a bird (*HSN*, 204 = *MW*, 308).

52 Whitehead and Boeseman, *A portrait of Dutch 17th century Brazil* (1989), 29.

53 Brienen, 'The zoological drawings of Eckhout and Marcgraf' (2007), 295.

54 Safier, 'The Editorial Itineraries of Marcgraf and Piso's *Historia Naturalis Brasiliae*' (2014).

55 Teixeira, 'Der Mythos der unberürthen Natur' (2008).

56 Du Rieu, *Album Studiosorum* (1875), 356: 'GUILELMUS PISO, inserviens illustriss. Comiti Mauritio. 34, Dr.M.'.

57 PISO to BARLAEUS, September 1645. Facsimile published in Pies, *Piso* (1981), 151-155. WILLEM PISO's father, the organ player HARMEN PIES, was buried in Leiden on 18 July 1645. See Pies, *Piso* (1981), 75.

58 Pies, *Piso* (1981), 151.

59 On 1 September 1648, in the Old Church of Amsterdam, WILLEM PISO, 37 years, married CONSTANTIA SPRANGER, '28 years' [but in fact 31 years old]. She was an orphan, being the daughter of the late cloth merchant GOMMERT SPRANGER (†1637) from Antwerp and MARIJA DE PICKER († c. 1621). She was probably raised by her aunt SUZANNA SPRANGER, who assisted her niece at the wedding. SUZANNA was the widow of the physician and botanist OUTGERT CLUYT (1578-1636), who, as PISO, was a friend of the well-known Dutch poet JOOST VAN DEN VONDEL. The marriage produced three children. Son HARMANUS PISO (†1658) was baptised there on 27 July 1649. Daughter MARIA was baptised on 11 December 1650, and daughter CORNELIA CATRINA PISO on 3 September 1659. Pies, *Piso* (1981), 114-117.

60 See for a discussion of PISO's addition to BONTIUS, Cook, *Matters of Exchange* (2007), 219–224.

61 Whitehead and Boeseman, *A portrait of Dutch 17th century Brazil* (1989), 29.

62 PISO to HEINSIUS, 28 September 1658. Facsimile published in Pies, *Piso* (1981), 159-162.

63 One of PISO's unnamed informers may have been the surgeon ABRAHAM DUURCOOP, head of the pharmacy of Recife, who in

the archives of the WIC is frequently named as working with Piso. Duurcoop was released from the service of the WIC on 19 june 1642, when the *Heeren XIX* ordered the disposal of many soldiers and related personnel. Wiesebron, *Brazilië in de Nederlandse Archieven*, 4 (2011), 178.

64 Piso, *Indiae utriusque* (1658), 107. See also Pickel, *Flora do nordeste do Brasil* (2008), 12.

65 Lichtenstein, 'Die Werke von Marcgrave und Piso' (1822), 270: 'Piso's Beschreibungen sind Durchgehends kürzer als Marcgrave's, und durch willkürliche Auslassungen werden sie gänzlich unvollständig. Oft ist auch, um den Schein eines wörtlichen Abschreibens zu vermeiden, die Stellung der Worte Marcgrave's verändert und so der Sinn der Rede nicht selten im Widerspruch mit sich selbst und der Abbildung. Piso's Werk verdient daher nur eine beiläufige Erwähnung, und kommt hauptsachlich nur da in Betracht wo es Irrtümer veranlasst hat, die widerlegt werden müssen'.

66 Moreira and Mendes, 'Sloths of the Atlantic Forest' (2016), 7. See Piso, *Indiae utriusque* (1658), 321. Eckhout's picture is in *Libri Picturati* A34.

67 Whitehead and Boeseman, *A portrait of Dutch 17th century Brazil* (1989), 31.

68 Dams, *Descriptions of Dutch Brazil* (2015), 279.

69 *Vitae* (1685), 7. Whitehead, 'Biography' (1979), 310.

70 Piso, *Indiae utriusque* (1658), 201, 288–289, 291–292.

71 Whitehead and Boeseman, *A portrait of Dutch 17th century Brazil* (1989), 30.

72 Rodrigues, 'Escôrço biobliográfico' (1957), xvii–xviii. A comparison of the meteorological tables in the *Historia* and the *Indiae utriusque* shows that the letterpress is identical, so the lead letters of the 1648 edition have been re-used in 1658.

73 *Lessus Funebris [...] Johannis Henrici Stechaviensis [...]* (Hafniae: Lamprecht / Altenburg n.d. [=1651]). The deceased theologian Johannes Heinrich (1621-1649) was the son of Michael Heinrich [=Henrici], Lutheran pfarrer in Grosstechau. See Thomas Walther (ed.), *Thüringer Pfarrerbuch: Band 6: Das Herzogtum Sachsen-Altenburg* (Leipzig, 2013), no. 759.

74 Du Rieu, *Album* (1875), 421 and University Leiden, ASF 10: '3 June 1652. Christianus Marckgravius, Lipstadio-Misnicus, ann 26, medicinae & mathesis studios, habitat in de Klocksteech, in de courvorstinne van Brandenborgch'. Two single pages from the lost *Album Amicorum* of Christian Marggrafe, with entries from Leiden professors, both dated July 1652, have survived. One entry is from Marcus Zuerius Boxhorn (1612-1653), the other from Antonius Thysius (1565-1640). Royal Library, The Hague, KW 73 B 40 and Middelburg, manuscript collection Koninklijk Zeeuwsch Genootschap, no. 6684.

75 See the introduction.

76 Christianus Marggravius, *Disputatio* medica περί του θείου [peri tou theiou] *sive De divino in morbis* (Leiden: J. and D. Elsevier, 1654).

77 ØLE WORM to VILLUM WORM, 15 July 1653: '... congratulations with the purchase of the *Herbarium vivum*, which may have belonged to MARCGRAVIUS and is described in DE LAET's Brazilian History'. ØLE WORM acknowledged the arrival of the herbarium in Copenhagen on 8 October 1653: 'I appreciate the Brazilian herbarium very much. It is extremely valuable and I will take great care of it until your return'. VILLUM WORM, enrolled at Leiden University on 7 July 1653. Shortly afterwards, VILLUM donated a 'Rijksdaalder' to the Lutheran Church in Leiden, where CHRISTIAN MARGGRAFE attended regularly. Both their names are on the same page of the Lutheran gift book, so it is plausible that the two gentlemen personally met there. See: UB Leiden, BPL 1450, fol. 85r ; Scheperlern, *Breve* (1965), letter nos. 1742 and1752 and hereafter footnote 86.

78 *Vitae* (1685), 7. Whitehead, (1979), 'Biography' (1979), 311. This is the only reference to such a collection of insects.

79 CHRISTIAN MARGGRAFE refers to the count's donation to the Danish king in his Latin letter to the trustees of Leiden University, dated 1 February 1655.

80 Whitehead and Boeseman, *A portrait of Dutch 17th century Brazil* (1989), 25; Wagener, *Thierbuch and Autobiography* (1979), 223.

81 For these donations, see at length Whitehead and Boeseman, *A portrait of Dutch 17th century Brazil* (1989).

82 CHRISTOPHER LINDENOV TIL LINDERSVOLD OG RESTRUP (CA 1612-1679) is named by JOHAN MAURITS as the person who told him that it would suit the Danish king ('*es nicht unangenehm sein wuerden*') to receive a gift of Brazilan paintings. See the letter from JOHAN MAURITS to FREDERIK III, 13 July 1654, printed in Larsen, *Frans Post* (1962), 251. See for this gift more in detail, Bencard, 'Fürstliche Geschenke' (2008). In Brazil, LINDENOV was responsible for the maintanance of the governement's buildings. In different monthly accounts (OWIC 69–70), he is named in relation to the 'Cloosterkerck' (14 November 1641); 'het Huys van de Hooge Raede' (12 June 1642), 't Raethuijs en Comptoir Generael opt Reciff' (3 January 1643), 'het Huijs daer de ambassadeurs van Angola sijn gelogeert' (27 March 1643) and 't Heerenhuijs, ten dienste van de Compagnie' (17 July 1643). His entry as 'captain' in the *Album Amicorum* of JOHANN PHILIPP MULHEISER, dated 31 August 1640, has not been identified before (Royal Library, The Hague). Later in life LINDENOV became admiral in the Danish fleet and ambassador to England. See Waldemar Westergaard (ed.), *The First Triple Alliance: The Letters of Christopher Lindenov, Danish Envoy to London, 1668-1672* (New Haven: Yale University Press, 1947).

83 CHRISTOFFEL LINDENOV was educated in 1629 at the military academy of Sorø. He donated several naturalia to the famous collector ØLE WORM. See: *Museum Wormianum* (1655), 192, 263, 308.

84 From the count's table in Recife, LINDENOV also knew the painter ALBERT ECKHOUT. After their return in the Netherlands in 1644, LINDENOV borrowed from ECKHOUT the sum of 3,000 guilders, for an investment in the Danish East India Company. See Van Gelder, 'Twee

Braziliaanse schildpadden door Albert Eckhout' (1960), 16.

85 In fact, at that time there were more siblings alive. From CHRISTIAN MARGGRAFE's last will and testament, made on 14 October 1671, we learn that there was also a brother, SAMUEL MARGGRAFE, apothecary in the German town of Spremberg in Niederlausnitz (currently near the border of Poland), and a sister BEATA MARGGRAFE, living with her brother SAMUEL. (ELO, notary archive JOACHIM TIERENS, invent. no. 1116, deed no. 92). In 1675 SAMUEL MARGGRAFE published a note about his invention of a *Wol-Approbirtes Vieh-Pulver* ('a well-approved cattle powder'), to be used in the cure against all kinds of cattle diseases. Sächsiche Landesbibliothek Dresden, Signatur Oecon.A.182,5.

86 JOHAN MAURITS to the University Trustees, 6 February 1655. This letter and all the others can be found in the Leiden University archive, ASF 290. Discussed by Meijer, 'De omstreden nalatenschap' (1972).

87 CHRISTIAN mentions this in his request to the University Trustees of 18 March 1655.

88 CHRISTIAN MARGGRAFE to ADOLPH VORSTIUS, 14 February 1655.

89 Perhaps this note gives a clue how MARGGRAFE's herbarium could be obtained by VILLUM WORM in 1653. A probable trajectory is the following: After the arrival in Leiden in 1644, GOLIUS found the herbarium in the chest that was delivered to him. A logical step was to give this collection of dried plants to the professor of botany, VORSTIUS. In turn, VORSTIUS gave the herbarium to DE LAET, to be used for the composition of the *Historia Naturalis Brasilae.* DE LAET ordered 15 woodcuts to be made from this herbarium, which book therefore had to go to the unknown woodcutter. Needless to say that the woodcutter was connected to the ELSEVIER publishing house. It makes sense to suppose that, because DE LAET died in 1649, the herbarium was never returned, but stayed with the Elseviers. In the spring of 1653 VILLUM WORM arrived in Amsterdam to negotiate with LODEWIJK ELZEVIER the printing of his father's famous book *Museum Wormianum.* Eventually t this book was published not by Lodewijk in Amsterdam, but in Leiden by his cousin and business partner JOHANNES ELZEVIER. So, it probably was during these negotiations that VILLUM WORM could buy the herbarium from the ELZEVIERS, who actually never had owned the herbarium. This route is all the more probable because DE LAET's assets were already sold in 1650. DE LAET's jewels and rarities were auctioned in Amsterdam in March 1650 and his large library was sold in Leiden in April 1650. In the catalogue of this sale no 'herbaria vivum' are mentioned. See: *Ordinaris dingsdaeghse courante,* 8 March 1650; *Catalogus bibliothecae ... De Laet* (1650); Hoftijzer, 'The library of Johannes de Laet' (1998); Andrade-Lima, 'Marcgrave's Brazilian Herbarium' (1977) and Wagner, 'Das Markgraf-Herbarium' (2008).

90 CHRISTIAN probably refers to the Greek expression Σύν θεω, used by his brother GEORG on his Wittenberg thesis and in his astronomical manuscripts, meaning 'with God', or 'with God's help'.

91 JOHAN MAURITS to the University Trustees, 19 March 1655. From his own memory the count could recall that in the chest had been

two skins of seahorses and an old rug (probably of native American origin).

92 *Ibidem.*

93 PISO to GOLIUS, 12 and 16 May 1655: 'alwaer ik bij sijn leven sijn kiste open sag met hem, weijnig te vooren eer hij na Africa vertrok, en daer in na mijn beste onthout, behalven sijn armlijk linden en wolle, wat students schriften gesien hebbe, een gestreept zeepaerts vell, wat out saet van Brasilse gewassen, ook gedroogde kruijden, en gedroogte ongedierties'.

94 *Ibidem,* 'gaf tot antwoort altijt, dat hij op geen gelt paste, en niet estimeerde als sijne astrologise en natuurlijke observatien waer te nemen, en te mogen seggen *Cantabit vacuus coram latrone viator,* en meer dergelijke beuselagtige woorden'. The expression is a citation from Juvenalis, *Satirae.*

95 Whitehead, 'Markgraf and Brazilian Zoology' (1979), 425 and Brienen, 'Georg Marcgraf'(2001), 91.

96 Statement by ABRAHAM DE VRIES before notary MARTIN BEECKMAN in The Hague, 1 June 1655 (Municipal archive Den Haag, Notary archive, inv. no. 267, deed 219). For CHRISTOPHER SCHETSKI's 'costgelt', see OWIC, 69, DN 8 March 1642.

97 Several observations by KECHEL are recorded in Pingré, *Annales Célestes* (1901).

98 Excerpt from the *Dingboeck* of the Academic *Vierschaar,* 19 June 1655: 'astronomische ende diergelijke mathematische observatien [..., welke] bequaem soude mogen sijn om in 't licht te brengen, mitsgaders oock noch eenighe brasiliaansche gewassen ende kleine natuurlijcke dingen al t' samen [...] noch leggende in een lade van heer verweerder [GOLIUS], waer bij oock een as van een globus met een copere meridiaen ende een kleijn [hoekmeet?] instrument, alles bij den nevensgaende inventaris breeder [gespecificeerd], [...]...uijtgesondert de electa (?) mathematica, tom 2, ende het planispherium van de suijder sterren, welcke twee den meergemelten prince aen den heer verweerder heeft belast aen sijn hooggemeltens handen te restitueeren'. Unfortunately, the inventory cited in the verdict is *not* preserved among the documents regarding the legal procedures. See also Meijer, 'De omstreden nalatenschap' (1972).

99 Schepelern, *Olai Borrichii itinerarium 1660-1665,* 1 (1983), 115 (26–27 April 1661).

100 *Catalogus van alle de principaelste Rarieteyten* (1671); English edition: *A catalogue of all the cheifest* [sic!] *rarities in the publick theater and anatomy hall* (1678).

101 The provenances given for some objects in the various catalogues published in different languages since 1671 are not very reliable. For example, 'the snout of an unknown fish from Brasil', noted as a gift from PISO (*Catalogus* 1671, 1678, no. 11) was in fact donated by the Leiden professor OTTO HEURNIUS before 1628. See: Kroon, *Bijdragen* (1910), 67, note 1. A large pod of the cassia tree PISO gave to the hortus in 1639 is first listed without provenance (1671, 1678:

H–39), but in 1699 this item is regarded as a gift of JOHAN MAURITS (L–39). From that date on, most Brazilian objects are listed in the catalogues as donated by the count. Gifts to the Leiden *Theatrum* by JOHAN MAURITS and PISO and are also mentioned in the notes of the Harderwijk magistrate ERNST BRINCK. Remarkable and horrible at the same time is his remark about PISO's donation of 'een grote slangenhuijt ... in welckers lichaem men gevonden heeft noch 2 vingers van een Negro'. See Roscam Abbing, *Brazilië* (2021), 163.

102 'Historiae Rerum Naturalium Brasiliae' (1648), 226; Almeida, *Historiae* (2016), 158.

103 Lemma 'Christiaan Marggraf' in *NNBW* 1 (1911). He obtained his degree as 'Medical Doctor' on 31 Juli 1659. As a token of gratitude for this promotion, CHRISTIAN MARGGRAFE donated a 'Rijksdaalder' to the Lutheran Church in Leiden. In 1653, a year after his arrival in Leiden, he had made a similar gesture. UB Leiden, BPL 1450, fol. 85r and 101r.

104 In 1665 CHRISTIAN MARGGRAFE lived at the corner of the Rapenburg and the Groenhazengracht. Later he moved to the Pieterskerckhof, where he held a *Collegium Chimicum* at least from 1674 until his death in 1687. The famous Irish physician THOMAS MOLYNEUX, who stayed with CHRISTIAN MARGGRAFE in 1684 and stood witness to the baptism of one of his children, called him 'a very noted physitian'. (*Correspondence of James Flamsteed*, 2 (1997), 165). During this stay with MARGGRAFE, MOLYNEUX visited several notable Dutch scholars, such as CHRISTIAAN HUYGENS in The Hague (7 August 1684) and the microscopist ANTONY VAN LEEUWENHOEK in Delft (13 February 1685).

105 Molhuysen, *Bronnen* 3 (1918), 142.

106 Morren, 'Een cholerisch geneesheer in de 18e eeuw' (1908). See also ELO, Crimineel vonnissenboek, inv. no. 3+18, fol. 38vs.

107 For her baptism, see Streekarchief Langstraat Heusden Altena, baptism records Babyloniënbroek, 7 Febr. 1655. Her father WILLEM ADRIAANSZ. VAN GAMEREN was *schout* ('sheriff') of Babyloniënbroek For the prenuptial agreement of the couple, see ELO, notary archive archive JOACHIM TIERENS, invent. no. 1118, 30 November 1674.

108 At their baptisms in the Leiden Lutheran Church, the children of the couple were named as follows: (1) ADRIANA ELISABETH (*1675); (2) GEORG (*1677); (3) THEODORA (*1679); (4) CHRISTIAEN (*1680); (5) WILHELMUS (*1683), (6) JOHANNES GODEFREDUS (*1684) and (7) JACOBUS (*1686). Three of them died young, one in May 1687. In March 1688, CHRISTIAN MARGGRAFE's widow enumerated the remaining 'Marckgraef' children as (1) ADRIANA ELISABET, (2) JURRIAEN (the renamed CHRISTIAEN), (3) WILHELMUS and (4) JACOBUS. Daughter ADRIANA ELISABET (†1709) married in 1702; Son JURRIAEN (†1726) first became a soldier employed by the VOC. He sailed to the Cape colony and the East Indies in 1708 and returned to Leiden in 1715. There he became a *velleplotter* (tanner). He married in 1718 and again in 1720. Son WILLEM († 1741) moved to Amsterdam, where he

became a grocer and married a German widow. They baptised three children between 1709 and 1714. The youngest son JACOBUS is last named in 1705, in the last will of his aunt DINA VAN GAMEREN, as being employed in the service of the country, indicating a career as a soldier.

109 Chr. Marggrafe, *Jacobi Le Mort, pseudochemici et ratiocinatoris dupondiarii ignorantia circa chemiam et universam scientiam naturalem* (Leiden 1687).

110 The widow ADRIANA VAN GAM[M]EREN (1655-1699) remarried on 11 March 1688 to ARNOLDUS KERCKHOVEN (1661-1691), a merchant in wood, and again on 7 August 1692 to GERRIT CRENA a *lakenhandelaar* (draper) from Dordrecht.

111 Advertisement in the *Oprechte Haerlemsche Courant*, 24 May 1687: 'Op den 28 Mey sal tot Leyden by PIETER DE GRAEF verkocht werden de nagelate Bibliotheecq van de Hr. Dr. MARGGRAEF, bestaende in seer schoone Medici en Historici Boecken. De Catalogi zijn tot Amsterdam by van Dalen, 's Hage, Beeck, Rotterdam, Santbergen, &c.'. During his career CHRISTIAN MARGGRAFE wrote two medical handbooks: *Prodromus medicinae practicae dogmaticae et vere rationalis* (Leiden: Boutesteyn, 1673; 1685[2]); *Materia medica contracta exhibens simplicia & composita medicamenta* (Leiden: Doude, 1674; Amsterdam: Wetstein, 1682[2]; Leiden: Boutesteyn, 1716[3]).

112 Terwen, 'The Buildings of Johan Maurits' (1979).

113 ELO, sign. LB 7000-1. These documents can be consulted online: https://www.erfgoedleiden.nl.

114 *De Navorscher* 5 (1855), 331.

115 Rammelman Elsevier, *Inventaris* (1864), 23.

116 Stokvis, *La médicine colonial* (1883).

117 Crommelin, 'Eine astronomische Handschrift' (1939).

118 Whitehead, 'Markgraf and Brazilian Zoology' (1979), 425.

119 ELO, sign. LB 18677.

120 Our transcriptions (as far as succeeded) are deposited with the original drafts in ELO, sign. LB 18677.

121 Royal Library in Copenhagen, MS Ny Kgl. Saml. 2007, 4°. This letter is also mentioned in MARGGRAFE to FROM, 26 June 1638.

122 Mentioned in MARGGRAFE to FROM, 26 June 1638.

123 Mentioned in MARGGRAFE to GOLIUS, 4 October 1638.

124 Mentioned in MARGGRAFE to GOLIUS, 26 June 1638.

125 The feather was hidden in a small manuscript booklet with the title *Tabula Anguli pro Prosthaphaeresibiq[ue]. Orbis in quinque Planetis. Propositio lateralum seu intervallorum.* (ELO, North no. 104).

126 A note with excerpted eclipse data is among the FLAMSTEED manuscripts in Greenwich (National Maritime Museum, Greenwich, P.R.O. vol. 41, fol. 26r). At the top of this paper FLAMSTEED states that he received this information from EDMOND HALLEY, who got it from THÉVENOT in 1668. We have not seen this document, which is referred to by North, 'Markgraf' (1979/1989), 397, footnote 17.

127 *Bibliotheca Thevenotiana* (Paris 1694), 210: 'Observationes Astronomicae a MARGGRAVIO in Brasilia factae cum earum usu'.

128 Lalande, *Astronomie*, 1 (1764), 537.
129 Lalande, *Astronomie*, 3d ed. (1792), vol. 1, 157. This multi-volume manuscript (*Tychonis Brahé Thesaurus observationum astronomicarum*) is now also kept in the Observatoire de Paris, sign. B 4 15–20.
130 Couplet, 'Extrait de quelques lettres écrite [...] du Brésil' [1700] (1706). See also Moreira, 'A expedição de Couplet à Paraíba' (1991).
131 Bigourdan, *Inventaire* (1895), 2–3.
132 *Ibidem.*
133 Lalande, in: *Histoire de l'Academie Royale des Sciences, Année 1766* (Paris 1769), 450.
134 'Collection d'observations astronomiques pour l'usage de l'Academie par De Lisle le cadet', Bibliothèque Nationale, Paris,, MS. n.a.f. 5156, pp. 63–64, cited in Saunders, 'The Archives of the Académie des Sciences' (1978), 696. Courtesy to Dr. Rob van Gent, who pointed us to this passage.
135 The transfer of Marggrafe's ready-to-press manuscript from Leiden to Paris, reconstructed here, based on factual evidence, as well as the copying process in Paris by J.N. De l'Isle, turns out to be completely different from the hypotheses previously formulated by, respectively, North (1979) or Matsuura (2011). North assumed that Ismael Bouillieu had played a major role in this, while Matsuura attributed this role to Erasmus Bartholin. See North, 'Markgraf' (1979/1989), 397 and Matsuura, *O observatório no telhado* (2011), 139-140.
136 See Godin's 'éloge' in the *Mémoires* of the Académie Royale des Sciences.
137 De l'Isle had been able to buy this correspondence from the heirs of Hevelius in 1726, on his way to his new assignment at the Russian Academy of Sciences in Saint Petersburg. Godin obtained an 18th-century copy made at De l'Isle's request. This copy of the Hevelius correspondence is now in the Bibliothéque Nationale in Paris.
138 Godin possessed also the astronomical legacy of the French astronomer Jacques-Eugene Louville (1671-1731), which in the 1780s was also bought by Lalande (*Astronomie*, 3rd edition, vol. 2 (1792), 180).
139 Lalande, *Astronomie*, second edition, 2 (1772), 160.
140 An Ulloa family archive seems not to exist. At least it is not mentioned in the *Actas del II centenario de don Ulloa*, the book commemorating Ulloa's life and work, published in 1995. Inquiries to colleagues in Spain also did not yield any result.
141 Joseph Townsend, *A Journey through Spain in the Years 1786 and 1787* (London, 1792), II, 440–441.
142 We have made unsuccessful attempts to recover the original manuscript at the library of the Real Instituto y Observatorio de la Armada in San Fernando and in the Archivo General de Simancas in Valladolid, heir institutions of the documents from Cadiz Observatory. See also Sellés, *Observatorio de Cádiz* (1988).
143 De l'Isle donated his copy of the manuscript to the *Dépôt de la Marine*. After the French revolution it went to the *Bureau des Longitudes*, and from there to the *Observatoire de Paris*, where at present it is preserved.

144 North, 'Markgraf' (1979), 397, footnote 14.
145 Observatoire de Paris, Mss B4–5; Biblioteca Nacional de Portugal, Lisbon, Mss 6, n. 37.
146 Personal communication by Luís Tirapicos, Lisbon.

Chapter 7

1 See Chapter 3.
2 Dekker, 'Early Explorations of the Southern Celestial Sky' (1987), 449. See also Knobel, 'On Frederick de Houtman's catalogue of the southern stars' (1917).
3 Dekker, 'On the dispersal of knowledge of the southern celestial sky' (1987). See also Verbunt and Van Gent, 'Early star catalogues of the southern sky' (2011).
4 See Chapter 3.
5 Eichstädt, *De confectione alchermes in Pomerania* (1634).
6 Omodeo, 'The scientific culture of [...] Laurentius Eichstadt' (2017).
7 Eichstädt, *Ephemeridum*, 2 (1636, 639[2]), n.p. [= 427]. Other predictions of eclipses visible in South America were for the solar eclipse of 3/13 November 1640 and the lunar eclipse of 1641 and the lunar eclipse of 24 March [= April n.s.] 1643. Eichstädt, *Ephemeridum*, 1 (1634), n.p. [= 257]; *Ephemeridum*, 2 (1636), n.p. [= 395, 455].
8 Eichstädt, *Ephemeridum*, 1 (1634), n.p. [= 16]; Omodeo, 'Eichstadt' (2017).
9 Schepelern, *Olai Borrichii itinerarium*, 1 (1983), 115.
10 The preface in Eichstädt, *Ephemeridum*, 2 (1636) is dated December 1635.
11 Johannes Bering matriculated at Franeker University on 20 October 1630 and at Leiden University on 14 December 1632. In December 1636 he became a professor at the University of Greifswald, first of mathematics and in 1645 of theology. Details about his study in the Netherlands can be found in his necrology by Heunius, *Ad Exequias [...] Johanni Beringio* (1658), 7. See also Friedländer, Ältere Universitätsmatrikeln. II. Universität Greifswald (1893), 559, 564, 634.
12 Leiden University Library, AC–I, inv. no. 22, fol. 90: 'omme aldaer de studenten te demonstreren de loop des hemels ende sterren'.
13 AC–I, inv. no. 315, fol. 264: 'voor dat hij een een jaer langh des Prof Goolius en de studenten heeft bedient op het nieuw gemaeckte plat int doen van de astronomische demonstratien'. Hereafter Pieter Vincentsz van Heemskerck requested a permanent appointment for this job, which he combined with a position as typographer in the adjacent printing shop of Elzevier, a position he had held for twenty years. He had also served Golius 'in his Arab works'. Indeed he obtained this position, in which he was obliged 'to serve Mr. Jacobo Golio and his students of astronomy', especially in the opening and closure of the windows, the lightning of the candles on the stairs of the Academy Building and in the small room below the newly made platform, and all other things that the Professor will order'. See AC–I,

inv. no. 315, fol. 148-149. See also Enschedé, 'De Leidsche Elseviers en hun meesterknecht P. Heemskerk' (1912).

14 Cf. Van Miert, 'Philology and Empiricism' (2013), 107-108.

15 De Wreede, *Snellius* (2007), 144, 147.

16 *Ibidem*, 47–48. In Prague SNELLIUS also made his acquaintance with JOHANNES KEPLER.

17 Snellius, *Descriptio cometae* (1619). Wreede, *Snellius* (2007), 100, 163.

18 This concerned the design of a celestial globe issued by the Dutch map maker JACOB AERTSZ. COLOM (1600-1673) from Amsterdam, constructed about 1635. This was the earliest European celestial globe with labels in Arabic script as well as in Latin and Greek. Only one – incomplete – copy of this globe survived. The constellation designs and the star positions on this globe by COLOM are in every respect identical to those on the 1603-edition of the celestial globe designed by WILLEM JANSZ. BLAEU. See Savage-Smith and Wakefield, 'Jacob Golius and Celestial Cartography' (1994).

19 Pingré, *Annales Célestes* (1901), 104, 106.

20 Van Berkel, 'De illusies van Martinus Hortensius' (1997). HORTENSIUS's first astronomical observation was the solar eclipse of 20 May 1621, when he was 16 years of age. In February 1625 he also observed a conjunction of the moon and Venus during a visit to Leiden. Pingré, *Annales Célestes* (1901), 55, 63.

21 Pingré, *Annales Célestes* (1901), 77.

22 This 'bon quart-de-cercle de cuivre' was also used in February 1632. Pingré, *Annales Célestes* (1901), 83, 88.

23 Pingré, *Annales Célestes* (1901), 90.

CHAPTER 8

1 Leiden University Library, AC–I, inv. no. 22, fol. 91–94. The construction was made by the carpenter PIETER JANS VAN HEININGEN, 'meester timmerman' for 2100 guilders.

2 Leiden University Library, AC–I, inv. no. 22, fol. 151. 'Noch soo sal den aennemer gehouden wesen te stellen het instrument onder op een stijl met vier carbeels, en boven met een klamp onder aen de binten dat het onbeweechlijck ende recht in 't loot staet. Ende dat men bequamelijck mach draeijen om aen alle die sijden gebruijckt te worden'.

3 'Ende dat men bequamelijck mach draeijen om aen alle die sijden gebruijckt te worden'. The carpenter PIETER JANS VAN HEININGEN, who had made the platform in 1633, also made the dome, this time for 550 guilders and under the condition that the work was finished in August 1634. AC–I, inv. no. 22, fol. 153.

4 Polman, *Markgraf e o Recife de Nassau* (1984).

5 Mota Menezes, *Atlas Arqueológico do Recife* (1998).

6 Specifications for the contractor of the platform, dated December 1632. Cf. Univ. Library Leiden, AC I, inv. 22, fol. 91–95. The original contract with signatures is in AC–I, 43–3.

7 The 'Rhijnland foot' was the common standard for measuring length

in the Dutch Republic. One Rhijnland foot (containing 12 'duim' or inches) had a length of 31.4 cm.

8 'Noch soo sal den aennemer gehouden wesen te stellen het instrument onder op een stijl met vier carbeels, en boven met een klamp onder aen de binten dat het onbeweechlijck ende recht in 't loot staet. Ende dat men bequamelijck mach draeijen om aen alle die sijden gebruijckt te worden'. Specifications for the building contractor of the turret, dated 14 June 1634. Cf. Univ. Library Leiden, AC I, inv. 22, fol. 151-153.

9 'Noch te maecken veertien luijcken ofte veijnsters, acht boven om op te setten, en zes neervallende en die soo groot te maecken om het instrument te mogen gebruijcken op die 80 en 20 graeden, gelijck in de teeckeninge te sien is'. *Ibidem.*

10 This room below the platform (a 'solderken om te stellen de globen') was made in 1633, a year before the construction of the dome, 'opdat de voorsz. Gool sal mogen ghebruiken de twee grote glooben die nu op de publique bibliotheque staen'. See Resolution 9 August 1633 (Univ. Library Leiden, AC I, inv. 22).

11 The globes MARGGRAFE brought along did not survive the journey back to the Netherlands. In 1659, GOLIUS possessed only the axis of one of the globes, with a brass meridian circle.

12 For more details on the still existing Blaeu quadrant see: Haasbroek, *Frisius, Brahe and Snellius and their Triangulations* (1968), 22–23 and Zuidervaart, *Telescopes from Leiden Observatory* (2007), 24.

13 This portable quadrant with a diameter of one foot was used to observe the moon eclipse of 20 December 1638.

14 How the horizon itself could be used as an azimuthal circle using the Snel quadrant is explained by the Leiden professor of astronomy, JOHAN LULOFS, in his *Inleiding tot eene natuur- en wiskundige beschouwing des aardkloots* (1750), 476–480. LULOFS fixed the quadrant at night in a certain position and measured at daylight the angle between the meridian and the point at the horizon towards which the quadrant was aiming. See for the procedure for measuring with such a quadrant, Chapman, 'Reconstructing the angle-measuring instruments of Pierre Gassendi' (1994), 105-110.

15 In 1656 an azimuthal circle with a radius of 20 inches (52 cm) was added to the quadrant, ordered from the Leiden instrument maker HENDRICK SNEEWINS. Since this artisan died before the azimuthal circle was finished, the work was completed by another Leiden instrument maker, JAN DAVIDT (the younger). Cf. Zuidervaart, *Telescopes from Leiden Observatory* (2007), 24.

16 See about the *pinacidium tychonicum*, Brahe's clever sighting device: King, *History of the telescope* (1955), 22. See also Straker, 'Kepler, Tycho, and the Optical Part of Astronomy' (1981).

17 Zuidervaart, *Telescopes from Leiden Observatory* (2007), 11.

18 Van Helden, 'Gassendi and the telescope'(1994), 330.

19 Willach, 'The Development of Lens Grinding and Polishing Techniques' (2001); idem, 'The Development of Telescope Optics' (2001).

20 Only SNELLIUS's quadrant was university property. The first telescope bought by Leiden University was the telescopic sight that the new professor of astronomy, CHRISTIAAN MELDER, adapted to the large quadrant in about 1669. See Zuidervaart, *Telescopes from Leiden Observatory* (2007), introduction, 11.
21 Schottmüller, 'Reiseeindrücke aus [...] Holland 1636' (1910), 265.
22 MARGGRAFE to GOLIUS, 26 June 1638.
23 In 1637 the only telescope makers known to be active in the Netherlands were (1) EVERT HARMANSZ STEENWIJCK, in Delft, between 1609 and 1654; (2) PAULUS RUYSCH in Utrecht, probably the same as the *tourneur d'Utrecht*, recommended in 1635 by GOLIUS to RENÉ DESCARTES; (3) ARIEN *de slyper* in Dordrecht, consulted in 1635 by ISAAC BEECKMAN and (4) JOHANNES SACHARIASSEN in Middelburg, who in 1634 taught BEECKMAN how to grind telescope lenses. (5) Finally, there was *Den Engelschen brilslyper by den Dam tot Amsterdam*, who in 1634 delivered two telescopes to MARTINUS HORTENSIUS. He is probably the same as *MR. GOETHAM, Engelsman, brillemaecker, wonende alhier op den Vijgendam*, who in 1665 is named as telescope maker in the last will of the Amsterdam theologian ADAM BOREEL. See Zuidervaart, 'The invisible technician made visible' (2012).
24 In the last part of his letter to GOLIUS, MARGGRAFE tells that he has three tubes, 'but I don't like any of them', because it concerned 'old tubes, which let light through at the sides'. With this remark he is probably referring to the sight tubes of the quadrant.
25 See vol. 2, manuscript page 7.
26 ELO, North no. 27vs. The note lists the following books: (1) JOHANNES BAYER's *Uranometria* (Augsburg, 1603); (2) JOHANNES KEPLER, *Tabulae Rudolphinae* (Ulm, 1627); (3) JOHANNES KEPLER, *Epitome astronomiae Copernicanae* (Lintz, 1618-1621); (4) WILLEBRORD SNEL, *Tiphys Batavus* (Leiden, 1624); (5) An edition of Archimedes; (6) A book of a certain 'De Castor'. Not clear which author is meant here. JOHN NORTH suggested a medical work of the Italian scholar GIROLAMO FRACASTORO (c. 1476-1553). In our view, the Portuguese scholar JOÃO DE CASTRO (1500-1548) is also a good candidate; (7) ISAAK HABRECHT, *Planiglobium coeleste ac terrestre* (Argentorati [Strasbourg], 1628); (8) Three 'tabulai quadratas' (unidentified works) by JACOB BARTSCH, printed also in Argentorati [Strasbourg]. (9) GUILLAUME DE NAUTONIER, *Mecometrie de Leymant* (Venice, 1603).
27 MARGGRAFE to GOLIUS, 26 June 1638. (ELO).
28 North, 'Markgraf' (1979), 401.
29 ELO, North, no. 111. Loose leaf (page 50) identified as coming from Philippe Garnier and Luca Donati, *Gemmulae Linguarum Latinae, Gallicae, Italicae & Germanicae* (Leiden: Ex officina Elseviriorum, 1637).

CHAPTER 9

1 ELO, North no. 66, dated 9 October 1636.
2 ELO, North no. 64. In this document MARGGRAFE addresses 16 numbered subjects (without caption) and seven 'problemata'. The dates

mentioned are: 20 May, 14 June, 23 September, 16-18 and 28–30 November and 24 December 1636.

3 *Ibidem*, '17 November 1636, Lugdunum h 8:53 vesperij'. ('vesperij' = 'in the evening').

4 North no. 22. This document contains draft observations, made between 13 January and 5 February 1637, mostly copied later in the Leiden observation register, North no. 52. One observation bears the date '8/18 Jan. 1635', but this seems a slip of the pen for '1637').

5 Nikitinski, 'Petrus Cunaeus über Astrologische Wahrsagerei' (2014), 252.

6 ELO, North no. 22, fol. 5recto and North no. 52, fol. 1r. (19 Jan 1637).

7 Van Helden, 'The astronomical telescope' (1976), 28.

8 Calculated according to the approximation: focus objective ≈ diameter image (here 0.009 meters) divided by the angular diameter of the Sun in radians (= 33 arcminutes, or 9.6 E-03 rad.), makes 0.94 meter. Cf. V. Greco *et al*, 'The optical principles of telescopes', in: Van Helden, *Catalogue of Early Telescopes* (1999), 99-111, esp. 108.

9 Calculated according to the approximation: focus objective ≈ diameter image (here 0.158 meter) divided by the angular diameter of the Sun in radians (= 33 arcminutes, or 9.6 E-03 rad.), makes 16.5 meters. Cf. *Ibidem*.

10 Drake, *Discoveries and Opinions of Galileo* (1957), 115.

11 Scheiner, *Rosa Ursina* (1630). Although SCHEINER eventually used a Kepler configuration, most of his projections were made with Galilean optics.

12 This platform for meteorological and astronomical purposes was erected at BEECKMAN's request in the summer of 1628. Cf. De Waard, *Journal tenu par Isaac Beeckman*, vol. 3 (1945), 112-113. See also: Van Berkel, *Isaac Beeckman* (1983), 104, note 21.

13 De Waard, *Journal*, vol. 3 (1945), 153.

14 HORTENSIUS to GASSENDI, June 1634, cited by Van Berkel, 'De illusies van Martinus Hortensius' (1997), 67.

15 Vaquero e.a., 'Revisited sunspot data' (2011).

16 Calculated according to the approximation: diameter objective ≈ average wavelength (of 550 nm) divided by the angular resolution in radians, (here ≈ 11 arc seconds, or 5.3 E-05 rad.), makes 10 mm. Cf. Greco *et al*, 'The optical principles of telescopes' (1999), 106.

17 Bolt & Korey, 'The world's oldest surviving telescopes' (2010).

18 In these calculations we followed the formula for the Galilean telescope developed by John North: $\Theta = 3438*(p \pm D/m)/(L + e*m)$. With Θ = field of view; p = diameter eye pupil; D = diameter objective; L = length of tube; e = effective pupil distance from ocular; m = magnification. The minus stands for the clear unvignetted field, and the plus for the partially illuminated field. When used visually, the eye pupil acts as a field stop, limiting the field of view. Assuming that (at night) p = 8 mm; D = 2 cm; L = 0.94 m and e = 1.0 cm, we derive that at night the unvignetted view Θ = 21 arc minutes. See: North, 'Thomas Harriot and the first telescopic observations of sunspots'

(1989), 128-133 and 142-144. See also: Casanovas, 'Early observations of sunspots' (1997), B, 3–20.

19 ELO, North no. 52, observation on 3 July 1637 of a Sco in the Heart of the Scorpion and the Moon's limb. The angular distance between these celestial bodies was at the time 18 arc minutes.

20 ELO, North no. 52, observations of Jupiter's moons recorded on 19 January, 22 February and 15 March 1637.

21 HORTENSIUS to GALILEI, 26 January 1637, in: *Le opere di Galileo Galilei*, 17 (1937) no. 3421, p. 18. For a Dutch translation by AD DAVIDSE, see https://adcs.home.xs4all.nl/hortensius/briefw-v.html. (Consulted December 2020). See also Keil, *Von Ocularien, Perspicillen und Mikroskopen* (2003), 48, citing a letter by GALILEO GALILEI to ELIA DIODATI, dated 6 June 1637. Similar complaints about the bad quality of telescopes in Holland are made in GASSENDI's letter to GALILEI of 19 January 1634 (idem, 40). See also Van Helden. 'Gassendi and the telescope' (1994), 330.

22 Davids, *Zeewezen en wetenschap* (1986), 73, 414. See about this proposal also GALILEI to HORTENSIUS, 15 August 1636, in: De Waard, *Journal tenu par Isaac Beeckman*, 4 (1953), 246 and HORTENSIUS to ELIA DIODATI, 24 November 1636, in: *Le opere di Galileo Galilei*, 16 (1936), 521.

23 Hortensius to Elia Diodati, 27 April 1637, in: *Le opere di Galileo Galilei*, 17 (1937), no. 3470.

24 Van Helden, 'Gassendi and the telescope' (1994), 333–336.

25 *Ibidem*, 333. See also Van Helden, 'The importance of the Transit of Mercury of 1631' (1976).

26 Gassendi, *Mercurius in sole visus et Venus invisa* (Paris: S. Gramoisy, 1631).

27 Hortensius, *Dissertatio de Mercurio in Sole viso et Venere invisa* (1633). See also the letters to and from HORTENSIUS on the website of AD DAVIDSE, https://adcs.home.xs4all.nl/hortensius/index.html (consulted December 2020).

28 FROM to LONGOMONTANUS, 1 May 1637 (Letter book GEORG FROM, Copenhagen, draft letter no. 54).

29 The crossings of the meridian consisted of 106 solar measurements (from 17° 37' to 61° 20') and 183 altitudes of a fixed star, the Moon or the planets (from 3° 33' to 70° 9').

30 Cf. Marriott, 'SkyMap Pro11' (2005). In the calculations, the following coordinates for the Leiden Academy Building are used: 52° 09' 25" North and 4° 29' 07" East.

31 Wesley, 'The Accuracy of Tycho Brahe's Instruments' (1978), 51.

32 Haasbroek, *Frisius, Brahe and Snellius and their Triangulations* (1968), 29.

33 As all observations were corrected for refraction, this potential cause must be declined.

34 Wesley, 'The Accuracy of Tycho Brahe's Instruments' (1978), 44, table 1. For individual stars, the mean error with BRAHE's revolving wooden quadrant (of 5 feet radius) varied between –67" and +100". Our results corroborated WESLEY's finding that multiple measurements of the

same star (for example, b Leo 7 times and a Aql 10 times) reduced the standard deviation of error to ~45". For strictly solar observations, the mean error was -1' 2" (standard deviation 1' 2") and for only non-solar observations the mean error was 49" (standard deviation 1' 33").

35 Haasbroek, *Frisius, Brahe and Snellius and their Triangulations* (1968), 29.

36 The clock tower on the academy building was erected in 1671. For the installation in 1636 of a timepiece with a bell at the academy, see: AC I, 22, fol. 193, 195.

37 See MARGGRAFE's description of the observatory in Volume 2.

38 ELO, North no. 51vs.

39 For details, see North, 'Georg Markgraf' (1979), 416; (1989), 228.

40 The smallest discrepancy was 20 s and the largest 65 min 39 s. For the remaining set of 45 points the error is roughly distributed symmetrically around the zero error. The average error was –4 s with standard deviation of 3.9 min. The minimum error was –9.95 min and the maximum 18.7 min.

41 MARGGRAFE to GOLIUS, 26 June 1638.

42 Ferrão & Soares, *The Voyage of Caspar Schmalkalden*, 2 (1998), 180.

43 MARGGRAFE to KECHEL, 18 May June 1639. 'Ich habe alfür in Astronomie noch wenig observiren können'.

44 *Progymnastica Mathematica Americana*, goals no. I–ii and I–iii.

45 See Volume 2, note of 7 September 1640.

46 Van Helden, *Measuring the Universe* (1985), 178.

47 North, 'Markgraf' (1979), 417.

48 *Progymnastica Mathematica Americana*, goal no. II–i.

49 ELO, North no. 54, last page. See for preliminary calculations of Recife's latitude from September 1639 also North no. 36.

50 *Progymnastica Mathematica Americana*, goal no. I–ii.

51 At the dawn of 25–26 June 1638, an earlier total lunar eclipse had happened, but that one was only partially visible from Recife. When leaving the umbra, the Moon had already set. Moreover, MARGGRAFE noted in his journal that at the time that 'it rained all night with strong winds', so that he saw the Moon only occasionaly 'leaping out from the clouds'.

52 These 'Fortunate Isles' or 'Isles of the blessed' were legendary islands in the Atlantic Ocean (positioned west of the Canary Islands), described in the Greek mythology as an earthly paradise. CLAUDIUS PTOLEMY introduced these imaginary islands as a reference point for the measurement of geographical longitude. This practise was continued through the Middle Ages until the seventeenth century. Compared to the current Greenwich meridian, a correction of 24° 03' West is needed.

53 See for MARGGRAFE's calculation, ELO, North doc. no. 10, discussed by Pimentel and Pereira, 'A astronomia e o mapa Brasilia qua parte paret Belgis' (2014), 220–223.

54 A comparison with other observers in Europe shows that this result of 3% was equalled by only one observer, namely WILLIAM CRABTREE in Broughtown near Manchester, who measured a duration of 3 hours

and 41 seconds. See for these other observations, Pingré, *Annales* (1901), 122-124.

55 The information about the lunar eclipse of 20 December 1638 was also shared with Kechel and From, who evidently also did not return any information on the Leiden observation.

56 Marggrafe to Golius, 16 February 1640 (ELO).

57 Eichstadt, *Ephemeridium* (1634) gave as the longitude of Leiden 29° 00' with respect to the Fortunate Islands, which in current coorinates would be 4° 56'. The real value is 4° 29', a difference of 0° 27'.

58 Hortensius to Pierre Gassendi, 12 October 1636, in: Gassendi, *Opera omnia* 6 (1658), 432. See the Dutch translation by Ad Davidse, https://adcs.home.xs4all.nl/hortensius/briefw-v.html#12-10-1636. (Consulted December 2020).

59 Marggrafe to Kechel, 18 May 1639 (ELO).

60 Barlaeus, *Rerum* (1647), 196-198; idem, *Brasilianische Geschichte* (1659), 551–552; idem, *History of Brazil* (2011), 190-192.

61 ELO, North no. 7.

62 Daxecker, 'Scheiner's main work "Rosa Ursina sive Sol"' (2005), 136.

63 Hortensius to Gassendi, 2 June 1634, in: Gassendi, *Opera omnia*, 6 (1658), 422.

64 Hortensius to Gassend, 12 October 1636, in: Gassendi, *Opera omnia* 6 (1658), 432.

65 Gassendi to Hortensius, 13 February 1637, in: Gassendi, *Opera omnia* 6 (1658), 92.

66 ELO, North no. 79.

67 ELO, North no. 33.

68 Value provided by SkyMap. For Marggrafe's draft calculations on this subject, see also ELO, North no. 36.

69 Courtesy to Ingrid Pot, André van Noort and Piet de Baar of Erfgoed Leiden, who assisted in establishing that the ink of all the words on this document was the same.

70 OWIC 58, no. 206, April 1643, published by Caland, 'Hofhouding' (1898).

71 *Progymnastica Mathematica Americana*, goal no. I–i.

72 See Knobel, 'On Frederick de Houtman's catalogue of the southern stars' (1917); Verbunt & Van Gent, 'Early star catalogues of the southern sky (2011) and Marggrafe's observations in vol. 2.

73 *Progymnastica Mathematica Americana*, goal no. III.

74 *Progymnastica Mathematica Americana*, goal no. I–v.

75 ELO, North no. 54, last page.

76 Flamsteed, *Historiae Coelestis Britannicae*, 3 (1725) [Often referred to as *Prolegomena*], 123-124. See also Lalande, *Astronomie* (1771), 142.

77 See Volume 2, note of 10 December 1641.

78 North, 'Markgraf' (1979), 407.

79 *Ibidem*. The Rhineland rod is taken as 3.767 m.

80 See Volume 2, note of 10 January 1641.

81 ELO, North no. 41.

82 Gospel of Matthew 2:1-12.

83 North, 'Markgraf' (1979), 399.
84 *Ibidem*, 407.
85 Hughes, 'Why did other European astronomers not see the December 1639 transit of Venus?' (2004).
86 MARGGRAFE had obtained a value for the Sun's highest elevation of 76 39 40" = 76.7 on 29 November 1639, the last day he was able to take measurements.
87 See for the international efforts to observe the Venus transits of 1761 and 1769, Woolf, *The Transits of Venus* (1959). For these efforts in the Netherlands and Indonesia, see Zuidervaart, *Van 'Konstgenoten' en Hemelse Fenomenen* (1999) and Zuidervaart and Van Gent, 'A Bare Outpost of Learned European Culture on the Edge of the Jungles of Java' (2004).
88 MARGGRAFE to GOLIUS, 26 June 1638.
89 HORROCKS's observation occurred during the Transit of Venus on 4 December 1639. See for all known historical observations of sunspots, the 'Revised Collection of Sunspot Group Numbers (v1.21; Date: 2020–04-19), online at http://haso.unex.es/haso/wp-content/uploads/2020/04/Revised-collection_v1.21.txt. (Consulted September 2021). See also Arlt and Vaquero, 'Historical sunspot records' (2020).
90 See Volume 2, note of 23 September 1640. MARGGRAFE made a sketch of three Jupiter satellites whose configuration matches rather well with current calculations. See Matsuura, *O observatório no telhado* (2011), 127, fig. 71.
91 A check on GALILEI's method of the determination of the longitude using the satellites of Jupiter would also require prevision tables of the satellites' positions (such as published in later years in astronomical almanacs such as the *Connoissance des temps*, 1679-1794). These tables were not available at the time, so MARGGRAFE would have calculated these himself, for instance based on his Leiden observations. For such a check he also would need to know the local time, which was difficult to achieve at night without a reliable timepiece.
92 Van Helden, 'Saturn and his anses' (1974), 112-113.
93 MARGGRAFE's last observation of Saturn was on 19 January 1641.
94 *Nieuwe ongewoon-wonderlykke staert-sterre op 't recif in Brazil gezien, op den 16 december anno 1652: Aldus vertoont; en affgebeelt door N.N.* (Amsterdam, 1653). Knuttel, no. 7353. The Recife observation is also mentioned in the *Hollantsche Mercurius* 3 [voor 1652] (Amsterdam: Brouwer, 1659), 112. This observation in Recife coincided with the first observation in Cape Town, South Africa, by governor JAN VAN RIEBEECK. The Recife observation is *not* listed in Kronk, *Cometography*, 1 (1999), 346–347. See also Morize, 'O cometa de 1652' (1904); Van Groesen, *Amsterdam's Atlantic* (2016), 148.
95 Naber, 'Het dagboek van Hendrik Haecxs' (1925), 293–294. HAECXS followed the comet every day until its disappearance on 25 January 1653. On the next day he wrote a prayer in his diary, wishing that the comet's appearance didn't predict anything bad.
96 HAECX is mentioned as living in Recife as early as April 1642. (Wiesebron, *Brazilië in de Nederlandse archieven* 4 (2011), 486). See also

OWIC 69 (31 July 1642): delivery of 400 pound twisted wax candles to the WIC; idem (2 Sept 1642): HAECX is one of the deacons of the Reformed Church of Mauritsstad and Recife; OWIC 70 (29 April 1644): HAECX delivered for 475 guilders 'goods to his Excellency'. He left Dutch Brazil later that year, but returned in May 1646.

97 Kronk, *Cometography*, 1 (1999), 355–356.

98 Although this study of astronomy at Groningen University is most likely an exaggeration, it seems plausible that SCHMALKALDEN received some instruction in astronomy from a local mathematical practitioner. As a seafaring nation, every Dutch city had a number of teachers who offered instruction in navigation. As SCHMALKALDEN arrived in Groningen on 4 May 1642 and entered the service of the West Indian Company already on 22 September 1642, his stay there was only four and a half months. SCHMALKALDEN probably did not enter the university as a student. His name is absent in the *Album Studiosorum* of Groningen University. Further, Groningen's main astronomer JOHANNES BORGESIUS (1617-1652) – at that time a calculator of almanacs – was at that time abroad to finish his study in medicine. BORGESIUS became professor of mathematics and astronomy only in 1645. See Joost, 'Auf den Meeren der Welt. Das Leben des Caspar Schmalkalden' (2002), 7 and *NNBW*, lemma 'Borgesius'.

99 Ferrão & Monteiro Soares, *The Voyage of Caspar Schmalkalden from Amsterdam to Pernambuco in Brazil*, vol. 1 (1998), 17.

100 Pingré, *Annales Célestes* (1901), 138-139.

101 MARGGRAFE's measurements are noted with an angular accuracy of 5''.

102 *Period I*: From 15 September 1639 – 18 March 1640: 61 solar and 259 stellar and planetary observations (total 320): uncertainty ± 8'; systematic error 7' | *Period II*: From 11 June 1640 – 7 February 1641: 89 solar and 92 stellar and planetary observations (total 181): uncertainty ± 5'; systematic error 2'. | *Period III*: From 3 October 1642 – 22 June 1643: 5 solar and 97 stellar and planetary observations (total 102): uncertainty ± 5'; systematic error 10'.

103 See Chapter 5.

104 OWIC 69, minutes of the Political Council, 20 and 22 January 1643: 'De Heeren Schepenen vertoonen dat de oude wooninge van sijn Excellentie soo is vervallen dat voor gedane reparatie het selve voor geen raethuijs bij haer kan gebruijckt worden'.

105 North, 'Markgraf' (1979), 404. It was the Brazilian historian JOSE ANTONIO GONSALVES DE MELLO who had suggested to NORTH that the count's first residence had been used as MARGGRAFE's observatory.

106 The date of *Vrijburg*'s commissioning can be deduced from an account of the count's treasurer, JOHANNES GREVINGH, dated January 1644, in which he claims compensation for the rent of the count's new house, starting 1 July 1642 (OWIC 58, no. 205, fol. 4).

107 Barlaeus, [English translation] (2011), 142.

108 Ferrão & Monteiro Soares, *The Voyage of Caspar Schmalkalden from Amsterdam to Pernambuco in Brazil*, vol. 1 (1998), 72.

109 Nieuhof, *Gedenkweerdig Brasiliaense Zee- en Lant-Reize* (1682), 19:

Vrijburg 'pronkte met twee hooge toorens, die vijf of zes mylen verre uit zee konden gezien worden, en den zeeman een baek verstrekten'.
110 Pereira da Costa, 'Primeira observações meteorologicas no Brazil' (1900), 113.
111 Carvalho, 'Um Naturalista do Seculo XVII' (1908), 216–217.
112 Gudger, '*Marcgrave*' (1912), 253, footnote 5.
113 Ihering, 'George Marcgrave' (1914), 309.
114 Moreira, '*Marcgrave e Piso*' (1917), 663.
115 Oliveira Lima, '*Sabios estrangeiros*' (1920), 338.
116 Wätjen, *Das holländische Kolonialreich in Brasilien* (1921), 124. WÄTJEN located the count's *Sternwarte* in *Vrijburg*'s garden.
117 Taunay, 'Escorço Bibliográfico' (1942), X.
118 Moraes, 'A Astronomia no Brasil' (1955), 95.
119 Whitehead, 'Biography' (1979), 312, note 12.
120 Snedegar, Lemma 'Georg Markgraf' (2014).
121 JOHANNES MICHAEL ANTONIUS POLMAN (1927-1986) was born in Amsterdam and fought at the end of the National Revolution of Indonesia in 1949. But in 1952, he decided to take a new direction in his life and came to Recife where, in 1957, he was ordained priest of the Order of the Sacred Heart of Jesus and became known as 'Father Jorge Polman'. As a science teacher at Colégio São João, in the neighborhood of Várzea, in Recife, he used observational astronomy as a pedagogical instrument with rare skill and was liked and admired as an educator and amateur astronomer.
122 Polman, *Markgraf e o Recife de Nassau* (1984); idem, 'First observatory' (1984).
123 North, 'Astronomia no Brasil holandês' (1980).
124 See Volume 2, note of 10 December 1641.
125 OWIC 69, minutes of the Political Council, 20 and 22 January 1643.
126 Naber, 'Het dagboek van Hendrik Haecxs' (1925), 301.
127 Pereira da Costa, *Inventario das armas* (1940), no. 363 (main house) and the surrounding buildings no. 349, 353–362 and 368–371. See also Nederveen, *Recife* (1989), 164-165.
128 Menezes, 'A casa de sua excelencia' (2005).
129 Zuidervaart, 'The invisible technician made visible' (2012), 49.
130 Matsuura, *O observatório no telhado* (2011), 128: On 18 January 1641 at 6:30 PM MARGGRAFE observed a conjunction between Saturn and Venus with his telescope. The distance he measured between these planets was 2/3 of the telescope's field of view. According to current calculations, the distance between Saturn and Venus at the first observation was 7.5 arc minutes, from which it can be concluded that the viewer's field of view was 11.25 arc minutes.
131 MARGGRAFE used the clepsydra on 15, 16 and 18 September 1639.
132 A length of 75.7 cm would lead theoretically to a time period for one swing back and forth of $T = 1.745$ sec. Half of it would be 87.3 sec. PINGRÉ found in 1640 a mean value of 94 sec, which corresponds with a theoretical length of 87.8 cm. The difference of 12.1 cm can be attributed to the length of the weight.

133 Pingré, *Annales Célestes* (1901), 140-141.
134 North, 'Markgraf' (1979), 228. For determining the local time of a central observation (T), MARGGRAFE interpolated between the times obtained (T_1 and T_2) from observations of a familiar star. He counted the number of oscillations (n_1 and n_2) of his pendulum in the two time intervals ($T_1 - T$) and ($T_2 - T$). Then the moment T is simply found, as $T = (n_1.T_2 + n_2.\ T_1)/(\ n_1 + n_2)$.
135 See MARGGRAFE's description of the observatory in Volume 2.
136 We compared 71 records of the *Hora Urbica* in Recife, noted at the beginning of an observing session, with the calculated current local time of the first astronomical observation. We found a mean difference of 9.6 min with standard deviation of 8.9 min. The difference may represent a real delay between the start of the observational session and the taking of the first measurement, but a discrepancy of 15 min was never reached.
137 BOULLIAU's copy of MARGGRAFE's observations (15 September 1639–22 November 1642) is preserved in the Observatoire de Paris, mss. B12-13.
138 ISMAËL BOULLIAU to JOHANNES HEVELIUS, 14 March 1662 (Observatoire de Paris). See for his visit to SAMUEL KECHEL also: Zuidervaart & Beurze, 'Kechel' (2014), 33.
139 Pingré, *Annales célestes* (1901), 278.
140 North, 'Markgraf' (1979), 397, footnote 17.
141 Halley, *Catalogus stellarum australium, sive, supplementum catalogi Tychonici* (1679).
142 Thrower, *Three Voyages of Edmond Halley* (1982), 44, 100-104, 188-190, 307–308, and Cook, *Edmond Halley* (1998), 279–280. The 'Pallacio das duas tores' is still named as such on the map 'Planta genografica da Villa de S. António do Recife de Pernambuco' (1763), Arquivo Histórico Ultramarino, Lisbon, ahu-pe_929.
143 See for HALLEY's determination of Recife's coordinates, Thrower, *Three Voyages* (1982), 190. MARGGRAFE's determination of Recife's position is discussed earlier in this Chapter, see fig. 112.
144 Lalande, *Astronomie*, 1 (1764), 537.
145 De la Hire, *Tabularum Astronomicarum* (1687), 43. See also his later editions, *Tabulae astronomicae Ludovici Magni* (1702; 1727), [usus Tabularum], 3; *Tables astronomiques* (1735), 4.
146 Flamsteed, *Historiae Coelestis Britannicae*, vol 3 (1725) [Usually referred to as *Prolegomena*], 123-124.
147 Observatoire de Paris, Cote B4/5 (Ancienne cote Cote Delisle 76): 'Recueil des observations astronomiques de Georg Maggraf'. (Eclipses de Lune; Hauteurs et azimuts du Soleil; Observations de Mercure par ses hauteurs et son azimuth; Divers observations pour les refractions; Fin du crepuscule du Soir).
148 [Maskelyne], 'Remarks on the planet Mercury' (1785).
149 DE L'ISLE's attempt to claim the priority for the discovery of comet Halley in 1758 is outlined in Zuidervaart, 'Geschonden vertrouwen'(2017).
150 Pingré, *Annales célestes* (1901).

151 Vaquero a.o, 'A new scenario for the onset of the Maunder minimum' (2011).

CHAPTER 10

1 'Historiae Rerum Naturalium Brasiliae' (1648), 239. Similar snakes were caught on 7 August and 6 October 1638.
2 Whitehead, 'Georg Markgraf and Brazilian Zoology' (1979), 426.
3 Brienen, 'The zoological drawings of Eckhout and Marcgraf' (2007), 295.
4 Vermij, *The Calvinist Copernicans* (2002), 104. Zuidervaart. *Telescopes from Leiden Observatory* (2007), 10.
5 Matsuura, *História da Astronomia no Brasil*, vol. 1 (2014).
6 The terminology 'thin' and 'thick description' was introduced in the social sciences by the anthropologist CLIFFORD GEERTZ (1926-2006). A thin description lacks context. It describes something without explaining its cultural significance. A thick description describes not only the action, but its significance.
7 Cook, *Matters of Exchange* (2007), 40: 'The intellectual activities we call science emerged from the ways of knowing valued most highly by the merchant-rulers of urban Europe'. See above, Chapter 2.
8 'Historiae Rerum Naturalium Brasiliae' (1648), 140.
9 Safier, 'Beyond Brazilian Nature' (2014), 175.
10 Whitehead, 'Biography' (1979), 309.
11 Trindade Medeiros and Albuquerque, 'Food flora in 17th century northeast region of Brazil' (2014).
12 Wiesenbron, *Brazilië in de Nederlandse archieven* (2005), 12.
13 Meuwese, *Brothers in Arms* (2012), 152.
14 See Long, 'Trading Zones in Early Modern Europe' (2015).
15 Later in life, JACOB JENSSØN NORDMAND became a skilled ivory turner at the Danish court. See Nordmand, 'En ubenaevnt Dansk kunstners Levnets-historie' (1784) and Bering Liisberg, 'Jacob Jensen Nordmand' (1893). The assertion by MARIANA DE CAMPOS FRANÇOZO that NORDMAND is the sculptor of the ivory furniture of Count JOHAN MAURITS has no grounds at all. See Françozo, *De Olinda a Olanda* (2009), 236.
16 ELO, North, no. 111. Loose leaf (page 50) identified as coming from Philippe Garnier and Luca Donati, *Gemmulae Linguarum Latinae, Gallicae, Italicae & Germanicae* (Leiden: Ex officina Elseviriorum, 1637).
17 See Chapter 6.
18 See about the indigenous interpreters in Dutch Brazil more in Meuwese, *Brothers in Arms* (2012), Chapter 3.
19 This scheme is not preserved. See *HNB*, 274.
20 MANUEL DE MORAES (1596?-1651?) was a former Luso-Portuguese Jesuit priest who defected to the Dutch side in the late 1620s and who came to the Netherlands in the mid-1630s, before returning to Brazil, where eventually he rejoined the Portuguese side. There he was persecuted for his behavior by the Catholic Inquisition. See: Mariana

de Campos Françozo, *De Olinda a Olanda* (2009), 66, 152-153 and Vainfas, *Traição* (2008).

21 Nationaal Archief (Den Haag), archief 1.04.02, inv. nr. 5585, fol 215.

22 For details, see the The Minor Planet Center (MPC) of the International Astronomical Union (IAU): https://minorplanetcenter.net/db_search/show_object?object_id=11238, and https://minorplanetcenter.net/db_search/show_object?object_id=11239 (both consulted in December 2021).

Appendix 1

1 Adapted from Brienen, 'Marcgraf' (2001), 108–110.

Appendix 2

1 See for Marggrafe's calculation, ELO, North doc. no. 10, discussed in Chapter 9.

Appendix 3

1 Sint Nicolaas, 'Vrijburg, getekend en gesneden' (2008). This article is unfortunately loaded with unsubstantiated claims.

2 Examination conducted at the Rijksprentenkabinet of the Rijksmuseum in Amsterdam on 10 November 2010.

3 Lastman, *De schat-kamer des grooten see-vaerts-kunst* (1621. Reprinted in 1624 and 1629).

4 Davids, *Zeewezen en wetenschap* (1985), 38, 157-158.

5 Lastman, *Beschrijvinghe van de kunst der stuer-luyden* (1642).

6 *Ibidem*, 57: 'Het tweede hooftstuck: handelt van de beweginghe der hemels lichten: ende hoe men door de hooght-metinge der selve vinden magh op wat breedt men is'.

7 De Groot, 'De *Schatkamer* en de *Kunst der Stuer-Luyden* (2002), 91.

8 Sint Nicolaas, 'Vrijburg, getekend en gesneden' (2008), 213.

9 See the 'Nederlandse Voornamenbank' of the Meertens Institute on https://www.meertens.knaw.nl/nvb/naam/is/siert. (Consulted December 2020).

10 A certain Hendrick Wijndaal (†1718) from Amsterdam entered the service of the VOC as a soldier in 1694, and Arend van Wijndal, married in Elst in 1738.

11 In 2020 the only internet hit in the Netherlands concerns a professional football player Owen Wijndal (*1999). In 1947 the family name did not exist in the Netherlands. See: https://www.cbgfamiliennamen.nl. (Consulted December 2020).

12 *Surnames in the United States Census of 1790: An Analysis of National Origins of the Population* (Genealogical Publishing Comp. 1969), 377.

CONSULTED ARCHIVES

THE NETHERLANDS

AMSTERDAM

- City Archive (Stadsarchief)
  - *Notary archive.*
  - *Indices.*
  - *Kwijtscheldingen.*

THE HAGUE

- National Archive
  - *archive First West Indian Company (OWIC).*
  - *collection of maps (kaartencollectie).*
- City Archive (Gemeentearchief)
  - *Notary archive.*
- Royal Library (Koninklijke Bibliotheek)
  - *Album Mulheiser, sign. KW 129 F 6.*

HEUSDEN

- Regional archive (Streekarchief) Langstraat Heusden Altena,
  - *DTB registers Babyloniënbroek.*

LEIDEN

- Heritage Leiden and the surrounding areas (Erfgoed Leiden en Omstreken – ELO)
  - *Marggrafe papers, library sign. LB 7000–1.*
  - *(numbered in pencil by JOHN NORTH in 1978).*
  - *Marggrafe draft letters, library sign. LB 18677.*
  - *Notary archive.*
  - *DTB registers.*
  - *Judgment book (Crimineel vonnissenboek).*
- Leiden University Library
  - *Archive Curators (Resolutiën 1628–1637).*
  - *Archive Senate (ASF 9: Volumen inscriptorium; ASF 290).*

UTRECHT

- City Archive (Stadsarchief)
  - *DTB registers.*
  - *Notary archive.*

Rotterdam
- City Archive (Stadsarchief)
  - *DTB registers.*
  - *Notary archive.*

DENMARK
København
- Natural History Museum of Denmark
  - *Marggrafe's herbarium: https://samlinger.snm.ku.dk/en/dry-and-wet-collections/botany/general-herbarium/the-marcgrave-herbarium/ (consulted September 2021).*
- Royal Danish Library (Det Kongelige Bibliotek)
  - *Letter book of Georg From, MS Ny Kgl. Saml. 2007, 4°.*

FRANCE
Paris
- Observatoire de Paris
  - *'Observations faites au Brésil' by Georg Marggrafe, manuscript sign. B 4–5.*
  - *Notes by J.N. de l'Isle.*
  - *Correspondence of Johannes Hevelius.*

GERMANY
Bonn
- Universitäts- und Landesbibliothek Bonn
  - *Book with signatures of Georg Marggrafe & Johan Maurits, sign. O 4' 236/1.*

Dresden
- Sächsiche Landesbibliothek
  - *Note about the invention of a 'Wol-Approbirtes Vieh-Pulver' by Samuel Marggrafe (1675), Signatur Oecon.A.182–5.*

Halle
- Universitäts- und Landesbibliothek Sachsen-Anhalt
  - *Album Amicorum of Andreas Sennert (1606–1689), sign. Misc. 55 oct.*
  - *Marggrafe's disputation from Wittenberg, sign. 003 779 483.*

POLAND
Kraków
- Jagiellonian Digital Library
  - *Handbooks: Libri Principis [A 37]: https://jbc.bj.uj.edu.pl/dlibra/publication/193892/edition/183824/content.*
  - *Libri picturati A 38: https://jbc.bj.uj.edu.pl/dlibra/doccontent?id=197455. (both consulted September 2021).*

PORTUGAL
Lisbon
- Biblioteca Nacional de Portugal
  - *Astronomical observations by Georg Marggrafe, Mss 6, n. 37.*

ABBREVIATIONS

DTB	Doop, Trouw en Begraafregisters
ELO	Erfgoed Leiden en Omstreken
KHA	Koninklijk Huisarchief
OdP	Observatoire de Paris
VOC	Verenigde Oostindische Compagnie
WIC/GWC	Geoctroyeerde West Indische Compagnie
OWIC	Oude WIC (1621-1674)

LITERATURE

Works by Georg Marggrafe

Marggrafe, Georgius, *De Causa continente morborum* (Wittenberg: Georg Muller, 1634).

—, [Short Latin poem of praise], in: Eichstaedt, *Pars Prima Ephemeridum Novarum et Motuum Coelestium* (Stettin: David Rentius, 1634).

—, [Latin poem of praise], in: Lauremberg, Petrus [et al], ΕΠΑΙΝΕΤΙΚΑ. *Mitrae Doctoralis: [...] imponebat clarissimo & literatissimo viro Dn. M. Salomoni Fischero, Luptia Megapolitano, In Alma Rosarum d. 30. April. Anno 1635* ([Rostock]: Praelum Reusnerianum, 1635).

—, [Latin comfort poem], in: *Epicedia ad clarissimum, excellentissimumque virum Dn. Laurentium Eichstadium [...] praematurum quidem at beatum obitum filii sui ex quatuor, dum vivebat, superstitis & unici Laurentii, acerbissime lugentem. Qui natus anno aera Christianae MDCXXXI, [...] denatus die bisextili MDCXXXVI ... in Templo Jacobaeo calendis Martiis honorifice sepultus* (Stettin: D. Rhetii, 1636). [copy in the Wojewódzka Biblioteka Publiczna in Toruń, Poland, sign, 102513; See *Handbuch des personalen Gelegenheitsschrifttums*, Vol. 1:3 (Universität Osnabruck, 2001), 131.

—, 'Historiae Rerum Naturalium Brasiliae, libri octo', in: *Historia Naturalis Brasiliae* (1648), [second part with own pagination and index] 1–292.

—, 'Tractatus Topographicus & Meteorologicus Brasiliae, cum Eclipsi Solari', in: Piso, *Indiae Utriusque re Naturali et Medica* (1658), [own pagination] 1–26.

Marcgrave, George & Willem Piso, *Historia Naturalis Brasiliae... in qua non tantum plantae et animalia, sed et indigenarum morbi, ingenia et mores describuntur et iconibus supra quingentas illustrantur* [J. de Laet, ed], (Leiden: F. Hackius & Amsterdam: L. Elzevier, 1648). A digital version of a hand coloured version can be consulted via https://disc.leidenuniv.nl.

See also https://archive.org/details/historianaturali12piso/page/n5/mode/2up (both consulted September 2021)

Literature discussing Georg Marggrafe's life and work

Andrade-Lima, Dárdano de; Maule, Anne Fox; Pedersen, Troels Myndel and Rahn, Knud, 'Marcgrave's Brazilian Herbarium, Collected 1638–44', *Botanisk tidsskrift* 71 (1977), 121-160.

Boogaart, E. van den & R.P. Brienen, *Information from Ceará from Georg Marcgraf (June–August 1639)* (Rio de Janeiro, 2002). Also published in: C. Ferrão & J.P. Monteiro Suarez (eds.), *Dutch Brazil*, 7 (Petrópolis 2002).

Brienen, R.P., 'Georg Marcgraf (1610–c.1644): A German Cartographer, Astronomer, and Naturalist-Illustrator in Colonial Dutch Brazil', *Itinerario. Bulletin of the Leyden Centre for the History of European Expansion* 25 (2001) 85-122.

—, *Art and Natural History at a Colonial Court: Albert Eckhout and Georg Marcgraf in Seventeenth-Century Dutch Brazil* (Dissertation, Northwestern University, 2002)

—, 'From Brazil to Europe: the zoological drawings of Albert Eckhout and Georg Marcgraf', in: *Early modern zoology: the construction of animals in science, literature and the visual arts*. Special issue of: *Intersections: yearbook for early modern studies* 7 (2007), 273–314.

Carvalho, Alfredo de, 'Um naturalista do século XVII. Georg Markgraf (1610-1644)', *Revista do Instituto Archeologico e Geographico Pernambucano*, 13 (1908), 212–222.

Gudger, Eugene Willis, 'George Marcgrave, the first student of American Natural History', *The Popular Science Monthly* (September 1912), 250–274.

—, 'George Marcgrave, A Postscript', *Science. New Series*, 40, no. 1032 (Oct. 1914), 507–509.

Hantsch, Victor, 'Georg Marggraf', *Berichte* über *die Verhandlungen der Saechsischen Akademie der Wissenschaften zu Leipzig, Philologisch-historische Klasse* 22 (1896) 199–227.

Ihering, Rodolpho von, 'George Marcgrave. O primeiro sabio que veiu estudar a natureza do Brazil – 1638 a 44', *Revista Museu Paulista*, 9 (1914) 307–315.

Jacob, Ernst Gerhard, 'Georg Markgraf (1610-1644), ein unbekannter deutscher Forscher in Brasilien und Angola', *Actas do V Colóquio internacional de estudos luso-brasileiros* ; 2 (1965), 1-11.

Laet, Johannes de, 'Ad Benevolos Lectores' in: 'Marcgravi de Liebstad, Historiae Rerum Naturalium Brasiliae, libri octo' (1648), p. 2–3.

Lichtenstein, Martin, 'Die Werke von Marcgrave und Piso über die Naturgeschichte Brasiliense, erlautert aus den wiederaufgefundenen originalzeichnungen', *Abhandlungen der Königlichen Preußischen Akademie der Wissenschaften zu Berlin*, 1814/15 (1818), 201–222; 1816/17 (1819), 155-178;1820/21 (1822), 237–257; 1826,49–66.

MacBryde, Bruce, 'Rediscovery of G. Marcgrave's Brazilian Collections (1638-1644)', *Taxon* 19 (1970), 349.

Marggrave, Christian, 'Vita Georgii Marggravii', in: idem, *Prodromus medicinae practicae dogmaticae & verè rationalis*. Second edition (Leiden: C. Bouteseyn, 1685), 8 unnumbered pages after the 'Praefatio'.

Matsuura, Oscar Toshiaki, *O observatório no telhado* (Recife: Cepe, 2011).

—, 'Um observatório de ponta no Novo Mundo', in: idem, *História da Astronomia no Brasil* (2014), 152-196.

Matsuura, Oscar T. & Huib J. Zuidervaart, 'America's earliest (European-style) astronomical observatory, founded and used by Georg Marggrafe in Dutch colonial Brazil (1638-1643)', in: Marcus Granato & Marta C. Lourenço (eds.), *Scientific Instruments in the History of Science. Studies in transfer, use and preservation* (Rio de Janeiro: MAST, 2014), 33–52.

Moreira, Juliano, 'Marcgrave e Piso', *Revista do Instituto Histórico e Geográfico Brasileiro*, 82 (1917), 791–808.

Morize, H., 'As mais antigas observações meteorológicas na America do Sul', *Revista do Instituto Archeologico e Geographico Pernambucano* (no. 54) 38 (1900), 107-112. [About two tables published by G. Hellmann in the *Meteorologische Zeitschrift* (1893).]
North, John D., 'Georg Markgraf, an astronomer in the New World', in: Van den Boogaart (ed.), *Johan Maurits Van Nassau-Siegen* (1979), 394–423. Reprinted with a new addendum in: idem, *The universal frame* (1989), 215–234.
—, 'Astronomia no Brasil holandês', *Ciência da Holanda*, 79 (1980) 12-15.
Pereira Da Costa, F.A. , 'Primeira observações meteorologicas no Brazil', *Revista do Instituto Archeologico e Geographico Pernambucano* (no. 54) 38 (1900), 113-119.
Pickel, Bento José, *Flora do nordeste do Brasil segundo Piso e Marcgrave no século XVII* (Recife: Edufrpe, 2008).
Pimentel, Jorge; Pereira, Levy, 'A astronomia e o mapa Brasilia qua parte paret Belgis, de Jorge Marcgrave', in: Matsuura, *História da Astronomia no Brasil* (2014), 197–228.
Polman, Jorge, 'First observatory in the Southern Hemisphere', *Sky and Telescope* (1984), 388.
—, *Markgraf e o Recife de Nassau* (Recife: Clube Estudantil de Astronomia, 1984).
Prazeres, Audemário, 'George Marcgrave e o desenvolvimento da astronomia moderna na America Latina, na cosmopolita Recife de Nassau', *Revista macroCosmo.com. A primeira Revista Eletronica Brasileira exclusiva de astronomia* 1:4 (March 2004), 40–55.
Roscam Abbing, Michiel, *Brazilië zien zonder de oceaan over te steken. De wandtapijten van Johan Maurits* (Amsterdam: Lias, 2021).
Safier, Neil F., 'Beyond Brazilian Nature. The Editorial Itineraries of Marcgraf and Piso's *Historia Naturalis Brasiliae*', in: Van Groesen (ed.), *The Legacy of Dutch Brazil* (2014), 168-186.
Snedegar, Keith, Lemma 'Georg Markgraf', in: Thomas R. Williams, *Biographical Encyclopedia of Astronomers* (New York: Springer, 2014), 737–738.
Storms, Martijn, 'De kaart van Nederlands Brazilië door Georg Marcgraf', *Caert-thresoor. Tijdschrift voor de Geschiedenis van de Kartografie* 30:2 (2011), 37–46.
—, 'Brasilia qua parte paret belgis: de wandkaart van Nederlands Brazilië door Georg Marcgraf', in: Wiesebron, *Brazilië in de Nederlandse Archieven, 5* vols. (2000–2013), 28–46. [Text also in Portuguese].
Taunay, Affonso de E., 'Jorge Marcgrave, de Liebstadt (1610-1644): escorço Biográfico', in: *História natural do Brasil* (São Paulo: Imprensa Oficial do Estado-Museu Paulista, 1942).
Teensma, Benjamin N., 'Onherkend want anoniem en zonder jaar. Georg Marcgraf's Braziliaanse manuscriptkaarten uit 1643', in: Wiesebron, *Brazilië in de Nederlandse Archieven.* Vol. 4 (2011), 61–72.
Wagner, Peter , 'Das Markgraf-Herbarium', in: Brunn & Neutsch (eds.), *Sein Feld war die Welt: Johann Moritz von Nassau–Siegen* (2008), 233–245.
Whitehead, Peter J.P., 'The Biography of Georg Marcgraf (1610-1643/1644) by his brother Christian, translated by James Petiver', *Journal of the Society for the Bibliography of Natural History* 9:3 (1979) 301–314.
—, 'Georg Markgraf and Brazilian Zoology' in: Van den Boogaart, *Johan Maurits van Nassau–Siegen* (1979), 424–471.
—, 'The Marcgraf Map of Brazil', *The Map Collector* (June 1987), 17–20.

Printed sources

Bittcher, Carl Friedrich Heinrich, *Pförtner Album. Verzeichnis sämmtlicher Lehrer und Schüler der Königlichen Preußischen Landesfchule Pforta vom Jahre 1343 bis 1843* (Leipzig, 1843).

Bogun, Kurt, 'Die Stammbuchsammlung in der Stadtbibliothek zu Königsberg. Teil III', in: *Vierteljahrsschrift* für *Wappen-, Siegel- und Familienkunde* 37 (1909), 17-108.

Brommer, Bea and Henk den Heijer, *Grote Atlas van de West-Indische Compagnie / Comprehensive Atlas of the Dutch West Indian Company: I. De Oude WIC / The Old WIC, 1621-1674* (Voorburg: Asia Maior/Atlas Maior, 2011).

Caland, F., 'Hofhouding van Johan Maurits Graaf v.Nassau in Brazilie' ["Lijste van de domesticquen aant hoff van Sijn Extie. Johan Maurits, Grave van Nassou etc., Gouverneur capitein ende admirael Generael van Brasil op den 1 April anno 1643, genietende de vrije taeffel"], *De Navorscher* 48 (1898) 557–558.

Erler, Georg (ed.), *Die jüngere Matrikel der Universität Leipzig, 1559-1809. 1. Bd. Die Immatrikulationen v. Wintersemester 1559 bis zum Sommersemester 1634* (Leipzig: Giesecke & Devrient, 1909).

—, *Die Matrikel und die Promotionsverzeichnisse der Albertus-Universität zu Königsberg i. Pr. 1544-1829. Bd. 1. Die Immatrikulationen von 1544-1656* (Leipzig: Duncker & Humblot, 1910).

Ferrão, Cristina and José Paulo Monteiro Soares (eds), *Theatrum Rerum Naturalium Brasiliae. Brasil-Holandes; Dutch-Brazil*, 5 vols. (Rio de Janeiro: Editora Index, 1993).

—, *The Voyage of Caspar Schmalkalden from Amsterdam to Pernambuco in Brazil*, 2 vols. (Rio de Janeiro: Editora Index, 1998). English adaptation and translation of Joost, *Die wundersamen Reisen* (1983).

Freninger, Franz Xaver, *Das Matrikelbuch der Universitæt Ingolstadt-Landshut-München (1472-1872)*, (München: Eichleiter, 1872)

Friedländer, Ernst, a.o., Ältere *Universitätsmatrikeln. I. Universität Frankfurt an der Oder, Erster Band (1506-1648)* (Leipzig: S. Hirzel, 1887).

—, Ältere *Universitätsmatrikeln. II. Universität Greifswald, Erster Band (1456-1645)* (Leipzig: S. Hirzel, 1893)

Haecxs, Hendrik, 'Het dagboek van Hendrik Haecxs, lid van den hoogen raad van Brazilië (1645-1654)' [edited by S.P. l'Honoré Naber], *Bijdragen en Mededeelingen van het Historisch Genootschap*, 46 (1925), 126–311.

Herckmans, Elias, 'Generale beschrijvinge vande Capitanie Paraiba', *Bijdragen en Mededeelingen van het Historisch Genootschap*, 2 (1879), 318–367.

Joost, Wolfgang (ed.), *Die wundersamen Reisen des Caspar Schmalkalden. nach West- und Ostindien 1642-1652* (Leipzig: Brockhaus, 1983). Re-issued as: *Casper Schmalkalden. Mit Kompas und Kanonen. Abenteuerliche Reisen nach Brasilien und Fernost, 1642-1652* (Stuttgart: Erdmann, 2002).

Knot, Gustav C., *Die alten Matrikeln der Universität Strassburg, 1621 bis 1793* (Strassburg: Trübner, 1897).

Molhuysen, P.C., *Bronnen tot de geschiedenis der Leidsche Universiteit*, 7 vols. (Leiden: Nijhoff, 1913-1924).

Ozment, Steven. *Three Behaim Boys. Growing Up in Early Modern Germany* (New Haven/London: Yale University Press, 1990).

Pingré, Alexandre-Guy, *Annales célestes du dix-septième siècle* [edited by Guillaume Bigourdan] (Paris, 1901).
Pudsey, Cuthbert, *Journal of a Residence in Brazil* [1629-1640] (Pétropolis: Index, 2000).
Rieu, W.N. du, *Album Studiosorum Academiae Lugduno Batavae MDLXXV–MDCCCLXXV* (Leiden 1875).
Waard, Cornelis de, *Journal tenu par Isaac Beeckman de 1604 à 1634*, Vol. 3 (The Hague, 1945).
Wackernagel, Hans Georg, a.o. (eds.), *Die Matrikel der Universität Basel, III Band, 1602-1666* (Basel, 1962).
Wasch, C.J.,'Classicale Acta van Brazilië [1633-1644]', [reprint from] *Kroniek van het Historisch Genootschap gevestigd te Utrecht*, 29 (1873).
—, 'Een doopregister der Hollanders in Brazilië', *Algemeen Nederlandsch Familieblad*, 5 (1888), 141-144; 169-172; 197–200; 225–228; 253–256; *idem* 6 (1889), 1–4; 25–28; 49–52;73–77.
—, 'Request der voornaamste kooplieden van Fernambuco aan de Vergadering der 19 der West-Indische Compagnie', *Algemeen Nederlandsch Familieblad*, 6 (1889), 111-112.
Wagener, Zacharias, *The "Thierbuch" and Autobiography of Zacharias Wagener* [Dutch Brazil, Vol. II] (Rio de Janaeiro: Index, 1997).
Weissenborn, Bernhard, *Album Academiae Vitebergensis. Jüngere Reihe, Teil 1 (1602-1660)* (Magdeburg: Selbstverlag der Historischen Kommission, 1934).

Primary literature

[Anonymous], 'Sommier discours over den staet vande vier geconquesteerde capitanias Parnambuco, Itamarica, Paraiba ende Rio Grande, inde noorderdeelen van Brasil', *Bijdragen en Mededeelingen van het Historisch Genootschap*, 2 (1879), 257–317.
—, *Journael ofte kort discours nopende de rebellye ende verradelijcke desseynen der Portugesen, alhier in Brasil voorgenomen, 't welck in Junio 1645 is ontdeckt. Ende wat vorder daer nae ghepasseert is, tot den 28 April 1647. Beschreven door een Lief-hebber, die selfs int begin der Rebellye daer te Lande is gheweest, ende aldaer noch is residerende* (Arnhem: Jan Jacobsz, 1647).
Avé-Lallemant, Robert, *Des Dr. Joachim Jungius aus* Lübeck *Briefwechsel mit seinem Schülern und Freunden* (Lübeck, 1863).
Barlaeus, Caspar, *Rerum per octennium in Brasilia. Et alibi nuper gestarum, sub Praefectura Illustrissimi Comitis J. Mauritii Nassoviae* (Amsterdam: J. Blaeu, 1647).
—, German translation: *Brasilianische Geschichte den Achtjahriger in selbigen Landen gefuhreter Regierung Seiner Furstlichen Gnaden hern Johan Moritz Furstens zu Nassau* (Cleve: T. Silberling, 1659).
—, Dutch translation: L'Honoré Naber, *Nederlandsch Brazilië onder het bewind van Johan Maurits Grave van Nassau, 1637-1644* ('s-Gravenhage: Martinus Nijhoff, 1923).
—, English translation: Blanch T. van Berckel-Ebeling Koning, *Caspar van Baerle [Caspar Barlaeus], The History of Brazil under the Governorship of Count Johan Maurits of Nassau, 1636-1644* (Gainesville, FL: University Press of Florida, 2011).

Bartholin, Thomas, *Twee hondert getal seltsame ondervindingen ofte geschiedenissen omtrent de Anatomia ofte ontledinge* (Dordrecht: J. Savry, 1657).
—, *Oratio in obitum clarissimi viri D. Henrici Fuiren* (Copenhagen: Morsing, 1659).
Bekkers, Johannes A.F., *Correspondence of John Morris with Johannes de Laet (1634-1649)* (Assen: Van Gorcum, 1970). Bodecher Banning, Johannes, *Poemata* (Leiden: Heger, 1637).
Bodecher Benning, Johannes, *Satyricon in corruptae juventutis mores corruptus* (Leiden: Le Maire, 1631).
—, *Epigrammata Americana ad Excellentissimum Illustrissimumque Comitem I. Mauricium, Comitem Nassaviae, Cattimeliboci, ac Dietziae, Beilsteinii Dominum, & Brasiliae Terra Marique Imperatorum* (Leyda: Lopes de Haro, 1639) – (Univ. Libr. Leiden, sign. KW 298 A 9). [4], 15 p (folio).
—, *Catalogus insignium librorum D. Ioannes Bodecheri Benningii, olim professoris philosophiae in Academia Leydensi. Quorum auctio habebitur Lugduni Batavorum in aedibus Joannis Maire. Ad diem 26. septemb. anno 1642* (Leiden: Le Maire. 1642). [Royal Library, Brussels].
Brahe, Tycho, *Astronomiae Instauratae Mechanica* (Wandsbek 1598). Digital Edition (Royal Library of Denmark) at: www.kb.dk/en/nb/tema/webudstillinger/brahe_mechanica (Consulted June 2021).
Broeck, Matheus van den, *Journael ofte Historiaelse Beschrijvinge van 't geen hy selfs ghesien ende waerachtigh gebeurt is, wegen't begin ende Revolte van de Portugese in Brasiel, als mede de conditie en het overgaen van de Forten aldaer* (Amstelredam: Van Goedesbergen, 1651).
Buck, Friedrich Johann, *Lebensbeschreibing derer verstorbenen preussischen Mathematiker* überhaupt*, und des vor mehr denn hundert Jahren verstorbenen grossen preussischen Mathematikers P. Christian Otters insbesondere* (Königsberg und Leipzig 1764).
Calado, Manoel, *O Valeroso Lucideno e Triumpho da Liberdade* (Lisboa: Craesbeeck, 1648). *Catalogus [....] Willebrordi Snellii* (1629).
Catalogus van alle de principaelste Rarieteyten die op de Anatomie Kamer binnen der Stadt Leyden vertoont worden (Leiden: D. van der Boxe, 1671). English edition: *A catalogue of all the cheifest rarities in the publick theater and anatomy hall of the University Leiden* (Leiden: M. van Leeuwen, 1678). Many reprints in Dutch, English, French and Latin.
Couplet, Pierre,'Extrait de quelques lettres écrite [...] du Brésil', *Histoire de l'académie royale des sciences, année 1700* (Amsterdam: Kuyper, 1706), 222–229; (second edition, Paris: Coignard, 1761), 171-178.
Descartes, René, *Discours de la méthode pour bien conduire sa raison et chercher la vérité dans les sciences* (Leiden: Jean le Maire, 1637).
Eichstaedt, Lorenz, *Prognosticon de Coniunctione Magna Saturni et Jovis in Trigono Igneo Leonis* (Stettin: N. Bartholdi, 1622).
—, *Ephemerides Astro-Meteorologicae ad Annos Aerae Christianae 1630. 1631. 1632. 1633. Oder, Teutsche Allmanach und Special Prognostica auff folgende vier Jahr nach der Heylsamen Geburt Jesu Christi. Mit sonderbarem Fleiß nach der newen Tychonischen Astronomy und Alten Calender gerechnet* (Leipzig: G. Grossens, 1631).
—, *De confectione alchermes in Pomerania par ari coepta. [...] Dissertatio & Exercitatio Medica* (Stettin, D. Rheticus, 1634).

—, *Pars Prima Ephemeridum Novarum et Motuum Coelestium Quinquennalis ad Annos Aera Christianae 1636, 1637, 1638, 1639, 1640. In Luminarium motibus & Eclipsibus ex Tabulis Danicis Christiani S. Longomontani, in reliquis Planetis ex Tabulis Rudolphinis Joannis Kepleri juxta exquisitas Nob. Tychonis Brahei observationes constructis, debita diligentia elaborate & supputata* (Stettin: David Rentius, 1634).

—, *Pars Altera Ephemeridum Novarum et Motuum Coelestium Decennalis ab Anno Aera Christianae 1641, incipientium, & in annum 1650 desinentium. In Luminarium motibus & Eclipsibus ex Tabulis Danicis Christiani S. Longomontani, in reliquis Planetis ex Tabulis Rudolphinis Joannis Kepleri juxta exquisitas Nob. Tychonis Brahei observationes constructis, debita diligentia elaborate & supputata* (Stettin: David Rhetius, 1636).

Flamsteed, John, *Historiae Coelestis Britannicae*, vol 3 (London: H. Meere, 1725) [Usually referred to as *Prolegomena*]

Gassendi, Pierre, *Mercurius in sole visus et Venus invisa* (Paris: S. Gramoisy, 1631).

—, *De apparente magnitudine solis humilis et sublimis. Epistolae quator. In quibus complura physica, opticaque problemata pronuntur & explicantur* (Paris: Heuqueville, 1642).

—, *Opera Omnia*, 6 volumes (Lyon: Anison & Devenet, 1658)

Golius, Jacob (ed,), *Muhammedis Ferganensis qui vulgo Alfraganus dicitur, Elementa astronomica, arabice et latine, cum notis ad res exoticas sive orientales quae in iis occurrunt* (Amstelodami: Jan Jansz. van Waesberge, 1669).

—, *Catalogus Bibliothecae Jacobi Golii* (Leiden: Hackius & Loopes de Haro, June 1668)

Halley, Edmond, *Catalogus stellarum australium: sive, supplementum catalogi Tychonici, exhibens longitudines et latitudines stellarum fixarum, quae, prope polum Antarcticum sitae, in horizonte Uraniburgico Tychoni inconspicuae fuere, accurato calculo ex distantiis supputatas, & ad annum 1677 completum correctas* (London: T. James, 1679).

Hedraeus, Bengt, *Nova et accurata astrolabii geometrici structura [...] quadrantis astronomici azimuthalis [...] claris & perspicuis exemplis illustrato* (Leiden: Van der Boxe, 1643).

Heunius, Johannes, *Ad Exequias Viro Pl. Reverendo ... Dn. Johanni Beringio* (Greifswald: Neuman, 1658).

Hire, Philippe de la, *Tabularum Astronomicarum, Pars Prior* (Paris: S. Michallet, 1687).

—, *Tabulae astronomicae Ludovici Magni* (Paris: Académie royale des sciences, 1702; 1727).

—, *Tables astronomiques, dresse'es et mises en lumiere par les ordres et par la magnificence de Louis le Grand* (Paris: Montalant, 1735).

Hortensius, Martinus, *Dissertatio de Mercurio in Sole viso et Venere invisa* (Leiden: Commelin, 1633).

[Kechel ab Hollensteyn, Samuel C]., *Catalogus Rarissimorum Librorum & Instrumentorum Mathematicorum, [...] Samuelis Caroli Kechelii ab Hollensteyn* (Leiden: Driehuysen, May 1668).

Kirstenius, Georgius, *Adversaria et Animadversiones in J. Agricolæ [...] Darinnen der Falsche und betriegliche gebrauch der Chymischen Aertzeneyen .... wiederleget wird* (Stettin: Mamphrasen, 1648).

Kralitz, Heinricus, *Memoria vitae novissima, quam bigae virorum [...] Johanni-Friderico Schroetero [...] ad. 13. Novembr. Anni 1625. Genae uno sub cippo dormitantibus, [...] ac bigae foeminarum [...] Barbarae Vogeliae [...] Iohannis Friderici Schroteri [...] conjugi [...] ad d. 9. Februar. Genae* (Dantzig: Georgii Rhetii, 1631). [not seen].

—, *Catalogus plantarum Horti Academici Lugduno-Batavi: quibus is instructus erat Anno MDCXXXV: praefecto ejusdem Horti D. Adolfo Vorstio [...]acc. index plantarum indigenarum, quæ prope Lugdunum in Batavis nascuntur* (Leiden: Elzevier, 1636). Reprinted as a facsimile in 1996 in a limited edition with an introduction by C.G.G.J. van Steenis (Zug, Switzerland: IDC, 1966).

Küchenmeister, August Theodorus, *Etwas von Liebstadt* (Dresden: Krause, 1743).

Laet, Johannes de, *Historie ofte Iaerlijck verhael van de verrichtinghen der Geoctroyeerde West-Indische compagnie, zedert haer begin, tot het eynde van't jaer sesthien-hondert ses en dertich* (Leiden: Elzevier, 1644).

—, *Catalogus bibliothecae ... Joannis de Laet* (Leiden: Hackius, 1650).

Lalande, J.J. Le François de, *Astronomie*, 2 vols. (Paris 1764). Second edition, 3 vols. (Paris 1771).

Lastman, Cornelis Jansz, *De schat-kamer des grooten see-vaerts-kunst. Inhoudende de seeckere gront-regulen, ende het recht ghebruyck der voornemelicke dingen, diemen inde groote zee-vaert behoort te verstaen* (Amsterdam: Lastman, 1621. Reprinted in 1624 and 1629).

—, *Beschrijvinghe van de kunst der stuer-luyden* (Amsterdam: Lastman, 1642).

Liebergen, Arnout van, *Apologiae, ofte waerachtighe verantwoordinghe [...] waerin, als in eenen klaren spiegel gesien kan worden,'t geene in Brasil gepasseert is ...* (Amsterdam: voor den autheur, 1643).

Linden ,Johannes Antonides van der, *De scriptis medicis libri duo* (Amsterdam: J. Blaeu, 1637).

Lulofs, Johan, *Inleiding tot eene natuur- en wiskundige beschouwing des aardkloots* (Leiden: Verbeek & Van Hoorn, 1750).

Marggrave, Christian, *Disputatio medica peri tou theiou sive de divino in morbis quod magnus Hippocrates in prognosticis observare medicum jussit* (Leiden: Elzevier, 1654).

Metius, Adriaen, *De novis ad autore inventis instrumentis.* Appendix in: idem, *Institutionum astronomicarum* (Franeker: Radeus, 1608).

Moreau, Pierre, *Histoire des derniers troubles du Brésil entre les Hollandais et les Portugais* (Paris: Courbé, 1651). Dutch translation: *Klare en waarachtige beschrijving van de leste beroerten en afval der Portugezen in Brasil* (Amsterdam: Hendricksz & Riewertsz, 1652).

Nieuhof, Johan, *Gedenkweerdig Brasiliaense Zee- en Lant-Reize, behelzende al het geen op dezelve is voorgevallen, beneffens een bondige beschrijving van gansch Neerlants Brasil* (Amsterdam: Wed. van Meurs, 1682).

[Nordmand, Jacob Jensen], 'En ubenaevnt Dansk kunstners Levnets-historie', in: *Samlinger til den Danske Historie*, II-3 (Copenhagen: Gyldendals, 1784), 134-158.

Paulli, Simon, *Quod Deus Bene Vortat, Hanc De Arthritide Disputationem*, [...] *pro prensandis summis in Medicina honoribus, [...] ad diem 10. Iulii* (Wittenberg: S. Auerbach, 1630).

—, *Quadripartitum Botanicum De Simplicivm Medicamentorvm Facultatibus in usus Medicinæ Candidatorum* (Strasbourg: Paulli, 1667).
Piso, Gulielmi, 'De medicina Brasiliensi, libri quatuor', in: *Historia Naturalis Brasiliae* (1648), [first part with own pagination and index] 1-122.
—, *Indiae Utriusque re Naturali et Medica, Libri Quatuordecim* (Amsterdam: L. & D. Elzevier, 1658)
Plante, Franciscus, *Mauritiados Libri XII. Hoc est Rerum ab Illustrissimo Heroe Ioanne Mauritio Comite Nassauiae &c in Occidentali India gestarum descriptio poetica* (Leiden: J. Le Maire, 1647).
Quelen, Augustus de, *Kort verhael van den staet van Fernanduc: toe-ge-eygent de E. Heeren gecommitteerde ter vergaderinghe van de negenthiene, inde geoctroyeerde West-Indische compagnie, ter camere van Amstelredam* (Amsterdam: n.pr, 1640).
Scheiner, Christoph, *Rosa Ursina sive Sol* (Bracciano, 1630).
Schepelern, H.D., *Breve fra og til Ole Worm, 1607-1654.* 3 vols. (Copenhagen: Munksgaard, 1965).
—, *Olai Borrichii itinerarium 1660-1665. The journal of the Danish polyhistor Ole Borch,* 4 vols. (Copenhagen: The Danish Society of Language and Literature, 1983).
Simon, Johann, [pfarrern zur Liebstadt], *Christliche Leichpredigt bey dem trawrigen Begräbnus des edlen gestrengen und ehrenvesten Rudolphen von Bünaw* (Dreszden, Schultz, 1603).
Snellius, Willebrord, *Descriptio cometae, qui anno 1618 mense Novembri primum effalsit* (Leiden: Elzevier, 1619).
—, *Catalogus variorum & insignium, Willebrordi Snellii à Royen* (Elzevier, 1629).
Worm, Øle, *Museum Wormianum, seu historia rerum rariorum Tam Naturalium, quam Artificialium, tarn Domesticarum, quam Exoticarum, quae Hafniae Danorum in aedibud Authoris servantur* (Leiden: Joh. Elzevier, 1655).

Secundary literature

Alcàntara Rodríguez, Mireia, Mariana Françozo and Tinde Van Andel, 'Looking into the flora of Dutch Brazil: botanical identifications of seventeenth century plant illustrations in the Libri Picturati', *Scientific Reports* 11, 19736 (2021). Doi.org/10.1038/s41598-021-99226-8.
—, 'Nature portrayed in images in Dutch Brazil. Tracing the sources of the plant woodcuts in the *Historia Naturalis Brasiliae* (1648)', to be published in *Plos One* (forthcoming).
Alcántara Rodríguez, Mireia; Isabela Pombo Geertsma; Mariana Françozo & Tinde van Andel, 'Marcgrave and Piso's plants for sale: The presence of plant species and names from the *Historia Naturalis Brasiliae* (1648) in contemporary Brazilian markets', *Journal of Ethnopharmacology* 259 (2020), https://doi.org/10.1016/j.jep.2020.112911.
Almeida, Argus Vasconcelos de, *Historiæ Rervm Naturalivm. Ensaios histórico-culturais sobre as ciências biológicas,* 1 (Recife: Ed. Univ. UFRPE, 2016).
Almeida, Argus Vasconcelos de; Maria Adélia Borstelmann de Oliveira; Isabelle Maria Jacqueline Meunier, 'Animais e plantas do horto zoo-botânico do palácio de Friburgo (1639-1645) construído por Maurício de Nassau no Recife', *Filosofia e História da Biologia,* 6:1 (2011), 19–35.

Applebaum, Wilbur & Robert A. Hatch, 'Boulliau, Mercator, and Horrocks's "Venus in sole visa": three unpublished letters', *Journal for the History of Astronomy* 14 (1983) 166-179.

Arlt, Rainer and José M. Vaquero, 'Historical sunspot records', *Living Reviews in Solar Physics* 17:1 (2020), 1–60.

Ashworth, W., 'Natural history and the emblematic worldview' in: D.C. Lindberg and R.S. Westman (eds.), *Reappraisals of the Scientific Revolution* (Cambridge 1990), 303–323.

Bencard, Mogens, 'Fürstliche Geschenke', in: Brunn, *Sein Feld war die Welt* (2008), 159-177.

Berkel, Klaas van, *Isaac Beeckman (1588-1637) en de mechanisering van het wereldbeeld* (Amsterdam, 1983).

–, 'Alexandrië aan de Amstel? De illusies van Martinus Hortensius (1605-1639), eerste hoogleraar in de wiskunde in Amsterdam', in: Haitsma Mulier [e.a.], *Athenaeum Illustre. Elf studies over de Amsterdamse Doorluchtige School, 1632-1877* (Amsterdam, 1997), 200–224. Reprinted in: K. van Berkel, *Citaten uit het boek der natuur* (Amsterdam, 1998), 63–84.

–, 'The city of Middelburg, cradle of the telescope', in: Albert van Helden, Sven Dupré, Rob van Gent & Huib Zuidervaart (eds.), *The Origins of the Telescope* (Amsterdam: KNAW Press, 2010), 45–72.

–, 'Rediscovering Clusius. How Dutch commerce contributed to the emergence of modern science', *BMGN – Low Countries Historical Review* 123:2 (2008), 227–236.

Bering Liisberg, Henrik Carl, 'Jacob Jensen Nordmand, Frederik den Tredies Kunstdrejer og Rustmester: Med lidt om Brasiliens Erobring og Kjøbenhavns Belejring', *Museum: Tidsskrift for Historie og Geografi* (1893), 245–284.

Bertele, Hans von, and Erwin Neumann, 'Der Kaiserliche Kammeruhrmacher Christoph Margraf und die Erfindung der Kugellaufuhr', *Jahrbuch der Kunsthistorischen Sammlungen in Wien*, 59 (1963), 39–98.

Besselaar, José van den, 'Breve introdução a leitura dos *Epigrammata Americana*', in: in: *Revista do instituto archeologico historico e geographico Pernambucano* 57 (1984), 9–39.

Bick, Alexander, *Governing the free sea. The Dutch West India Company and commercial politics, 1618-1645* (Ph.D. dissertation, Princeton University, 2012).

Bierens de Haan, David, 'Josephus Scaliger J.C. Fil. als cirkel-quadrator', in: idem, *Bouwstoffen voor de geschiedenis der wis- en natuurkundige wetenschappen in de Nederlanden* (Amsterdam: 1878), 280–314.

Bigourdan, Guillaume, 'Inventaire général et sommaire des manuscrits de la Bibliothèque de l'Observatoire de Paris', in *Annales de l'Observatoire*, 21 (1895) 2–60.

Bolt, Marvin & Michael Korey, 'The world's oldest surviving telescopes', in: Albert van Helden, Sven Dupré, Rob van Gent & Huib Zuidervaart (eds.), *The Origins of the Telescope* (Amsterdam: KNAW Press, 2010), 231–256.

Boogaart, E. van den (ed.), *Johan Maurits van Nassau-Siegen 1604-1679: a humanist prince in Europe and Brazil. Essays on the occasion of the tercentenary of his death* (The Hague 1979).

–, 'A Well-Governed Colony. Frans Post's Illustrations in Caspar Barlaeus's History of Dutch Brazil', *The Rijksmuseum Bulletin* 59 (2011), 236–271.

—, *Het land van de suikermolen. Johan Maurits' Brazilië* (Zwolle: WBooks, 2021).
Boukema, Robbert, *"Tot handhavinghe van de Ware Religie". Het Leven van Theoloog-Bewindhebber Johannes De Laet (1581-1649)* (Master Thesis Colonial And Global History, Leiden University, 2018).
Bouwman, André (et al), *Stad van Boeken. Handschrift en druk in Leiden, 1260–2000* (Leiden, 2008).
Boxer, Charles R., *The Dutch in Brazil (1624-1654)* (Oxford: Clarendon Press, 1957). Dutch translation *De Nederlanders in Brazilië 1624-1654* (Amsterdam: Atlas, 1993).
Bremmer, R. & P. Hoftijzer (eds.), *Johannes de Laet (1581-1649): A Leiden polymath.* Special issue of *Lias* 25:2 (1998), 135–230.
Brendecke, Arndt and Susanne Friedrich, 'Introduction', in: Brendecke, Arndt; Susanne Friedrich and Stefan Ehrenpreis (eds.), *Transformations of Knowledge in Dutch Expansion* (2017), 1-18.
Brienen, R.P., *Visions of savage paradise. Albert Eckhout, court painter in Colonial Dutch Brazil* (Amsterdam, 2006).
Brunn, Gerhard & Cornelius Neutsch (eds.), *Sein Feld war die Welt. Johann Moritz von Nassau-Siegen (1604-1679). Von Siegen* über *die Niederlande und Brasilien nach Brandenburg* (Münster [etc.]: Waxmann, 2008).
Burnett, Charles, 'From Astronomica to Exotica: Jacob Golius's Edition of al-Farghānī's On the Science of the Stars in Comparison with the Earlier Versions', in: Jan Loop & Jill Adrian Kraye, *Scholarship between Europe and the Levant,* (Leiden: Brill, 2020), pp. 60–86.
Buvelot, Quentin (ed.), *Albert Eckhout. A Dutch Artist in Brazil* (Zwolle: Waanders 2004).
Casanovas, J., 'Early observations of sunspots: Scheiner and Galileo', in: B. Schmieder, J. C. del Toro Iniesta & M. Vásquez (eds.), *1st. Advances in Solar Physics Euroconference, Advances in the Physics of Sunspots, ASP Conference Series,* 118 (1997), B, 3–20.
Chapman, Allan, 'Reconstructing the angle-measuring instruments of Pierre Gassendi', in: W.D. Hackmann and A.J. Turner (eds.), *Learning, language and invention. Essays presented to Francis Maddison* (Aldershot: Variorum, 1994), 103-116. Reprinted in Allan Chapman, *Astronomical instruments and their users* (Aldershot: Variorum 1996).
Christianson, John Robert, *On Tycho's Island. Tycho Brahe and his assistents 1570-1601* (Cambridge: University Press, 2000).
Cook, Allan, *Edmond Halley. Charting the Heavens and the Seas* (Oxford: Clarendon Press, 1998).
Cook, Harold, *Matters of Exchange. Commerce, Medicine, and Science in the Dutch Golden Age* (New Haven & London, 2007).
Cocquyt, Tiemen, 'The Holland circle. Instrumental in establishing the Dutch surveyor and instrument maker's identity', in: Ulrike Gehring and Peter Weibel (eds.), *Mapping spaces : networks of knowledge in 17th century landscape painting* (Karlsruhe: ZKM / Münich: Hirmer, [2014]), 384–395.
Corrêa do Lago, Pedro and Bia, *Frans Post {1612-1680}. Catalogue Raissoné.* [English translation] (Milan: 5 Continents Editions, 2007).
Crommelin, C.A., 'Eine astronomische Handschrift aus dem 17. Jahrhundert', *Isis* 31:1 (1939), 48–50.

Dams, Britt, 'Elias Herckmans. A poet at the borders of Dutch Brazil', in: Huigen et al., *The Dutch Trading Companies as Knowledge Networks* (2010), 17–37.
—, 'Transcriptions and Descriptions of Exotic Fauna in Barlaeus's *Rerum per Octennium* and de Laet's *Historia Naturalis Brasiliae*', in: Harold John Cook and Sven Dupré (eds.), *Translating Knowledge in the Early Modern Low Countries* (Münster: LIT Verlag, 2012), 333–348.
—, *Comprehending the New World in the Early Modern Period: Descriptions of Dutch Brazil (1624-1654)* (Ph.D. diss. Universiteit van Gent, 2015).
Davids, Karel (C.A.), *Zeewezen en wetenschap. De wetenschap en de ontwikkeling van de navigatie-techniek in Nederland tussen 1585 en 1815* (Amsterdam/ Dieren: Bataafsche Leeuw, 1986).
Davidson, Ake, 'Deutsche Stammbücher in swedischen Sammlungen', in Jörg-Ulrich Fechner, *Stammbücher als kulturhistorischen Quellen* (Munchen/ Braunschweig, 1981), 89–95.
Daxecker, Franz, 'Christoph Scheiner's main work "Rosa Ursina sive Sol"', *Acta Universitatis Carolinae. Mathematica et Physica*, 46 (2005), Suppl, 127-140.
Dekker, Elly, 'Early Explorations of the Southern Celestial Sky', *Annals of Science* 44 (1987), 439–470.
—, 'On the dispersal of knowledge of the southern celestial sky', *Der Globusfreund* 35/37 (1987), 211–230.
Deiman, Jan C., 'Het instrumentarium van de Utrechtse Sterrenwacht', *Tijdschrift voor de Geschiedenis der Geneeskunde, Wiskunde, Natuurwetenschappen en Techniek*, 10 (1987), 174-189.
Diedenhofen, Wilhelm, 'Johan Maurits and his garden', in: Van den Boogaart (ed.), *Johan Maurits Van Nassau-Siegen* (1979), 197–236.
Dijksterhuis, Fokko Jan, 'Moving around the ellipse. Conic sections in Leiden, 1620-1660', in: Lüthy, Christoph and Sven Dupré, *Silent Messengers. The Circulation of Material Objects of Knowledge in the Early Modern Low Countries* (Berlin: LIT Verlag, 2011), 89-124.
Dijkstra, Arjen F.B., *Between academics and idiots: a cultural history of mathematics in the Dutch province of Friesland (1600-1700)* (Enschedé: Ph.D. diss Univ. Twente, 2012).
Drake, Stillman, *Discoveries and Opinions of Galileo* (New York, 1957).
Dreyer, J. L.E., *Tycho Brahe. A picture of scientific life and work in the sixteenth century* (Edinburgh, Black, 1890; reprinted New York, etc. 1963).
Eekhout, R.A., 'The Mauritias. A neo-Latin epic by Francis Plante', in: Van den Boogaart (ed.), *Johan Maurits Van Nassau-Siegen* (1979), 377–393.
Enschedé, J.W., 'De Leidsche Elseviers en hun meesterknecht P. Heemskerk (overl. in 1654)', *Het Boek* (1912), 25–31.
Findlen, Paula, *Possessing Nature. Museums, Collecting, and Scientific Culture in Early Modern Italy* (1994).
Françozo, Mariana de Campos, *De Olinda a Olanda. Johan Maurits van Nassau e a circulação de objetos e saberes no Atlântico holandês (século XVII)*, (Thesis Universidade Estadual de Campinas, 2009).
—, 'Global Connections: Johan Maurits of Nassau-Siegen's Collection of Curiosities', in: Van Groesen (ed.), *The Legacy of Dutch Brazil* (2014), 105-123.

Gelder, H.E. van, 'Twee Braziliaanse schildpadden door Albert Eckhout', *Oud Holland* 75 (1960), 5–30.

Gonsalves de Mello, José Antonio, 'Johan Bodecher Banning: um professor da universidade de Leiden no Brasil Holandes (1638-1639)', in: *Revista do instituto archeologico historico e geographico Pernambucano* 57 (1984), 169-176. Reprinted in: Hans Bots and Maxim Kerkhof (eds.), *Forum Litterarum: Miscelanea de estudos literarios linguisticos e historicos oferecido a J.J. van den Besselaar* (Amsterdam and Maarssen. 1984), 81–85.

—, *Tempo dos Flamingos. Influência da ocupaçâo holandesa na vida e na cultura do norte Brasil* (Rio de Janeiro, 1947). Translated and adapted into Dutch by G.N. Visser and B.N. Teensma, as *Nederlanders in Brazilië. De invloed van de Hollandse besetting op het leven en de cultuur in Noord-Brazilië* (Zutphen: Walburg pers, 2001).

Graham Burnett, D., *Descartes and the Hyperbolic Quest: Lens Making Machines and their Significance in the Seventeenth Century* (Transactions of the American Philosophical Society, 2005).

Grell, Ole Peter; Cunningham, Andrew; Arrizabalaga, Jon (eds.), *Centres of Medical Excellence? Medical Travel and Education in Europe, 1500-1789* (Ashgate Publishing, 2010).

Greco, V. *et al*, 'The optical principles of telescopes', in: Van Helden, *Catalogue of early telescopes* (1999), 99-111.

Grafton, Anthony, *Athenae Batavae: The Reseach Imperative at Leiden, 1575-1650* (Leiden, 2003).

Groesen Michiel van, 'Van vaandrig in Brazilië tot dichter in Dordrecht. Het album amicorum van Johann Philipp Mulheiser (ca. 1603–ca. 1677)', *De zeventiende eeuw* 24 (2008), 196–209.

—, *Amsterdam's Atlantic. Print Culture and the Making of Dutch Brazil* (Univ. of Pennsylvania Press, 2016).

—, 'Abraham Willaerts: Marine Painter in Dutch Brazil and the Atlantic World', *Oud Holland* 132(2019), 65–78.

Groesen, Michiel van (red.) *The Legacy of Dutch Brazil* (New York: Cambridge University Press, 2014).

Groot, S.J. de, and B.J. de Groot, 'De *Schatkamer* en de *Kunst der Stuer-Luyden* (1621-1714). De navigatieboeken van Cornelis Jansz. Lastman', in: *Schatkamer. Veertien opstellen over maritiem-historische onderwerpen aangeboden aan Leo M. Akveld* (Franeker: Van Wijnen, 2002), 82–99.

Haasbroek, N.D., *Gemma Frisius, Tycho Brahe and Snellius and their Triangulations* (Delft, 1968).

Harmsen, A.J.E., 'Barlaeus' descriptions of the Dutch Colony in Brazil', in: Zweder von Martels (ed.), *Travel Fact and Travel Fiction: Studies on Fiction, Literary Tradition, Scholarly Discovery, and Observation in Travel Writing* (Leiden: E.J. Brill, 1994), 158-169.

Heidhues, Mary Somers, 'An Early Traveler's Compendium: Caspar Schmalkalden 's Images of Asia', *Archipel* 70 (2005), 145-184. See: https://www.persee.fr/doc/arch_0044-8613_2005_num_70_1_3976. (Consulted September 2021).

Heineman, Karen M., *Het Academiedak. De Sterrenwacht, het uurwerk van 1589, de academietoren,* Unpublished paper, Faculty of Art History, Leiden, n.d. [c. 1979]. Copy in library Rijksmuseum Boerhaave, Leiden.

Heijer, Henk den, 'Bewindhebbers, gouverneurs en raden van bestuur (het bestuur van de West-Indische Compagnie in de Republiek en in Brazilië)', in: Wiesenbron, *Brazilië in de Nederlandse archieven,* 2 (2005), 17–41.
–, *De geschiedenis van de WIC* (Walburg Pers, 2002).
Heijer, Henk den; Teensma, Ben, *Nederlands-Brazilië in kaart. Nederlanders in het Atlantische gebied 1600-1650* (Zutphen: Walburg pers, 2011).
Helden, Albert van, 'Saturn and his anses', *Journal for the History of Astronomy,* 5 (1974), 105-121.
–, 'The astronomical telescope, 1611-1650', *Annali dell'Istituto e Museo di Storia della Scienza di Firenze,* 1:2 (1976) 13–36.
–, 'The importance of the Transit of Mercury of 1631', *Journal for the History of Astronomy,* 7 (1976), 1-10.
–, *Measuring the Universe. Cosmic dimensions from Aristarchus to Halley* (Chicago: University of Chicago Press, 1985).
–, 'Gassendi and the telescope: toward a research community', *Actes du Colloque International Pierre Gassendi,* Vol. 2 (Digne les Bains, 1994), 329–339.
–, *Catalogue of early telescopes* (Florence, 1999).
Helk, Vello, *Stambogs-skikken id et Danske monarki indtol 1800* (Odense: Universitetsforlag, 2001).
Henderson, James, *A History of the Brazil, Comprising its Geography, Commerce, Colonization, Aboriginal Inhabitants* (London: Longman a.o., 1821).
Heuvel, C. van den, *'De Huysbou' A reconstruction of an unfinished treatise on architecture, town planning and civil engineering by Simon Stevin* (Amsterdam: KNAW-press, 2005).
Hoftijzer, Paul, 'Sic transit gloria ...The end of the officina Hackiana', *Quaerendo* 26:4 (1996), 258–273.
–, 'The library of Johannes de Laet' (1581-1649), *Lias* 25:2 1998), 201–216.
Howse, Derek, 'The Greenwich List of Observatories: a world list of astronomical observatories, instruments and clocks, 1670-1850', *Journal for the History of Astronomy* 17:4 (1986).
Hughes, David W., 'Why did other European astronomers not see the December 1639 transit of Venus?', in; D.W. Kurtz, (ed.), *Transits of Venus: New Views of the Solar System and Galaxy* [Proceedings IAU Colloquium No. 196], (Mabridge: International Astronomical Union, 2004)
Huigen, Siegfried; Jong, Jan de; and Kolfin, Elmer, *The Dutch Trading Companies as Knowledge Networks* (Leiden: Brill, 2010) [Intersections, Volume 14].
Huisman, Tim, *The Finger of God. Anatomical practices in 17th-century Leiden* (Leiden, 2009).
Hulsman, Lodewijk, 'Gisberth de With en Anna Paes. De geschiedenis van het huwelijk van een Dordtenaar en een Braziliaanse in de 17e eeuw', *Oud Dordrecht* 23:2 (2005), 52–62; 23:3 (2005), 36–45 and 24:1 (2006), 61–70.
Joost, Wolfgang, 'Auf den Meeren der Welt. Das Leben des Caspar Schmalkalden (1616-1673)', in: idem, *Casper Schmalkalden. Mit Kompas und Kanonen.* (2002), 7.
Joppien, R., 'The Dutch vision of Brazil. Johan Maurits and his artists', in: Van den Boogaart (ed.), *Johan Maurits Van Nassau-Siegen* (1979), 297–376.
Jorink, Eric, *Reading the book of nature in the Dutch Golden Age* (Leiden-Boston: Brill, 2010). Translation by Peter Mason of *Het 'Boeck der Natuere'.*

Nederlandse geleerden en de wonderen van Gods schepping 1575-1715 (Leiden, 2006).
Juynboll, Wilhelmina M.C., 'Jacob Golius', in: *Zeventiende-eeuwsche beoefenaars van het Arabisch in Nederland* (Utrecht: Kemink, 1931), 119-183.
Kaiser, Frederik, 'Geschichte und Beschreibung der Sternwarte in Leiden', *Annalen der Sternwarte in Leiden*, 1 (1868), I–LXVIII.
Kallinen, Maija, 'Lectures and Practices. The Variety of Mathematical and Mechanical Teaching at the University of Uppsala in the 17th Century', in: M. Feingold and V. Navarro-Brotons (eds.), *Universities and Science in the Early Modern Period* (2006), 111-125.
Keil, Inge, *Von Ocularien, Perspicillen und Mikroskopen, von Hungersnöten und Friedensfreuden, Optikern, Kaufleuten und Fürsten Materialien zur Geschichte der optischen Werkstatt von Johann Wiesel (1583-1662) und seiner Nachfolger in Augsburg* (Augsburg, 2003).
King, Henry C., *History of the telescope* (London, 1955).
Knobel, E.B., 'On Frederick de Houtman's catalogue of the southern stars', *Monthly Notices of the Royal Astronomical Society* 77 (1917). 414–432.
Kragh, Helge, *Jørgen From (1605-1651): Et Billede fra Dansk Astronomi- og Lærdomshistorie* (Aarhus: Centre for Science Studies, University of Aarhus, Denmark, Research group: History and philosophy of science, July 2014) [Research Publications on Science Studies, no. 22].
—, 'Georgius Frommius (1605-1651) and Danish Astronomy in the Post-Tychonian Era', *Acta Baltica Historiae et Philosophiae Scientiarum* 3:1 (Spring 2015), 45–68.
Kreuszig, August Hermann, *Album der evangelisch-Lutherischen Geistlichen im Konigreiche Sachsen von der Reformationszeit bus zum Gegenwart* (Crimmitschau: Rober Raab, 1898).
Krempel, Léon, 'Biographische und stilistische Notizen zu Frans Post', in: Idem (ed.): *Frans Post (1612-1680): Maler des verlorenen Paradieses* (Petersberg, 2006), 19–28.
Krohn, Thomas, *Christoph Nothnagels Lehr- und Forschungstätigkeit an der Universität Wittenberg – Mathematisch-astronomische Weltsicht des 17. Jahrhunderts im Spiegel universitärer Lehre* (Ph.D. diss. Wittenberg, 2014).
Kroon, J.E., *Bijdragen tot de geschiedenis van het geneeskundig onderwijs aan de Leidsche Universiteit 1575—1625* (Diss. Leiden, 1910).
Larsen, Erik, *Frans Post. Interprète du Brésil* (Amsterdam/ Rio de Janeiro: Colibris Editora, 1962).
Long, Pamela O., *Artisan/Practitioners and the Rise of the New Sciences, 1400-1600* (Oregon: State University Press, 2011).
—, 'Trading Zones in Early Modern Europe', *Isis* 106:4 (2015), 840–847.
Lunsingh Scheurleer, Th.H. & G.H.M. Posthumus Meyjes (eds.), *Leiden University in the seventeenth century : an exchange of learning* (Leiden, 1975).
Lunsingh Scheurleer, Th.H.; C. Willemijn Fock & A.J. van Dissel, *Het Rapenburg. Geschiedenis van een Leidse gracht.* 6 vols. (Leiden: n.p., 1986-1992).
Marriott, Chris, 'SkyMap Pro11', *Software for Astronomers* (Devon, 2005).
[Maskelyne, Nevil], 'Remarks on the planet Mercury', *The Edinburgh Magazine, or Literary Miscellany* 1 (1785), 29–31; Translated into Dutch as 'Aanmerkingen over de planet Mercurius', *Algemeene Vaderlandsche Letter–Oefeningen*, 7:2 (1785), Mengelwerk, 194-196.

Matsuura, Oscar Toshiaki, *História da Astronomia no Brasil, Vol. 1* (Recife: Cepe, 2014)

Meijer, Theodorus J., 'De omstreden nalatenschap van een avontuurlijk geleerde', *Leids Jaarboekje* (1972), 63–76.

Menezes, José Luiz Mota, *Atlas Arqueológico do Recife, Módulo 7, Presença Holandesa* (Recife 1998).

–, 'A casa de sua excelência, o príncipe João Maurício de Nassau', *Clio Arqueológica*, no 18 (2005). [Universidade Federal de Pernambuco – UFPE].

Meuwese, Mark, *Brothers in Arms. Dutch-Indigenous Alliances in the Atlantic World, 1595-1674* (Leiden- Boston: Brill, 2012).

Monteiro Carolina and Erik Odegard, 'Slavery at the Court of the "Humanist Prince". Reexamining Johan Maurits van Nassau-Siegen and his Role in Slavery, Slave Trade and Slave-smuggling in Dutch Brazil', *Journal of Early American History* 10 (2020) 3–32.

Moraes, Abrahão de, 'A Astronomia no Brasil', Capítulo II, in: F. Azevedo (ed.). *As Ciências no Brasil* (São Paulo: Editora Melhoramentos, 1955), 84-161.

Moreira, Danielle; Mendes, Sérgio L., 'Sloths of the Atlantic Forest in the sixteenth and seventeenth centuries', *Anais da Academia Brasileira de Ciências*, 88:3 (2016), suppl., 1799-1808.

Moreira, I.C., 'A expedição de Couplet à Paraíba – 1698', *Revista Brasileira de História da Ciência* 5 (1991) 23–31.

Morize, Henrique, 'O cometa de 1652', *Revista do Instituto Archeologico e Geographico Pernambucano*, 11 (1904), 611–613.

Morren, Th., 'Een cholerisch geneesheer in de 18e eeuw', *Leidsch Jaarboekje* (1908), 85–89.

Mostert, Ilja, 'Cornelis Goliath (1617-1660), wereldberoemd kaartmaker in Zeeland en Zuid-Amerika', *Caert-Thresoor, Tijdschrift voor de Geschiedenis van de Kartografie* 37:4 (2018), 149-159.

Moulin, D. de, 'Medizinische und naturwissenschaftliche Aspecte der Regerungszeit des Grafen Johann Moritz von Nassau als Gover neur in Brasilien 1637-1644', in: De Wer, *Soweit der Erdkreis reicht* (1979), 33–46.

Mout, M.E.H.N., 'The youth of Johan Maurits and aristocratic culture in the early seventeenth century', in: Van den Boogaart (ed.), *Johan Maurits Van Nassau-Siegen* (1979), 12–38.

Muller, E. & K. Zandvliet, *Admissies als landmeter in Nederland voor 1811. Bronnen voor de geschiedenis van de landmeetkunde en haar toepassing in administratie, architectuur, kartografie en vesting- en waterbouwkunde* (Alphen a/d Rijn, 1987).

Nikitinski, Oleg, 'Petrus Cunaeus uber Astrologische Wahrsagerei und über die wurde der Exakten Wissenschaften', *Humanistica Lovaniensia*, 63 (2014), 249–268.

Noordegraaf, Leo & Gerrit Valk, *De gave Gods. De pest in Holland vanaf de late middeleeuwen.* (Amsterdam 1996).

North, John D., 'Thomas Harriot and the first telescopic observations of sunspots', in: idem, *The Universal Frame* (1989), 109-144.

–, *The universal frame: historical essays in astronomy, natural philosophy and scientific method* (London / Ronceverte, 1989).

Odegard, Erik L.L., *Colonial Careers. Johan Maurits van Nassau-Siegen, Rijckloff Volckertsz. Van Goens and Career-Making in the Early Modern Dutch Empire* (Ph.D. Dissertation Leiden University, 2018).

Oers, Ron van, *Dutch Town Planning Overseas during VOC and WIC Rule, 1600-1800* (Zutphen: Walburg pers, 2000).

Oliveira Lima, Manoel, 'Sabios estrangeiros', *Revista do instituto archeologico historico e geographico Pernambucano*, 21 (1920), 336–345.

Omodeo, Pietro Daniel, 'The Scientific Culture of the Baltic Mathematician, Physician, and Calendar-Maker Laurentius Eichstadt (1596-1660)', *Journal for the History of Astronomy* 48:2 (2017), 135-159. published in German as: 'Die wissenschaftliche Kultur des Mathematikers, Arztes und Kalender-machers Lorenz Eichstaedt (1596-1660)', in: Klaus-Dieter Herbst und Werner Greiling (eds.): *Schreibkalender und ihre Autoren in Mittel-, Ost- und Ostmitteleuropa (1540-1850)* (Bremen, 2018), 109-135.

Ormeling,Ferjan,'Het gebruik en de effectiviteit van symbolen op oude kaarten', *Geo-Info* 9:3 (2012), 20–27.

Otterspeer, Willem, *Groepsportret met Dame. I: Het bolwerk van de vrijheid. De Leidse universiteit 1575-1672* (Amsterdam, 2000).

Pereira da Costa, A. (ed.), *Inventario das armas e petrechos belicos que os holandêses deixaram em Pernambuco e dos prédios edificados ou reparados até 1654* (Recife: Impensa Official, 1940).

Pies, Eike, *Willem Piso (1611-1678)* (Düsseldorf: Interna-orb, 1981).

Pouls, H.C., *De landmeter Jan Pietersz. Dou en de Hollandse Cirkel* (Delft, 2004).

Rammelman Elsevier, Willem J.C., *Inventaris van het Archief van de Gemeente Leiden*, Vol. 2 (Leiden, n.d. [1864]).

Raeder, H., E. Strömgren & B. Strömgren (eds.), *Tycho Brahe's Description of his Instruments and Scientific Work* (København, 1946).

Ramminger, Johann, *Neulateinische Wortliste. Ein Wörterbuch des Lateinischen von Petrarca bis 1700* (NLW): www.neulatein.de/neulateinische_wortliste.htm.

Ratelband, Klaas, *Nederlanders in West-Afrika, 1600-1650. Angola, Kongo en Sao Tomé* (Zutphen: Walburg pers, 2000). [Publication of a manuscript, written in 1945].

Rodrigues, J.H., 'Escôrço biobliográfico', in G. Piso, *História natural e médica da* Índia *Ocidental* (Rio de Janeiro: Instituto Nacional do Livro, 1957), 3–67.

Rooseboom, Maria, *Bijdrage tot de geschiedenis der instrumentmakerskunst in de Noordelijke Nederlanden tot omstreeks 1840* (Leiden, 1950).

Ruestow, Edward, *Physics at seventeenth- and eighteenth-century Leiden. Philosophy and the new science in the university* (The Hague, 1973).

Savage-Smith, Emilie and Colin Wakefield, 'Jacob Golius and Celestial Cartography', in: in: W.D. Hackmann and A.J. Turner (eds.), *Learning, language and invention. Essays presented to Francis Maddison* (Aldershot: Variorum, 1994), 238–260.

Scharf, Cláudia Philippi, *Libri Principis e as ilustrações de fauna do Brasil Holandês: fatura, técnicas, materiais e autores* (Salvador: Ph.D. dissertation Universidade Federal da Bahia, 2019).

Saunders, E. Stewart, 'The Archives of the Académie des Sciences', *French Historical Studies* , 10:4 (1978), 696–702.

Schmidt, Benjamin, 'Space, time, travel: Hugo de Groot, Johannes de Laet, and the advancement of Geographic Learning', *Lias* 25 (1998), 177-199.

Schottmüller, Kurt, 'Reise-eindrücke aus Danzig, Lübeck, Hamburg und Holland 1636. Nach dem neuentdeckten II. Teil von Charles Ogiers Gesandtschaftstagebuch', *Zeitschrift des Westpreüssischen Geschichtsvereins* 59 (1910), 201–273.

Sellés, Antonio Lafuente y Manuel, *Observatorio de Cádiz (1753-1831)* (Cadiz: Instituto de Historia y Cultura Naval, 1988).

Silva, Maria Angelica da, and Melissa Mota Alcides, 'Collecting and framing the wilderness. The garden of Johan Maurits (1604–79) in North-East Brazil', *Garden History* 30:2 (2002), 153-176.

Sint Nicolaas, Eveline, 'Vrijburg, getekend en gesneden', *Bulletin van het Rijksmuseum* 56 (2008), 202–213.

Sluijter, Ronald G.H., *'Tot ciraet, vermeerderinge ende heerlyckmaeckinge der universiteyt': bestuur, instellingen, personeel en financiën van de Leidse universiteit, 1575-1812* (Hilversum: Verloren, 2004).

Smet, Ingrid A.R. De, 'Town and Gown in the Dutch Golden Age: The Menippean Satires of Jan Bodecher Benningh and "Amatus Fornacius"', in: G. Tournoy and D. Sacré (eds), *Myricae. Essays on Neo-Latin Literature in Memory of Jozef IJsewijn* (Leuven University Press/Peeters Press, 2000), 491–521.

Sousa-Leao, Joaquim de, *Frans Post 1612-1689* (Amsterdam: Van Gendt, 1973).

Souza Leão Vieira, Daniel de, *Topografias Imaginárias: A Paisagem Política do Brasil Holandês em Frans Post, 1637-1669* (Ph. D. Dissertation Leiden University, 2010).

Starcke, E., 'Die Coninxloo's', *Oud Holland* 16:3 (1898), 129-146.

Straker, Stephen, 'Kepler, Tycho, and the 'Optical Part of Astronomy': the Genesis of Kepler's Theory of Pinhole Images', *Archive for History of Exact Sciences,* 24:4 (1981), 267–293.

Stokvis, B.J., *La médicine coloniale et les médecins hollandais du 17e siècle* (Amsterdam 1883).

Teixeira, Dante Martins, 'Der Mythos der unberürthen Natur. Die Naturgeschichte in Hollandisches Brasilien (1624-1654) und ihr Beitrag zur Kenntnis der jungeren Geschichte der Fauna der Neuwen Welt', in: Brunn & Neutsch (eds.), *Sein Feld war die Welt. Johann Moritz von Nassau-Siegen* (2008), 197–232.

Terwen, J.J., 'The Buildings of Johan Maurits' in: Boogaart (ed.), *Johan Maurits van Nassau-Siegen, 1604-1679,* (1979), 87-104.

Thorndike, Lynn, *A History of Magic and Experimental Science: The seventeenth century* (Macmillan, 1958).

Thrower, N.J.W (ed.), *Three Voyages of Edmond Halley in the Paramore, 1698-1701,* 2 vols. (Farnhem: Ashgate, 1982). [Works issued by the Hakluyt Society, second serie, 156].

Tol, Joris J.S. van den, *Lobbying in Company: Mechanisms of political decision-making and economic interests in the history of Dutch Brazil, 1621-1656* (Ph.D. dissertation Leiden University, 2018).

Trindade Medeiros, Maria Franco and Ulysses Paulino Albuquerque, 'Food flora in 17th century northeast region of Brazil in *Historia Naturalis Brasiliae*', *Journal of Ethnobiology and Ethnomedicine* 10:50 (2014).

Vainfas, Ronaldo. *Traição. Um jesuíta a serviço do Brasil holandês processado pela Inquisição.* (São Paulo: Cia. das Letras, 2008).
Vaquero, José M.; M.C. Gallego, Ilya G. Usoskin, and Gennady A. Kovaltsov, 'Revisited sunspot data: A new scenario for the onset of the Maunder minimum', *The Astrophysical Journal Letters*, 731:2 (2011), 1–4.
Veldt, James van der, 'An Autograph Letter of John Maurice of Nassau, Governor of the Dutch Colony in Brazil (1636-1644)', *The Americas* , 3:3 (1947), 311–318.
Verbeek, Theo & Eric-Jan Bos, *Descartes en Leiden. Vrienden en vijanden, bewonderaars en bestrijders* (Leiden, 2003).
Verbunt, Frank and Robert H. van Gent, 'Early star catalogues of the southern sky. De Houtman, Kepler (second and third classes), and Halley', *Astronomy & Astrophysics* 530, A93 (ESA 2011), 1–26.
Vermij, Rienk, *The Calvinist Copernicans. The reception of the new astronomy in the Dutch Republic, 1575-1750* (Amsterdam, 2002).
Vinde, Lea van de (ed.), *Shifting image. In search of Johan Maurits* (The Hague: Mauritshuis/Waanders, 2019).
Vrijer, M.J.A de, *Henricus Regius, een cartesiaansch" hoogleeraar aan de Utrechtsche hoogeschool* ('s Gravenhage: Nijhoff, 1917).
Wätjen, Hermann, *Das holländische Kolonialreich in Brasilien. Ein Kapitel aus der Kolonialgeschichte des 17. Jahrhunderts* (Den Haag: Nijhoff / Gotha: Perthes, 1921). Translated into Portuguese as *O Dominio Colonial Hollandez no Brasil. Um capitulo da historia colonial do seculo XVII* (São Paulo: Companhia Editora Nacional, 1938).
Wer, Guido de (ed), *Soweit der Erdkreis reicht. Johann Moritz von Nassau-Siegen 1604-1679* (Cleve: Städtisches Museum, 1979).
Wesley, Walter G., 'The Accuracy of Tycho Brahe's Instruments', *Journal for the History of Astronomy*, 9 (1978) 42–53.
Whitehead, Peter J.P., Boeseman, M., *A portrait of Dutch 17th century Brazil. Animals, plants and people by the artists of Johan Maurits of Nassau* (Amsterdam / Oxford: North Holland Publishing, 1989).
Wiesebron, Marianne, *Brazilië in de Nederlandse archieven (1624-1654): inleiding op de Nederlanden in de zeventiende eeuw en de collecties in Nederlandse archieven en bibliotheken* (Research School CNWS, 2004).
—, *Brazilië in de Nederlandse archieven 1624-1654: de West-Indische Compagnie: overgekomen brieven en papieren uit Brazilië en Curaçao* (Research School CNWS, 2005).
—, *Brazilië in de Nederlandse Archieven (1624-1654): Documenten in het Koninklijk Huisarchief en in het Archief van de Staten-Generaal / O Brasil Em Arquivos Neerlandeses (1624-1654): Documentos no Arquivo da Casa Real e no Arquivo dos Estados Gerais* (Leiden: Research School CNWS, 2008).
—, *Brazilië in de Nederlandse archieven (1624-1654): Oude West Indische Compagnie. Vol. 4: Correspondentie van de Heren XIX en Notulen van de Hoge en Secrete Raad van Braziliё / O Brasil em arquivos neerlandeses (1624-1654): Companhia das* Índias *Ocidentais velha: Cartas enviadas pelos Diretores XIX Atas diárias do Alto e Secreto Conselho do Brasil* (Leiden: Leiden University Press, 2011).
Wiesenfeldt, Gerhard, *Leerer Raum in Minervas Haus. Experimentelle Naturlehre an der Universität Leiden, 1675-1715* (Amsterdam 2002).

Wigartho, *Des mit allerley Merckwurdigkeiten Gelehrter Leute aus Alt- und Neue Zeiten beschäftigten Mercurii, I: Enthält Nachrichten von Gelehrten, Die Gemeiner Leute Kinder gewesen*, Vol. 1 (Schwabach: Enderes, 1738).

Willach, Rolf, 'The Development of Lens Grinding and Polishing Techniques in the First Half of the 17th Century', *Bulletin of the Scientific Instrument Society*, no. 68 (2001), 10-15.

—, 'The Development of Telescope Optics in the Middle of the Seventeenth Century', *Annals of Science* 58 (2001), 381–398.

Winter, P.J. van, *Hoger beroepsonderwijs avant la lettre. Bemoeiingen met de vorming van landmeters en ingenieurs bij de Nederlandse universiteiten van de 17e en 18e eeuw* (Amsterdam, 1988).

Witkam, H.J., 'Jean Gillot, een Leids ingenieur', *Leids Jaarboekje* 59 (1967), 29–54 and *ibidem*, 61 (1969), 39–70.

Witkam, J.J., *Jacobus Golius (1596-1667) en zijn handschriften* (Leiden: Brill, 1980).

Woolf, Harry, *The Transits of Venus: A Study of Eighteenth-Century Science* (Princeton, N.J.: Princeton Univ. Press, 1959).

Wreede, Liesbeth C. de, *Willebrord Snellius (1580-1626). A humanist reshaping the mathematical sciences* (Diss. Utrecht 2007).

Zandvliet, Kees, 'Johan Maurits and the Cartography of Dutch Brazil, the South-West-passage and Chile', in: Boogaart, E. van den (ed.), *Johan Maurits van Nassau-Siegen* (1979), 494–516..

—, *Mapping for Money. Maps, plans and topgraphic paintings and their rol in Dutch overseas expansion during the 16th and 17th centuries* (Amsterdam, 1998).

Zoeteman, Martine, *De studentenpopulatie van de Leidse universiteit, 1575-1812. 'Een volk op zyn Siams gekleet eenige mylen van Den Haag woonende'* (Leiden: University Press, 2011).

Zuidervaart, Huib J., *Van 'Konstgenoten' en Hemelse Fenomenen. Nederlandse Sterrenkunde in de Achttiende Eeuw* (Rotterdam: Erasmus Publications, 1999).

—, *Telescopes from Leiden Observatory and other collections, 1656-1859. A Descriptive Catalogue* (Leiden, 2007).

—, 'The 'invisible technician' made visible. Telescope making in the seventeenth and early eighteenth-century Dutch Republic' in: Alison D. Morrison-Low, Sven Dupré, Stephen Johnston, Giorgio Strano (eds.), *From Earth-Bound to Satellite. Telescopes, Skills and Networks* (Leiden-Boston: Brill, 2012), 41-102.

—, 'Geschonden vertrouwen. De Franse astronoom Joseph-Nicolas Delisle en zijn relatie met Nederlandse collega's tijdens de zoektocht naar de komeet van Halley (1758-1759)', in: Jos Gabriëls, Ineke Huysman, Ton van Kalmthout, Ronald Sluijter (eds.), *In vriendschap en vertrouwen. Cultuurhistorische essays over confidentialiteit* (Hilversum: Verloren, 2017), 117-129.

Zuidervaart, Huib, and Veerle Beurze, 'Samuel Carolus Kechel ab Hollenstein (1611-1668). Wiskundige en astronoom in de marge van academisch Leiden', *Leids Jaarboekje* 106 (2014), 25–58.

Zuidervaart, Huib, and Rob van Gent, 'A Bare Outpost of Learned European Culture on the Edge of the Jungles of Java: Johan Maurits Mohr (1716-1775) and the Emergence of Instrumental and Institutional Science in Dutch Colonial Indonesia' in: *Isis: An International Review devoted to the History of Science and its Cultural Influences* 95 (2004) 1–33.

INDEX

Abrahamsen, Jacob, 330, 338
Acosta, José de, 219
Alcides, Melissa Mota, 158
Aldrovandi, Ulisse, 164
Al-Farghānī, Ahmad ibn Muhammad ibn Kathīr, 20-21, 185
Alvares, Pedro, 265, 328
Alvez Camello, Belchior, 265, 328
Alvez Sr, Belchior, 328
Anchieta, Joseph de, 285
Anderson, Douglas, 12
Andrade-Lima, Dárdano de, 154
Arciszewksi, Christoffel, 32
Arentsdr, Lysbeth, 315
Argenterio, Giovanni, 55
Arien de slyper, 358

Baar, Piet de, 362
Baijert, Cornelis, 328
Bailly, Griettie, 61, 63, 81, 112, 178, 315, 321
Banchem, Jan Jacobs van, 68
Baneke, David, 12
Barcquer, Jan Willemsz, 338
Barlaeus, Caspar, 40, 42, 44- 45, 105, 107, 110, 131-132, 134-136, 168, 246, 262
Barreto de Meneses, Francisco, 265
Bartholin, Barthol, 73, 75, 318, 319
Bartholin, Caspar, 73, 75, 319
Bartholin, Erasmus, 75, 164, 319, 354
Bartholin, Petrus, 229
Bartholin, Thomas, 73-75, 118, 318-319, 325
Bartsch, Jacob, 219, 228, 311, 358
Baudius, Anna, 63, 75, 318-319
Baudius, Dominicus, 75
Bayer, Johann, 196, 219, 358
Beeckman, Isaac, 79, 201, 223, 227, 280, 356, 358-359
Beeckman, Martin, 326, 329, 351
Beerensteyn. See Bernstein
Behaim, Stephan Carl, 325
Bennigsen, Maria Sophia von, 317
Bering, Johannes, 199, 200, 355
Berkel, Klaas van, 29
Bernegger, Matthias, 197
Bernstein, Georg Heinrich von, 72, 298, 317
Bick, Alexander, 323
Bicker, Gerrit, 327
Bicker, Maria, 327
Bierens de Haan, David, 330
Bignon, Jean-Paul, 187
Bigourdan, Guillaume, 187, 273
Blaeu, Johannes, 40, 125, 136, 138, 151, 285, 340, 342
Blaeu, Willem Jansz, 65- 66, 194-195, 210, 230, 251, 266, 268, 356
Bodecher Benning, Johannes, 109-111, 119, 185, 299, 323, 331-332
Bodecher Benning, Nicolaas, 331
Boeseman, Marinus, 38, 171
Bonjour, Henry, 108, 295, 300
Bonjour, Pierre, 108-109, 300
Bontius, Jacobus, 168-169
Boogaart, Ernst van den, 114, 136
Borch, Ole, 181
Boreel, Adam, 358
Borgesius, Johannes, 364
Böthführ, Heinrich Julius, 322
Boulliau, Ismael, 189, 268, 326, 354
Boxer, Charles, 40
Boxhorn, Marcus, 73, 78, 348
Brahe, Tycho, 18-19, 21-23, 58, 64-

66, 81, 89, 112-113, 143, 187, 193-194, 197, 199, 201, 208, 211-213, 215, 218-219, 230-231, 235, 240, 248, 266, 268, 272, 279-280, 283, 297, 315, 319, 360
Brandis, Henrick, 318
Brendecke, Arndt, 28
Breyne, Jacob, 343
Brienen, Rebecca Parker, 113-114, 154, 167, 178, 278, 344
Brinck, Ernst, 33, 352
Brochmand, Christian, 319
Brochmand, Jesper Rasmussen, 77-78, 81, 319
Brommer, Bea, 342
Bronckhorst, Everard, 315
Bronckhorst, Jan Gerritsz. van, 84
Brosterhuyzen, Jan van, 342
Brouwer, Hendrik, 162, 260
Bünaw, Rudolph von, 311
Burgerhuys, Michael, 32
Burgh, Albert Coenraads, 35, 86, 88, 109, 323, 324

Calado do Salvador, Manoel, 42, 44-45, 309
Call, Jan van, 35
Camerarius, Philipp, 312
Camp, Clement le, 337
Campen, Jacob van, 33, 345
Campos Françozo, Mariana de, 367
Carbone, Giovanni Battista, 189
Carvalho, Alfredo de, 263
Cassini, Jacques, 187
Cassini, Jean-Dominique, 187
Castelli, Benedetto, 223
Castro, João de, 358
Cathagen, Hendrick Christoffels, 325, 334
Ceulen, Mathias van, 88, 324
Claes, Barbara, 325
Claess, Maria, 108
Clerck, Elisabeth de, 336
Clusius, Carolus, 160-161, 164, 171, 345
Cluyt, Outgert, 347
Codde van den Burgh, Dirck, 102
Colom, Jacob Aertsz., 356
Coninxloo, Pieter van, 38, 309
Cook, Harald, 28-30, 281, 283
Coppijn, Joos, 130, 338
Correa do Lago, Pedro and Bia, 103
Couplet des Tortreaux, Pierre, 187-188, 268
Crabtree, William, 244, 256-257, 361
Cralitz. See Kralitz
Crena, Gerrit, 353
Crommelin, Claude August, 184
Crommeling, Joos, 76
Crüger, Petrus, 314

Dalen, Cornelis II van, 34
Dams, Britt, 154, 165, 171
David, Jacob, 63, 130
David Sr., Jan, 63, 130
David Jr., Jan, 357
Davidse, Ad, 321, 360, 362
Dekker, Elly, 194
Delisle. See Isle, de l'
Descartes, René, 65, 315, 319, 322, 358
Diedenhofen, Wilhelm, 42
Dietericus, Johannes Cunradus, 317
Donati, Luca, 358, 367
Dou, Jan Pietersz, 71
Douw de jonge, Johan, 62
Drewisch, Andreas, 95
Druyf, Dirk, 76
Duin, Marieke van der, 12
Dürer, Albrecht, 169, 192-193
Duurcoop, Abraham, 324, 329, 347-348
Duyfhuysen jr, Jacob, 327

Eckhout, Albert, 38, 116, 139, 171, 175, 328, 344, 346, 349
Écluse, Charles de l'. See Clusius
Eduwaer, Arent, 315
Eduwaer, Dirck, 61, 63, 315
Eduwaer, Eduart, 315
Eduwaer, Hillegond, 315
Eichstädt, Lorenz, 24, 59, 60, 197-199, 218-219, 226, 240, 245, 252, 279, 291, 297, 313
Elzevier, Abraham, 77, 163
Elzevier, Bonaventura, 163
Elzevier, Isaac, 77, 179
Elzevier, Johannes, 350

Elzevier, Lodewijk, 40, 163, 169, 347, 350
Elzevier, Lodewijk III, 151, 163
Erpenius, Thomas, 66

Fabricius, Elisabeth, 312-313
Fabricius, Jacob, 57-58, 197, 312-313
Fabricius, Johann, 60, 298, 314
Fabricius, Otto, 314
Falcoburgius (Van Valckenburg), Adrianus, 74, 318
Falloo, Andries, 108-109, 300, 330
Falloo, Frederick, 109, 295, 300
Falloo, Henricus, 109
Falloo, Maria, 109, 300
Fincke, Anna, 73
Fincke, Margrethe, 73
Fincke, Thomas, 73
Fischer, Salomon, 58, 313
Fischer, Thomas, 313
Flamsteed, James, 186, 252-253, 268, 270, 272-273, 353
Flinck, Govert, 34
Fontana, Francesco, 258
Fracastoro, Girolamo, 219, 358
Fraem. See Frommius
Frederik Hendrik, prince of Orange, 33, 308
Frederik III, King of Denmark, 175, 349
Friedrich Wilhelm I, Elector of Brandenburg, 175
Friedrich, Susanne, 28
Friis, Christen, 77
Friis, Georg Andreas (=Jørgen), 77, 320-321
Frommius (From), Georg (= Jørgen), 63, 75, 77-79, 81, 83, 96, 112, 184-185, 229, 319, 321, 332
Fuchs, Leonart, 55
Fuiren, Henrik, 72-73, 75, 83, 299, 317-319
Fuiren, Jørgen, 73
Fuiren, Thomas, 73, 299, 317-318

Gakelius, Petrus, 318
Galilei, Galileo, 65, 77, 223, 227, 258
Galison, Peter, 47
Gameren, Adriana Willemsdr van, 182, 296, 353
Gameren, Dina van, 353
Gameren, Willem Adriaansz. van, 352
Gariges, Anna, 109, 300
Gariges, Maria, 109, 300
Garnier, Philippe, 358, 367
Gassendi, Pierre, 199, 227, 228-229, 244-245, 247-248, 259, 315, 357
Geertz, Clifford, 367
Gehrels, Tom, 286
Gent, Rob van, 354
Gessner, Conrad, 164
Gesteira, Heloisa Meireles, 213
Gillot, Jean, 316
Göbel, Friedrich, 57, 64, 312, 313
Godin, Louis, 188, 268, 354
Goebelius. See Göbel
Goedaert, Johannes, 165
Goetham, Mr., 358
Goliath, Cornelis Bastiaensz, 105, 121, 139, 140, 265, 329-330
Golius, Jacob, 20-21, 63, 65-68, 71, 78, 89, 91, 105, 110-111, 151-152, 173, 176-179, 181-182, 184-186, 189, 200-202, 208, 216, 218, 221, 227, 236, 244-245, 257, 266, 268, 281, 283, 285, 292, 296, 315-316, 322, 325, 343, 350-351, 355, 357-358
Gomes de Sylveira, Duarte, 133-134
Gonsalves de Mello, José Antonio, 331, 364
Gool, Jacob. See Golius
Gool, Theodorus, 182
Graef, Pieter de, 353
Gravesande, Arent van 's, 319
Grevingh, Johannes, 44, 147, 310, 330, 335, 364
Grotelande, Cornelis Dircx van, 335
Gudger, Eugene, 21, 23, 263

Haasbroek, Nicolaas Dirk, 230, 233
Habrecht, Isaac, 53, 197, 219, 358
Hackius, François, 40, 63, 151, 162-163, 168-169
Haecxs, Hendrik, 259-260, 363-364
Haeften, Joost van, 324

Hagen, Adolfus Fredericus von, 72, 83, 298, 317
Hagen, Christiaen van, 62
Hahn, Elisabeth, 317
Halley, Edmond, 186, 268-270, 353
Hamel, Henrick, 331
Hansen, Mathias, 319
Hantzsch, Victor, 24
Heddense, Hessel, 328
Hedraeus, Bengt, 91, 213, 326
Heemskerck, Pieter Vincentsz van, 63, 200, 326, 355
Hegerus, Franciscus, 163
Heidhues, Mary Somers, 345
Heijer, Henk den, 124, 342
Heinfogel, Conrad, 192
Heiningen, Pieter Jans van, 356
Heinrich, Johannes, 348
Heinrich, Michael, 348
Heinsius, Nicolaas, 169
Helden, van. See Van Helden
Henderson, James, 269
Herckmans, Elias, 130-134, 162, 339, 340
Herman, Paulus, 204
Heurnius, Otto, 74, 318, 351
Hevelius, Johannes, 20, 173, 179, 188, 239, 296, 354
Hire, Gabriel-Philippe de la, 187
Hire, Philippe de la, 187, 268, 270-272
Holwarda, Johannes Phocylides, 244
Hondius, Jodocus, 193-194
Hondius, Wilhelmus, 95
Honinckhuijsen, Caspar, 128, 337-338
Horrocks, Jeremiah, 244, 256-257
Hortensius, Martinus, 79, 81, 110, 201-202, 225, 227-229, 244-245, 247-248, 280, 321-322, 356, 358, 360
Houten, C.J. (Kees) van, 286
Houten–Groeneveld, Ingrid van, 286
Houtman, Frederick de, 194, 251, 268
Hovius, Adrianus, 91, 326
Huyck, Antoine van, 184, 292, 332
Huyck, Jean van, 184, 292, 332
Huygens, Christiaan, 218, 259, 352
Huygens, Constantijn, 33, 109

Ihering, Rodolpho von, 263
Isle, Joseph-Nicolas de l', 187-189, 270, 272, 273, 319, 354

Jaccheus, Gilbert, 319
Jacobsz, Adriaen, 342
Jacopucci dos Reis, Filipe, 13
Jansen, Matheus, 147
Janss, Pieter, 108
Jansz, Cornelis, 305
Jentsch, Georg, 325
João IV, King of Portugal, 329
Johan Maurits van Nassau-Siegen, *Passim*
Jol, Cornelis, 147
Jonston, Jan, 167
Joost, Wolfgang, 260
Joppien, Rüdiger, 156
Julius Ceasar, Roman Emperor, 107
Jungius, Joachim, 326
Juweel, Jannetje Jansdr, 119

K, D. à, born near Citzâ (= Zeitz), 300
K, D., born Stettin, 300
K., S., born Prague, 301
Keblusek, Marika, 12
Kechel ab Hollesteyn, Anna Elisabeth, 75, 319
Kechel ab Hollesteyn, Samuel, 20, 61, 63, 75-76, 79-81, 91, 112-113, 145, 175, 179, 184-186, 237, 246, 268, 292, 296, 315, 318-319, 321, 326, 341, 351, 362
Kepler, Johannes, 22, 59, 64, 194, 197, 199, 216, 218-219, 228, 240, 251, 256-257, 311, 315, 356, 358
Kepler, Suzanna, 311
Kerckhoven, Arnoldus, 353
Ketelaer, Andries, 38, 309
Keyser, Pieter Dircksz, 194
Kistenius (Kirsten), Georg, 72, 83, 298, 317
Kragh, Helge, 78
Kralitz, Heinrich, 88-90, 237, 300, 325

Laet, Johannes de, 22-23, 40, 42, 63, 73, 83-86, 93, 112, 116-117, 136, 151-152, 154-155, 158, 160-165, 167-169, 171-172, 177, 181, 184, 253, 258, 277-278, 285, 293, 295, 318, 322-323, 326, 343, 346, 350
Lalande, Jêrome Le François de, 187-188, 270, 354
Lamberts, Aeltie, 327
Lange, Marcus, 299, 322
Langren, Michiel Florent van, 244
Lanoy, Philipp de, 313
Lansbergen, Philippus van, 78-79, 201, 321
Lastman, Cornelis Jansz, 305-306
Laurentius, Andreas, 320
Leandro da Costa, José Rodrigues, 13
Leeuwenhoek, Antony van, 352
Letting, Jacobus, 323
Lichtenstein, Martin, 171
Liesveld, Cornelia van, 323
Lieven, Bernhard von, 321
Lieven, Reinhold von, 321-322
Linden, Johannes Antonides van der, 90, 325
Lindenov, Christoffel, 175, 349
Lindenov, Henrik, 321
Lindenov, Jacob, 321
Linnaeus, Carolus, 172
Lipsius, Justus, 56-57, 299, 313
Lonck, Hendrick Corneliszoon, 31
Long, Pamela, 47, 284
Longomontanus, Christen Sørensen, 77-78, 81, 199, 229, 272, 319
Longorius, Marcus, 299, 322
Loo, Maria van, 319
Loth, Christoph, 60, 314
Louis XIV, King of France, 187, 270
Lourenço, Marta, 12
Louville, Jacques-Eugene, 354
Louwman, Peter, 12
Lucanus, Marcus Annaeus, 107
Lulofs, Johan, 357
Lus, Alida, 75, 319

Maire, Jean le, 40, 65, 163
Man, Cornelis de, 139-140, 340
Manson, Dirck, 324
Mansvelt, Hendrick Frederick van, 114, 120, 122, 333, 335-336
Marckgraeff, Ascanius, 327
Marckgraeff, Christoffel, 96, 327
Marggraf, Christoph, 311, 327
Marggrafe, Adriana Elisabeth, 352
Marggrafe, Beata, 350
Marggrafe, Christian, 20, 24, 53, 63, 105, 113, 119, 154, 171-173, 175-179, 181, 183, 193, 199, 281, 283, 286, 291, 295-296, 307, 322, 343, 348-350, 352-353
Marggrafe, Georg, *Passim*
Marggrafe, Georg, schoolmaster of Liebstadt, 51, 53
Marggrafe, Georg, son of Christian, 182, 296, 352
Marggrafe, Georg, vicar in Pappendorf, 51
Marggrafe, Gottfried, 51
Marggrafe, Jacobus, 352, 353
Marggrafe, Johannes Godefredus, 352
Marggrafe, Jurriaen (= Christian II), 286, 352
Marggrafe, Samuel, 350
Marggrafe, Theodora, 352
Marggrafe, Wilhelmus, 352
Marggravius. See Marggrafe
Maskelyne, Nevil, 272
Matham, Theodor, 159, 160, 345
Matsuura, Oscar, 23, 185, 212, 354
Matthews, Steven Paul, 29
Matthias, Johannes, 319
Maydell, Anna von, 321
Meckelborg, Hans, 319
Meijer, Theo, 23
Meisnerus, Eusebius, 326
Melder, Christiaan, 358
Metius, Adriaen, 200, 315
Michiels, Jan, 333
Mieris, Willem van, 204
Milanen, Wilhelm van, 85, 87-88, 322-324
Molenius, Johannes, 320-321
Mols, Hans, 341
Molyneux, Thomas, 352
Mombers, Grietje, 317

Monte, Guidobaldo del, 321
Moraes, Manuel de, 263, 285, 345, 367
Moreau, Pierre, 44
Moreira, Juliano, 263
Moritz, Landgraf of Hessen-Kassel, 38
Morris, Guideon, 339
Morris, John, 163
Mort, Jacob le, 182
Mota Menezes, José Luiz, 100-101, 205, 265, 328
Moucheron, Cosmo de, 119, 335
Moucheron, Hendrick de, 118-119, 335
Moucheron Jr., Hendrick de, 335
Mouffet, Thomas, 346
Mule, Hans Jørgensen, 82-83, 320-321
Mulheiser, Johann Philipp, 108-109, 331, 349
My, Hieronymus van der, 67

Napier, John, 199
Nassau-Dietz, Hendrik Casimir van, 308
Nassau-Siegen, Johan Maurits, *Passim*
Nassau-Siegen, Johann Moritz von. See: Nassau-Siegen, Johan Maurits van
Nassau-Siegen, Johann VII von, 33
Nassau-Siegen, Wilhelm von, 33
Nautonier, Guillaume le, 200, 219, 358
Nerée, Richard Johannes de, 77
Nicolai, Anna, 314
Nieuhof, Johan, 262, 342
Nieulandt, Cornelis, 341
Nieuwland, Ilja, 12
Noort, André van, 362
Nordmand, Jacob Jenssøn, 284, 333, 367
North, John, 21-23, 72, 189, 205, 219, 234, 239, 253, 255, 262-264, 267, 273, 305, 311-312, 317, 354, 358-359, 364
Nothnagel, Christoph, 57-58, 197, 298, 313-314

Odegard, Erik, 107
Oers, Ron van, 96
Oeseler, Philippus, 317
Olearius, Adam, 167
Oliveira Lima, Manoel de, 263
Ostendorf, Christoph, 12
Otterius (Otter), Christian, 82, 322
Outheynen, Henrick Willemss, 324

Paes de Azevedo, Anna Gonsalves, 336
Paulli, Simon, 54, 57-58, 197, 291, 312-313
Peeters, Gillis, 38-39, 96-97, 328
Pelt, Anthonis, 324
Pereira da Costa, Francisco Augusto, 263
Pereira, Levy, 142-143
Pergaeus, Apollonius, 71
Picard, Jean, 187, 319
Pickel, Bento José, 174, 347
Picker, Marija de, 347
Pies. See Piso
Pies, Harmen, 323
Pieters, Heijlien, 322
Pimentel, Jorge, 142, 143
Pingré, Alexandre Guy, 261-262, 264, 267, 273, 281
Piso, Cornelia Catrina, 347
Piso, Harmanus, 347
Piso, Maria, 347
Piso, Willem, 41, 63, 86-88, 93, 100, 103, 105, 111, 115, 117-119, 147, 152, 154, 164, 168-172, 178-179, 181, 237, 257, 278, 286, 293, 296, 300, 315, 323-325, 327, 329, 334-335, 346-348, 351-352
Plancius, Petrus, 193-194, 196
Plante, Franciscus, 40, 119, 136
Platter, Thomas, 54
Poel, Cornelius van der, 132, 339
Pol, Milo van de, 12
Polman, Jorge, 23, 205, 263-264, 365
Post, Frans, 38, 100, 102-103, 125, 132, 136, 139, 204, 303, 328-329, 337
Post, Pieter, 33
Pot-Noordman, Ingrid, 12, 184, 362
Pots, Elisabeth, 324

Ptolemy, Claudius, 192-193, 254, 272, 297, 361
Pudsey, Cuthbert, 327

Quelen, Augustus de, 309

Rammelman Elzevier, Willem, 183-184
Ray, John, 167
Reaal, Laurens, 110, 227
Rebhun Sr., Nicolaus, 311
Rebhun, Nicolaus, 311
Rhodius, Ambrosius, 314
Riccioli, Giovanni Battista, 272
Richer, Jean, 268
Riebeeck, Jan van, 363
Rivet, Claude, 69
Rochefort, Charles de, 167
Röchlin, Christina Elisabeth, 317
Rodrigues, J.H., 13, 172
Roscam Abbing, Michiel, 328, 343, 345
Rubbens, Marritgen Pieters, 324
Rudolph II, Holy Roman Emperor, 22, 311
Ruys, L.J., 184
Ruysch, Paulus, 358
Ryff, Petrus, 54

Sachariassen, Johannes, 358
Safier, Neil, 155, 167, 282
Savry, Jacob, 69
Scaliger, Josephus Justus, 106, 200-201, 219, 330
Scharf, Claudia, 154-155, 158, 160, 343
Scheiner, Christoph, 130, 223, 225, 248, 359
Schetski, Christopher, 178-179, 351
Schickard, Wilhelm, 197
Schmalkalden, Caspar, 136, 260, 262, 294-295, 344-345, 364
Schooten the elder, Frans van, 66, 71, 316
Schooten the younger, Frans van, 316
Schoppe, Sigismund von, 32, 265
Schouten, Claes Simonsz, 326
Schröter, Johann Friedrich, 89, 325
Sennert, Andreas, 56-57, 299, 307
Sennert, Daniël, 54-55, 57, 59, 197, 312-313
Sherard, William, 343
Silnaer, Elisabeth, 108, 292, 299, 330
Silnaer, Hans, 108, 292, 299
Silva, Maria Angelica da, 13, 158
Simon, Elisabeth, 51
Simon, Johann, 50- 53, 311
Sint Nicolaas, Eveline, 305-306
Sloane, Hans, 154
Smient, Elbert Gerritsz, 339
Sneewins, Hendrick, 357
Sneewins, Johannes, 326
Snel van Royen. See Snellius
Snellius, Willebrord, 65-66, 68, 78-79, 200-201, 219, 230, 283, 315-316, 356-358, 360-361
Sousa-Leao, Joaquim de, 102
Sporen, Jeremias, 38, 309
Spranger, Constantia, 347
Spranger, Gommert, 347
Spranger, Suzanna, 347
Steenwijck, Evert Harmansz, 358
Stevin, Simon, 96, 328
Stokvis, Bernard, 184
Storms, Martijn, 126, 141
Struys, Cornelia, 119
Symoensz, Arent, 178

Taube, Berendt von, 321
Taube, Wilhelm von, 321
Taunay, Affonso, 263
Teensma, Ben, 124, 126, 141
Teixeira, Dante Martins, 168
Thévenot, Melchisédech, 186-187, 268, 270, 353
Thielemanss, Elsjen, 108
Thysius, Antonius, 348
Tierens, Joachim, 350, 352
Tirapicos, Luís, 12, 355
Torm, Erik Olufsen, 79, 321
Townsend, Joseph, 188

Uexküll, Gertrude Elisabeth von, 321
Uexküll, Johann von, 321
Uexküll, Konrad von, 81, 321-322
Uexküll, Otto von, 63, 81-82, 321

Uexküll, Sophia von, 321
Ulloa, Antonio de, 17, 188-189, 354
Ursinus, Christophorus, 82-83, 321
Ursinus, Gabriel, 321
Usingen, van. See Mansvelt, van

Vaez, Antonio, 327
Valet, Joan, 147
Van Helden, Albert, 12
Varenius, Bernard, 326
Vasconcelos de Almeida, Argus, 45
Vaz de Llacerda, Antônio, 94
Veenhuisen, Thijs Teunissen van, 339
Verhochen, Leendert Reijnsen, 330
Vingboons, Johannes, 98, 148, 342
Visscher, Claes Jansz., 140
Vogel, Barbara, 325
Vondel, Joost van den, 347
Voorde, Balthasar van de, 110, 331, 332
Vorstius (Vorst), Adolph, 63, 68-69, 73-74, 89-90, 110, 151, 173, 176-177, 316, 325, 350
Vos, Jan de, 67
Vossius, Gerard, 110
Vries, Abraham de, 178-179, 326, 329, 351

Wagener, Zacharias, 99-100, 103, 157, 158, 175, 207, 260, 327-328
Walbeeck, Johannes van, 339
Wätjen, Hermann, 36, 41, 263
Wendelin, Govaert, 244
Wesley, Walter, 231, 360
Whitehead, Peter, 23-24, 38, 105, 141, 154, 156, 171, 178, 184, 263, 278, 345
Wiesebron, Marianne, 284
Wijnhoff, Anna, 63, 77
Wilhelm IV, Landgraf of Hessen-Kassel, 38, 64
Willaerts, Abraham, 38
Winnenvelt, Nicasio, 122
With, Gijsbert de, 122, 124, 336-337
With, Witte de, 327
Worm, Øle, 73-74, 118, 162-164, 167, 278, 321, 325, 349
Worm, Søster, 321
Worm, Villum, 174-175, 296, 347, 349-350
Wy(n)dal, Syrt Egbers, 306

Zandvliet, Kees, 121, 130-131, 134, 136, 141, 328
Zeidler, Johannis, 313
Ziller Camenietski, Carlos, 12

For Product Safety Concerns and Information please contact our EU representative GPSR@taylorandfrancis.com Taylor & Francis Verlag GmbH, Kaufingerstraße 24, 80331 München, Germany

Batch number: 10347037

Printed by Printforce, the Netherlands